ANALOG INTEGRATED CIRCUIT DESIGN

MICROELECTRONICS SERIES

Electronic Integrated Circuits and Systems, Franklin C. Fitchen

Thick-Film Microelectronics: Fabrication, Design and Applications, Morton L. Topfer

Electronic Integrated Systems Design, Hans R. Camenzind

MOS Integrated Circuits: Theory, Fabrication, Design, and Systems Applications of MOS/LSI, Engineering Staff of American Micro-systems, Inc.

Analog Integrated Circuit Design, Alan B. Grebene

ANALOG INTEGRATED CIRCUIT DESIGN

Alan B. Grebene

Exar Integrated Systems, Inc.
Sunnyvale, California

MICROELECTRONICS SERIES

Van Nostrand Reinhold Company Regional Offices:
New York Cincinnati Chicago Millbrae Dallas

Van Nostrand Reinhold Company International Offices:
London Toronto Melbourne

Library of Congress Catalog Card Number: 72-3869
ISBN: 0-442-22827-9

Manufactured in the United States of America

Published by Van Nostrand Reinhold Company
450 West 33rd Street, New York, N. Y. 10001

Published simultaneously in Canada by Van Nostrand Reinhold Ltd.

15 14 13 12 11 10 9 8 7 6 5 4

Library of Congress Cataloging in Publication Data

Grebene, Alan B 1939-
Analog integrated circuit design.

(Microelectronics series)
Includes bibliographical references.
1. Integrated circuits. 2. Electronic analog computers—Circuits. I. Title.
TK7874.G69 621.3819'57'35 72-3869
ISBN 0-442-22827-9

PREFACE

The advent of integrated circuit technology has altered many of the established circuit design techniques and principles. This is particularly evident in the field of analog integrated circuits where the designer finds himself faced with a new set of design constraints and ground rules. In writing this book, it is my intention to educate the practicing electronics engineer in the fundamental design principles, capabilities, and applications of monolithic analog circuits.

The contents and the organization of this book is aimed primarily at the practicing engineer in the solid state circuits field. However, the subject matter is treated rigorously, and from a fundamental viewpoint, to make it suitable as a text for graduate study in semiconductor circuits.

In the preparation of the text, it is assumed that the reader is familiar with the basic theory and the principles of solid state devices and circuits. Therefore, the solid state device theory, which is already well-covered elsewhere in the literature, is reviewed only briefly, and most of the space is devoted to the novel circuit design approaches unique to monolithic circuits. The overall field of analog integrated circuits is far too wide to be covered efficiently in a single book. Consequently, in planning this book, I have deliberately limited myself to the area of monolithic analog circuits. Hybrid integrated circuits, which represent an area of overlap between the discrete and the monolithic circuits, will not be covered explicitly.

In the area of microelectronics, and particularly in monolithic integrated circuits, it is necessary for a circuit designer to assimilate the knowledge of adjoining fields, namely, process technology and solid state device design. To satisfy this need, this book is composed of two separate parts which complement each other. The first part of the text, Chapters 1 through 3, provides a comprehensive survey of integrated circuit process technology as well as device design and layout principles. These chapters are intended to familiarize the designer with the physical structures, advantages, and limitations of monolithic components. This knowledge is imperative to an analog integrated circuit designer since a successful design is one which efficiently utilizes the advantages of monolithic devices while avoiding their shortcomings.

The second part of the text, Chapters 4 through 11, is aimed at the design and application of different classes of analog integrated circuits. Chapter 4 provides a transition between the two parts of the text by exploring some of the basic circuit building blocks which, in turn, serve as the starting point for the design of larger and more complex functions. In Chapters 5 through 9 specific classes of analog circuits such as operational amplifiers, regulators, multipliers, and active filters are discussed individually. Chapter 10 covers the field of digital and analog interface circuits. Design of highly specialized circuits intended to operate under radiation environment, or at micropower or high voltage levels, are discussed in Chapter 11. Consumer and communication circuits, which represent a significant class of analog ICs, overlap in part with the classification of circuits given in Chapters 5 through 11. Therefore, they are not treated as a separate topic but covered implicitly in each of these chapters.

Part of the material covered in this book is patterned after a sequence of graduate level courses in integrated electronics which I taught at Santa Clara University. Therefore, when preceded by courses on solid state circuits and semiconductor electronics, this book will be well suited for a senior or graduate level course.

I am indebted to many people who have contributed directly or indirectly to the preparation of this book. In particular, I would like to thank my wife, Karen, who typed and retyped the draft of the manuscript several times over. I would also like to extend my gratitude to my colleagues and associates at Signetics Corporation and specifically to Messrs. G. Kelson and H. Stellrecht for their assistance in the organization and the technical accuracy of the book, and to Mrs. K. Koivisto for her assistance in typing the manuscript.

Saratoga, California A. B. G.

CONTENTS

ANALOG INTEGRATED CIRCUIT DESIGN

1 Integrated Circuit Fabrication Processes

The field of microelectronics represents an overlap of a wide variety of technical disciplines, ranging from physical chemistry and metallurgy to solid state device and circuit theory. Therefore, in the area of microelectronics and particularly in monolithic integrated circuits, it is necessary for a circuit designer to assimilate the knowledge of his adjoining fields, namely process technology and solid state device design.

The aim of this chapter is to familiarize the reader with the fabrication processes for analog integrated circuits. As a designer, one does not need to know the specific details of each and every process step, since often each of these fabrication processes become by themselves areas of specialization and rapid technological development. However, it is imperative that a successful design engineer at all times be familiar with the capabilities and the constraints of each of the major process steps in the fabrication of integrated circuits.

1.1 THE PLANAR PROCESS

It has taken less than a decade in time to convert integrated circuit technology from a laboratory curiosity to a major branch of the electronics industry. The key to this rapid growth of the integrated circuit field has been the development of process technology. Even though the specific nature of monolithic circuit fabrication processes is well diversified, the bulk of the fabrication steps associated with present-day integrated circuit technology can be grouped into a single package under the name "The Planar Process."[1] Prior to the evolution of this process, the solid state electronics field was dominated by germanium devices. The introduction of the planar process in 1960 revolutionized the microelectronics field almost overnight; and silicon, rather than germanium, emerged as the predominant semiconductor material.

As applied to semiconductor devices, the term *planar* does not mean two dimensional in the strict definition of the word. Instead, in this context, the term is used to mean that the monolithic components fabricated by this process extend below the surface of the silicon substrate; however the plane surface of the semiconductor remains relatively flat and unaltered through the sequence of different fabrication steps. Planar process technology is comprised of five independent processes: epitaxy, surface passivation, photolithography, diffusion, and thin film deposition. A schematic illustration of these process steps is given in Fig. 1.1, in terms of the cross-sectional view of a silicon wafer.

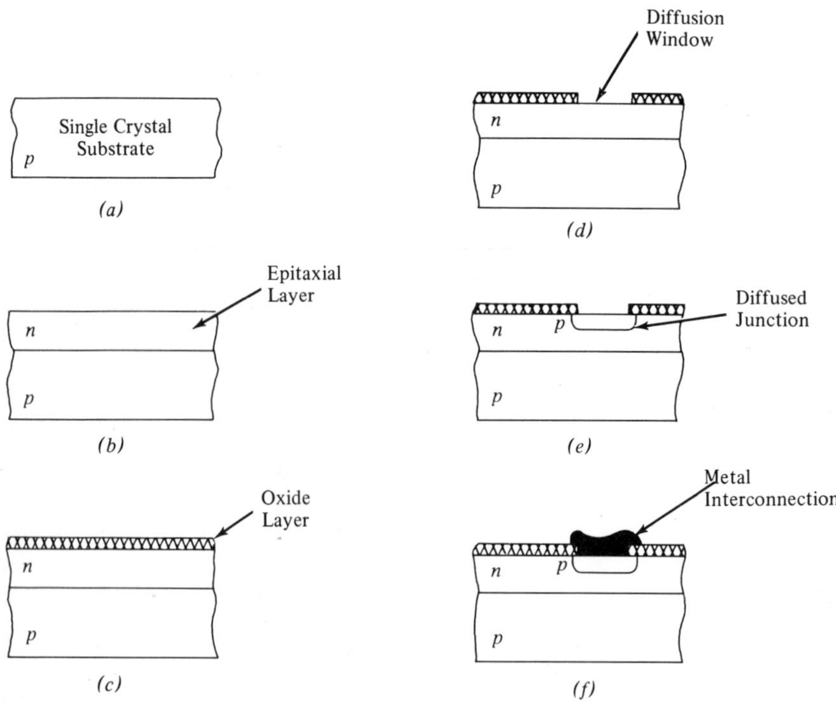

Figure 1.1 Basic process steps in planar technology: Epitaxy; surface passivation; photolithography; diffusion; and metallization.

Epitaxy, derived from the Greek word meaning "arranged upon" is a growth technique wherein the single crystal structure of a silicon substrate can be extended by vapor-phase deposition of additional atomic layers of silicon. By controlling the deposition rates and introducing controlled amounts of impurities into the carrier gases utilized during the epitaxial growth process, the thickness and the resistivity type of the deposited layer can be accurately controlled. In Fig. 1.1(*b*), this is illustrated for the case of an *n*-type epitaxial layer grown on a single crystal *p*-type substrate.

Thermal growth or deposition of a dielectric layer, such as SiO_2, on top of the silicon wafer provides a convenient means of electrically passivating the silicon surface, and forming a protective mask against contamination of the surface by undesired impurities. This is illustrated in Fig. 1.1(*c*).

Photolithographic techniques are used to etch selective openings on the passivating oxide layer. These openings serve as diffusion windows from which controlled amounts of impurity doping can be introduced into the single crystal silicon substrate or the epitaxial layer, as shown in Fig. 1.1(*d*). Using photographic techniques, the diffusion windows can be greatly reduced in size, without sacrificing the accuracy of their alignment or lateral dimensions.

Controlled amounts of dopant impurities can be introduced into the preselected parts of the silicon surface through the diffusion windows in the oxide layer. Solid state diffusion of these dopants into silicon at high temperatures results in the formulation of a *pn* junction within the single crystal silicon, as shown in Fig. 1.1(*e*). Since the diffusion of impurities proceeds sideways as well as downward from the diffusion windows at the surface, the resulting junction edge is not exposed to air on the surface, but is protected by the surface oxide layer.

Electrical contact to the semiconductor regions can be formed by depositing a thin metal film of high electrical conductivity, such as aluminium, over the windows cut in the oxide. These conductive films can then be etched into desired interconnection patterns on the surface of the silicon wafer thus completing the monolithic circuit structure.

As compared with earlier solid state device fabrication techniques, the planar process offers two unique advantages which make it ideally suited for integrated circuits:

1. The semiconductor junctions are protected on the surface by an oxide layer and are not exposed to air. This results in low leakage currents and high reliability.
2. Photographic reduction, masking, and etching techniques are used to determine device geometries, thus making possible very small dimen-

sions and simultaneous fabrication of a large number of devices and circuits on the same silicon surface.

Monolithic integrated circuit technology owes its rapid growth and acceptance to these two unique features of the planar technology.

1.2 DIFFUSION

The diffusion process is by far the most widely used method of introducing controlled amounts of impurities into the silicon substrate. It is a relatively well understood and highly reproducible process step which readily lends itself to the batch-processing advantages of the planar technology, since a large number of silicon wafers can be processed simultaneously.

Diffusion, as a general process, is the mechanism by which different sets of particles confined to the same volume tend to spread out and redistribute themselves evenly throughout the confining volume. In the case of solid state devices and integrated circuits, the relevant diffusion process is the one dealing with the movement and the distribution of impurity atoms in a crystalline lattice structure. In crystalline solids, the diffusion process is significant only at elevated temperatures where the thermal energy of the individual lattice atoms has a statistical chance of overcoming the interatomic forces which hold the lattice together.

In the single crystal silicon lattice, the impurities can move through the lattice by any one or a combination of the two dominant diffusion mechanisms. These physical mechanisms can be briefly outlined as follows:

Substitutional Diffusion: The impurity atoms propagate through the lattice by replacing a silicon atom at a given lattice site.
Interstitial Diffusion: The impurity atoms do not replace the silicon atoms at the regular lattice sites, but instead move into the interstitial voids in the three-dimensional lattice structure.

In the fabrication of analog integrated circuits, substitutional diffusion is by far the most important, since all the impurities intentionally introduced into the silicon substrate to form the junction and the device structures diffuse substitutionally. This is not necessarily the case for digital integrated circuits where controlled amounts of interstitial impurities such as gold, copper, or nickel can be introduced into the silicon lattice to reduce the minority carrier lifetime.

The impurities which are diffused into silicon to determine the type and resistivity of various regions of the semiconductor material are elements from Groups III or V of the Periodic Table, for *p*- or *n*- type doped regions, respectively. Some of these dopants are listed in Table 1.1. For

TABLE 1.1 *p*- AND *n*-TYPE DOPANTS FOR SILICON DEVICE FABRICATION

p-Type		*n*-Type	
Boron	(B)	Phosphorus	(P)
Aluminum	(Al)	Arsenic	(As)
Gallium	(Ga)	Antimony	(Sb)
Indium	(In)		

integrated circuit fabrication, B, P, As, and Sb are the most commonly used dopants.

Diffusion Theory

Although the diffusion process proceeds in all three dimensions simultaneously, for the analysis of the fundamental properties of that process, it is sufficient to consider only one dimension. As will be described in Chapter 2, the geometry and dimensions of most semiconductor devices fabricated by the planar process justify this one-dimensional assumption.

The fundamental physical property of the diffusion process is that the particles tend to diffuse from a region of high concentration to that of a lower one, at a rate proportional to the concentration gradient between the two regions. This is known as Fick's first law, and can be mathematically expressed as

$$F = -D\partial N/\partial x \tag{1.1}$$

where F is the net particle flux density, i.e., the net number of particles flowing through a unit surface area normal to the direction of flow per unit time, N is the number of particles per unit volume, and x is the distance measured parallel to the direction of flow. The term D is called the diffusion coefficient and has units of $(\text{length})^2/\text{time}$. The magnitude of D gives a measure of the relative ease or difficulty with which the diffusing particle can move about in its environment. The negative sign appears in (1.1) to indicate that the particle flow is directed from a region of high concentration toward one of lower concentration.

In all diffusion problems, one is interested in the variations of the impurity concentration with time, as well as with distance. The fundamental law of diffusion which relates the time rate of change of concentration to the spatial coordinates of the region in question can be derived from

Fick's law by applying the continuity principle to Eq. (1.1) as follows: Consider a region or a volume in the material enclosed by an area A normal to the flow, and a width dx paralellel to the flow. The net flow of particles into this volume can be written as

$$FA - \left[F + \frac{\partial F}{\partial x}\,dx\right]A = -\frac{\partial F}{\partial x}\,dxA \tag{1.2}$$

This net flow into the volume can be related to the concentration change within the volume as

$$\frac{\partial N}{\partial t}\,dxA = -\frac{\partial F}{\partial x}\,dxA \tag{1.3}$$

The expression for $(\partial F/\partial x)$ can be obtained by differentiating (1.1). Substituting the result into (1.3), one obtains

$$\frac{\partial N}{\partial t} = D\,\frac{\partial^2 N}{\partial x^2} \tag{1.4}$$

This equation is known as Fick's second law, and it defines the impurity distribution at any given point within the semiconductor crystal as a function of time.

The diffusion coefficient D is a property of the particular doping impurity, and is an exponential function of temperature. Figure 1.2 shows the diffusion coefficients of various dopants in silicon, as a function of temperature.[2]

Since (1.4) is a second order partial differential equation, its particular solution applicable to a given diffusion problem depends on the boundary conditions associated with the given diffusion process. Most of the diffusion processes encountered in the fabrication of integrated circuits fall into one or the other of the following two classes of boundary conditions: constant source (complementary error function) diffusion or limited source (Gaussian) diffusion. The properties of the diffusion profiles associated with each of these classes of diffusions will now be considered separately.

Constant Source (Complementary-Error Function) Diffusion. In this type of diffusion, the impurity concentration at the semiconductor surface is maintained at a constant level throughout the diffusion cycle, i.e., $N(0,t)$ = constant = N_0. The constant impurity level on the surface is determined by the temperature and the carrier gas flow rate of the diffusion furnace. In most constant source diffusion systems, it is convenient to let the surface concentration, N_0, be determined by the solid solubility concentration limit of the particular dopant in silicon. As shown in Fig. 1.3, solid solubility limit is also a strong function of the temperature.

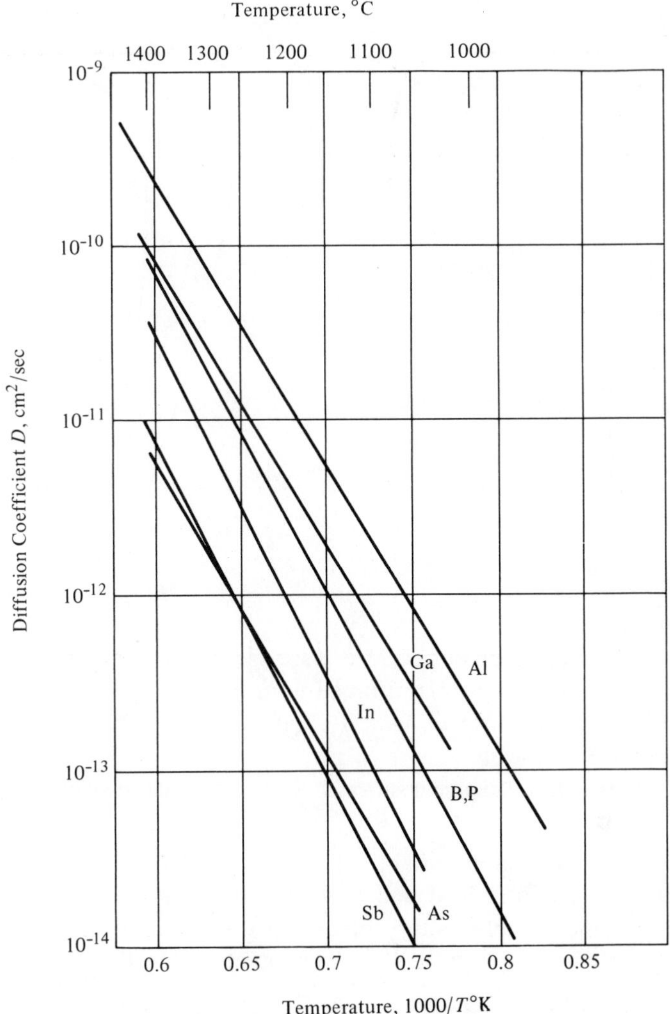

Figure 1.2 Diffusion coefficients of various dopants in silicon.[2]

The constant source diffusion process can be described by the solution of (1.4) satisfying the boundary conditions:

$$N(0,t) = N_0 = \text{const} \qquad \text{and} \qquad N(x,o) = 0 \tag{1.5}$$

The resulting particular solution of Eq. (1.4) can be expressed as

$$N(x,t) = \text{N}_0 \left(1 - \frac{2}{\sqrt{\pi}} \int_0^{x/2\sqrt{Dt}} e^{-\lambda^2}\, d\lambda\right) \tag{1.6}$$

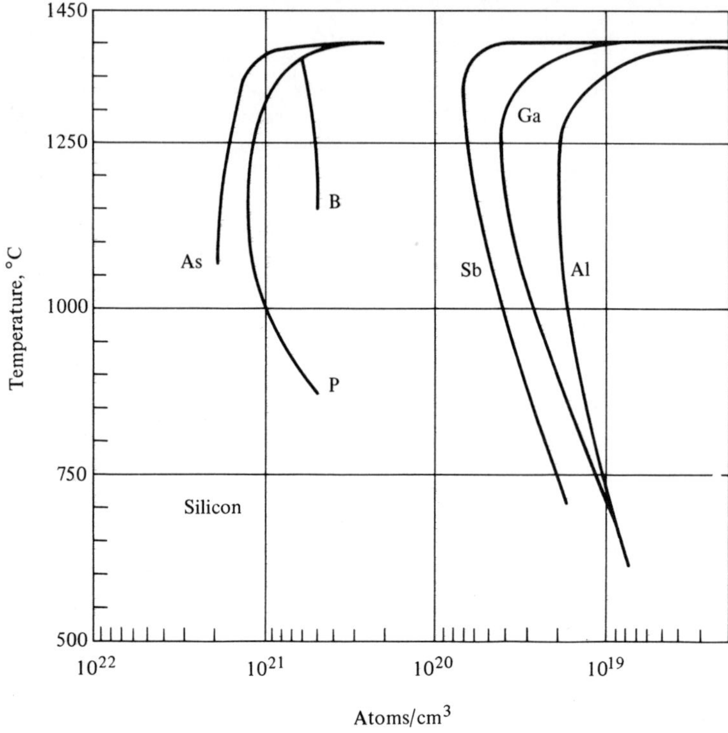

Figure 1.3 Solid solubilities of various dopants in silicon.[3]

where λ is an integration variable. The portion of the solution inside the parentheses is a well defined and tabulated function of its argument, and is known as the complementary error function (abbreviated erfc). Thus, the resulting concentration at any given point within the silicon material can be written as

$$N(x,t) = N_0 \operatorname{erfc}\left(\frac{x}{2\sqrt{Dt}}\right) \tag{1.7}$$

A graph of the complementary error function is shown in Fig. 1.4 for a range of values of its argument.

A typical set of impurity distribution profiles formed by complementary error function diffusions is shown in Fig. 1.5. Note that in this type of diffusion, the total amount of impurities diffused into silicon increases indefinitely with time. Similarly, the impurity concentration at any point within the material (except at the surface) is a monotonically increasing function of time; therefore, as t goes to infinity, the entire wafer would

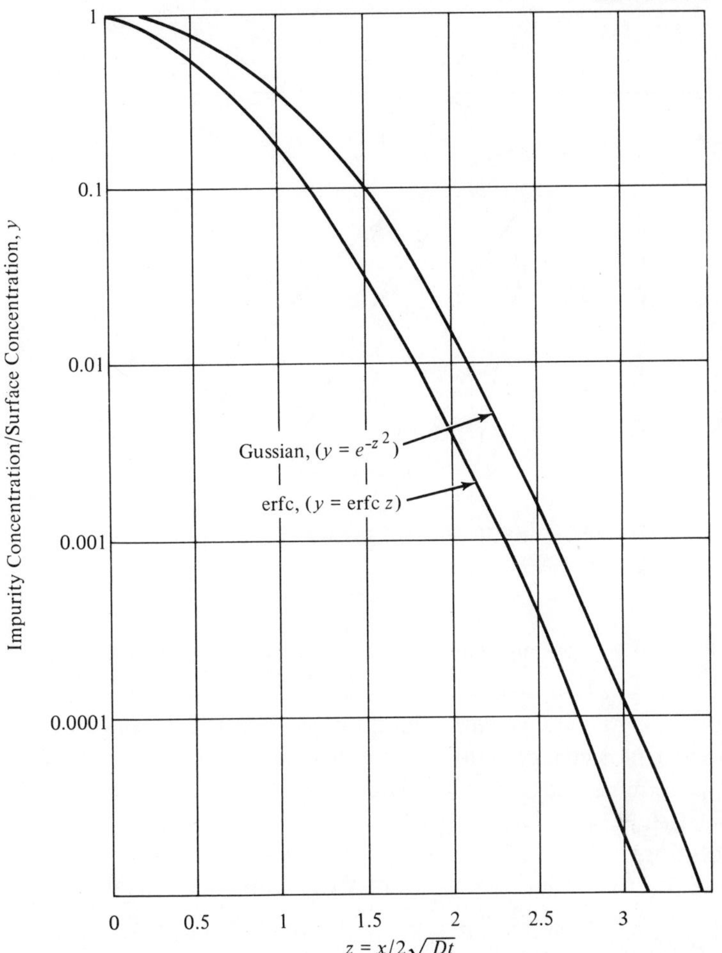

Figure 1.4 Gaussian and complementary error-function distributions.

have a uniform concentration of N_0. If the diffused impurity type is different from the resistivity type of the substrate material, a junction is formed at the points where the diffused impurity concentration is equal to the background concentration already present in the substrate. These junction depths are shown as points x_1, x_2, and x_3, respectively for the diffusion profiles of Fig. 1.5. In the fabrication of monolithic circuits, constant source diffusion is commonly used for the isolation and the emitter diffusion steps.

Example: A p-n junction is to be formed by a constant source boron diffusion into a silicon substrate, having an initial n-type impurity con-

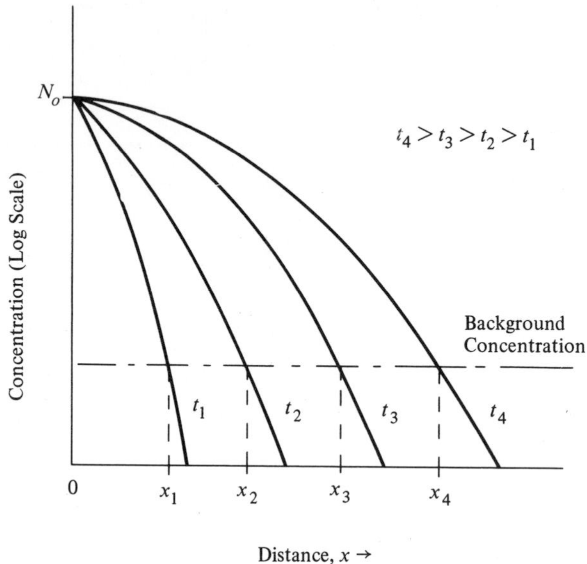

Figure 1.5 Constant source diffusion profiles as a function of time.

centration of 10^{16} atoms/cm^3. The diffusion step is performed at 1250°C, with the surface concentration of boron maintained at the solid solubility limit. Determine the diffusion time t to reach a junction depth of 9 microrons (μ).

Solution: The surface concentration is determined from Fig. 1.3 as $N_0 = (5.5)(10^{20})$ atoms cm^3. At 1250°C, diffusion coefficient of boron is $(5)(10^{-12})$ cm^2/sec (see Fig. 1.2). The junction is formed when $N(x,t)$ equals the background concentration. Thus:

$$\frac{N(x,t)}{N_0} = (1.82)(10^{-6}) = \text{erfc}\,\frac{x}{2\sqrt{Dt}}$$

where $x = 9\ \mu$. Solving for t from Fig. 1.4, one finds $t = 4500$ sec $=$ 66.5 min.

Limited Source (Gaussian) Diffusion. In this type of diffusion, the total amount of impurity introduced into the semiconductor during the diffusion step is limited. This is achieved by depositing on the silicon surface a fixed number of impurity atoms per unit of exposed surface area, during a short "predeposition," step prior to the actual diffusion. This predeposition is then followed by a "drive" cycle where the impurity source is turned off, and the amount of impurity already deposited during the "predeposition" step is diffused into the silicon substrate. With this type of diffusion, the depth of penetration of the impurities during the predeposition step is assumed to be negligible as compared with the final junction depth achieved after "drive" cycle. Thus, the initial impurity distribution $N(x,0)$ is assumed to be a delta function on the semiconductor surface. Then the basic diffusion equation (1.4) is solved with the boundary condition,

$$\int_0^\infty N(x,t)\,dx = Q = \text{const} \tag{1.8}$$

where Q is the initial deposited impurity concentration on the surface (in atoms/cm²). The corresponding solution to the diffusion equation is

$$N(x,t) = \frac{Q}{\sqrt{\pi\,Dt}}\,e^{-x^2/4\,Dt} \tag{1.9}$$

The impurity profile given by (1.9) is known as the Gaussian distribution. A normalized plot of this distribution is shown in Fig. 1.4 as a function of its argument. Figure 1.6 shows a sketch of the limited source diffusion profile for increasing values of time. Note that in this type of diffusion, the surface concentration N_0 is inversely proportional to the square root of the diffusion time. In integrated circuit fabrication, the limited source diffusion is commonly used in forming the transistor base regions.

Example: 10^{16} boron atoms/cm² are deposited on the silicon surface prior to a limited source diffusion step. The silicon substrate has an initial n-type impurity concentration N_B of 10^{16} atoms/cm³. Determine the junction depth after a diffusion of one hour at 1100°C.

Solution: The diffusion coefficient of boron at 1100°C is found from Fig. 1.2 to be $(3)(10^{-13})$. Then (1.9) can be rewritten as

$$e^{-x^2/4\,Dt} = \frac{N_B\sqrt{\pi\,Dt}}{Q} = (5.8)(10^{-5})$$

From Fig. 1.4, one obtains

$$\sqrt{\frac{x^2}{4\,Dt}} = 3.12\,\mu \qquad \text{or} \qquad x = 2.05\,\mu$$

Basic Properties of the Diffusion Process

In the design and layout of monolithic integrated circuits, the following three fundamental properties of the diffusion process must be considered:

1. All diffusions proceed simultaneously. The impurities introduced in an earlier diffusion step continue to diffuse during subsequent diffusion

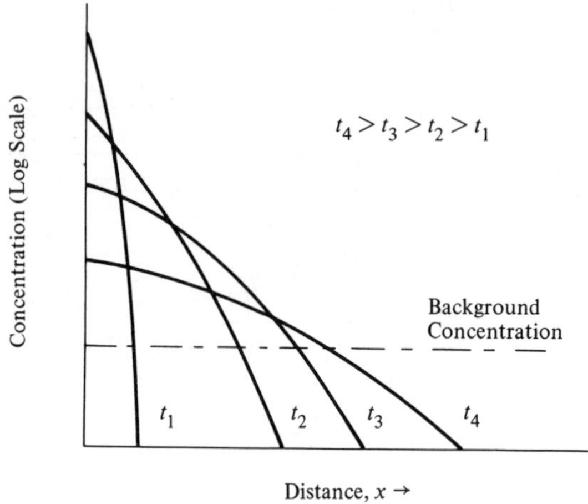

Figure 1.6 Limited source (Gaussian) diffusion profiles as a function of time.

cycles. Therefore, when calculating the total effective diffusion time for a given impurity profile, one must often consider the effects of subsequent diffusion cycles. The effects of the subsequent diffusions on a given impurity profile can be estimated by defining an effective (Dt) product for the particular impurity profile as

$$(Dt)_{\text{eff}} \approx D_1 t_1 + D_2 t_2 + D_3 t_3 + \cdots \tag{1.10}$$

where t_1, t_2, t_3, etc. are the different diffusion times and D_1, D_2, D_3, etc. are the corresponding diffusion coefficients as determined by the respective temperatures of the diffusion cycles. Thus, for example, in the planar device fabrication, the emitter region of a bipolar transistor is formed by a diffusion process which succeeds the base diffusion step. Therefore, the effective Dt product of the base region contains a finite contribution from the emitter diffusion step.

2. The diffusion profiles of (1.7) and (1.9) are both functions of $(x/\sqrt{Dt})$. Therefore, for a given surface and background concentration, the junction depths x_1 and x_2 associated with the two separate diffusions having different times and temperatures, can be related as

$$\frac{x_1}{x_2} = \sqrt{\frac{D_1 t_1}{D_2 t_2}} \tag{1.11}$$

3. The diffusions proceed sideways from a diffusion window as well as downward. In considering the lateral dimensions of the planar devices, particularly in the case of lateral *pnp* transistors (see Section 2.1) these lateral diffusion effects need to be considered. Figure 1.7 shows the lateral

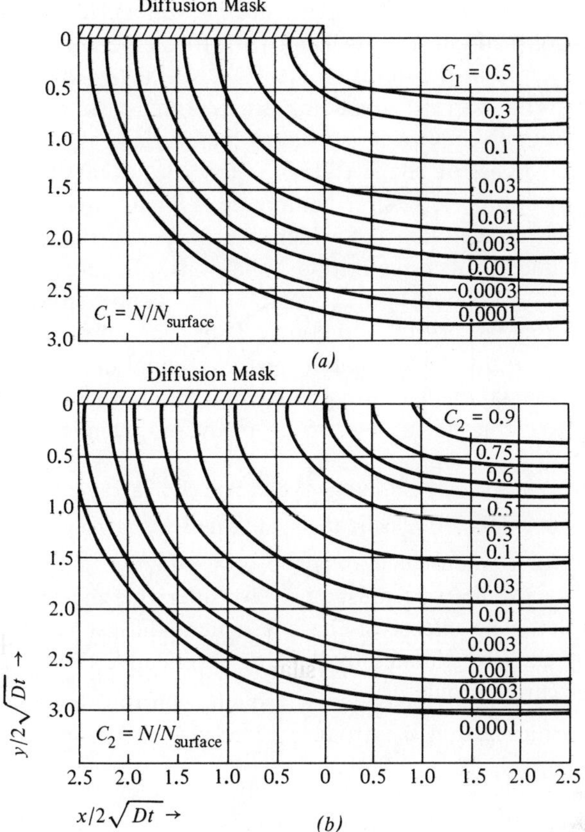

Figure 1.7 Lateral diffusion profiles for (*a*) constant source; (*b*) limited source diffusions.[4]

diffusion effects for various concentration ratios.[4] Figure 1.7(*a*) gives the constant concentration contours for the case of a constant source diffusion. The diffusion contours of Fig. 1.7(*b*) correspond to the limited source case. In both cases the side diffusion is about 75 to 80 percent of the vertical penetration for contours which are more than an order of magnitude below the surface concentration.

1.3 EPITAXY

Epitaxy is a growth technique where the single crystal structure of a silicon substrate can be extended by vapor-phase deposition of additional atomic layers of silicon. Epitaxial growth, or deposition, is carried out in a special furnace called a "reactor" where silicon wafers having a clear and chemically polished surface are heated up to temperatures comparable to those encountered in the diffusion step (i.e., 1000 to 1200°C). During the epitaxial growth, vapors containing silicon are passed over the heated substrate. Normally hydrogen is used as the carrier gas, with either silicon tetrachloride ($SiCl_4$) or silane (SiH_4) as the source of silicon. During the epitaxial growth process, source compound is chemically reduced, resulting in free silicon atoms, some of which are deposited on the single crystal substrate. Under proper deposition conditions, the interatomic forces of the single crystal silicon lattice constrain the deposited silicon atoms to follow the original crystal structure. Thus, structurally, the deposited epitaxial layer forms a continuation of the original crystal structure.

When $SiCl_4$ is used as the source of silicon, the hydrogen gas also serves as a reducing agent, leading to an overall chemical reaction:

$$SiCl_4 + 2\,H_2 \rightleftharpoons Si + 4\,HCl \tag{1.12}$$

In the case of silane, free silicon is obtained by pyrolytic decomposition of SiH_4:

$$SiH_4 \rightarrow 2\,H_2 + Si$$

Normally the $SiCl_4$ process requires somewhat higher temperatures than silane decomposition and also has a slower growth rate.

During the process of epitaxial growth, controlled amounts of *p*- or *n*-type impurities are also introduced into the carrier gas to control the

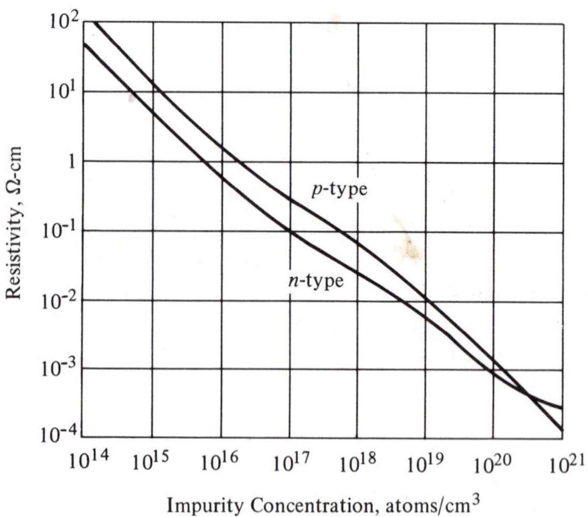

Figure 1.8 Resistivity of silicon as a function of uniform impurity concentration at 300°K.[5]

type and the resistivity of the deposited layer. Unlike the diffusion process, epitaxial growth proceeds by uniform addition of atomic layers onto the substrate. Thus, the dopant impurities are uniformly distributed through the epitaxial layer, and do not show a concentration gradient. Furthermore, epitaxial layers can be grown over diffused regions or over other epitaxial layers.

Since the impurity distribution within the epitaxial layer is uniform, its resistivity, ρ, can be expressed as

$$\rho = \frac{1}{qN\mu} \tag{1.13}$$

where q is the electronic charge, and N is the uniform impurity concentration within the layer; and μ is the carrier mobility. Since μ is an implicit function of concentration, the evaluation of the actual resistivity is largely emperical.[5] Figure 1.8 shows the resistivity of uniformly doped *p*- or *n*-type silicon substrate or epitaxial layers as a function of impurity concentration.

Redistribution of Impurities During Epitaxy

Since epitaxial growth is a high temperature process, the impurities at the episubstrate interface tend to redistribute themselves via the diffusion process. For example, in the case of an *n*-type epitaxial layer grown on a *p*-type substrate, the epi-substrate interface no longer represents a step junction but becomes graded due to the diffusion of impurities from the epitaxial layer into the substrate and vice verse. Consequently, the impurity distribution at the epi-substrate interface may look as shown in Fig. 1.9.

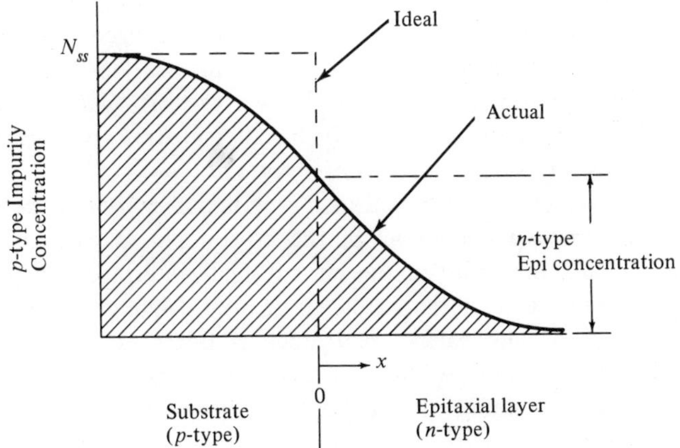

Figure 1.9 Impurity redistribution at epitaxial layer-substrate interface.

The dotted line in the figure shows the ideal p-type impurity distribution at the interface and the shaded area corresponds to the actual distribution.

For relatively rapid rates of epitaxial growth (i.e. $> 0.2\ \mu$/min), the impurity distribution $N(x,t)$ across the interface can be written as

$$\frac{N(x,t)}{N_{ss}} \approx \frac{1}{2} \operatorname{erfc} \frac{x}{\sqrt{Dt}} \tag{1.14}$$

where N_{ss} is the impurity concentration within the bulk of the substrate. The total amount of interdiffusion across the episubstrate interface during the entire device fabrication cycle can be estimated by defining an effective (Dt) product for the out-diffusion process, in accordance with (1.10).

In terms of minimizing impurity redistribution during epitaxial growth, the silane (SiH_4) process is preferred over $SiCl_4$ because of its higher growth rate and lower deposition temperature. This results in a very sharp epi-substrate transition which is desirable for high frequency devices.

Crystal Defects in Epitaxial Layers

During the epitaxial deposition process, a number of crystal defects may occur. Depending on their nature and density, these defects and imperfections may affect the overall electrical characteristics of the epitaxial layer, and of the device junctions formed in it. When present in large numbers, these crystal defects reduce the minority carrier lifetime within the epitaxial layer, increase the junction leakages, and cause localized voltage breakdowns.

The most commonly encountered imperfections in the epitaxial structures are "dislocations" and "stacking faults." The dislocation defects are caused by imperfect arrays of atoms within the localized regions of the substrate lattice, and by mechanical stress. The dislocations tend to appear in clusters, and be oriented along lines known as "slip planes." They can be reduced by taking additional care during the substrate preparation, i.e., etching and polishing steps, prior to epitaxy.

Another prevalent crystal defect in the epitaxial regions is the stacking fault caused by improper stacking of the crystal planes over a localized region. Such a fault may start at the epi-substrate interface and propagate through the deposited layer thickness; or it may originate entirely within the epitaxial region. This latter occurrence is often associated with excessively high growth rates. Additional types of crystal defects in epitaxial layers are pits or pyramids on the epitaxial layer surface which are in general due to either excessive or parasitic impurities present in the epitaxy system.

Polycrystalline Silicon Growth

If silicon is epitaxially grown on a noncrystalline substrate, such as an SiO_2 layer, the deposited layers of silicon tend to be oriented in random directions, thus leading to a polycrystalline structure. Polycrystalline silicon does not have the electrical properties associated with single crystal semiconductors, but can be used for a variety of special applications in integrated circuit fabrication. One particular application of polycrystalline silicon growth will be described later in this chapter, in connection with dielectrically isolated device structures.

1.4 SURFACE PASSIVATION

Passivation of the silicon surface by an inert dielectric layer is one of the basic features of the planar process. In almost all cases, surface passivation is provided by a thermally grown layer of silicon dioxide (SiO_2). This passivating layer performs three fundamental functions:

1. It serves as a diffusion mask and allows selective diffusions into silicon through the windows etched into the oxide.
2. It protects the junctions from exposure to the moisture and other contaminants in the atmosphere.
3. It serves as an insulator on the device surface on which the metal interconnections can be formed.

In addition to these three functions, the SiO_2 layer is often utilized as the dielectric region of monolithic capacitances. The silicon dioxide layers can be formed on the silicon surface by any one of several methods. Some commonly used techniques consist of thermal oxidation, pyrolytic deposition, and anodic or gas plasma oxidation. Among these, thermal oxidation is by far the most commonly used growth technique.

Thermal Oxidation of Silicon

During the thermal oxidation step, an oxide layer is formed on the silicon surface through the basic chemical reaction,

$$Si + O_2 \rightarrow SiO_2 \tag{1.15}$$

In the presence of some water vapor, the oxidation process is significantly accelerated, and proceeds in accordance with the chemical reaction

$$Si + 2\,H_2O \rightarrow SiO_2 + 2\,H_2 \tag{1.16}$$

Except for the initial oxidation step (see Fig. 1.1*c*), the thermal oxide growth is not generally performed as a separate fabrication step. Instead it is often incorporated into the diffusion cycle by providing an oxidizing atmosphere within the diffusion furnace during the latter part of each diffusion cycle. This in turn provides sufficient oxide on the previous diffusion windows for the next masking step.

Thermal oxidation normally proceeds at a temperature range of 900 to 1200°C. During oxidation, a carrier gas containing the oxidizing agent (normally oxygen gas or water vapor) is passed over the heated wafer substrate. Some of the silicon on the surface of the wafer is used up during the oxide growth process. From density and molecular weight considerations, it can be shown that to form a thermally grown SiO_2 layer of thickness x_0, one uses up approximately 0.42 x_0 of silicon from the wafer surface.

Kinetics of thermal oxide growth are well understood and covered in the literature.[7,8] Oxidation proceeds by an inward motion of the oxidizing species toward the silicon–SiO_2 interface. Therefore, as the oxidation process proceeds, it is necessary for the oxygen molecules to diffuse through a thicker layer of SiO_2 to get to the silicon surface where the chemical reactions of (1.15) or (1.16) can take place. Consequently the time rate of oxide growth decreases rapidly with increasing oxide thickness. It can be shown[8] that for very thin layers of SiO_2, the growth rate is linear with time. However, as the oxide thickness x_0 increases, the growth rate becomes proportional to $\sqrt{t}$. Figures 1.10*(a)* and *(b)* show

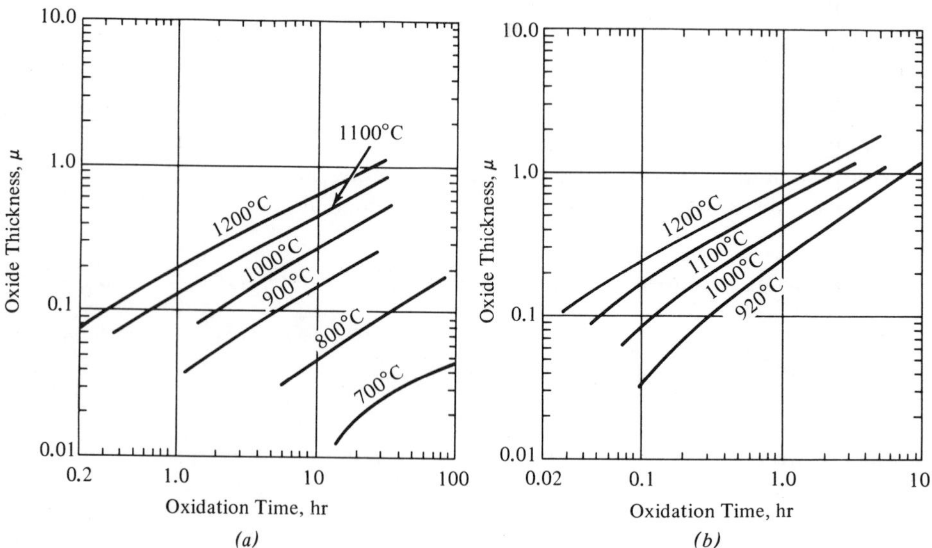

Figure 1.10 Typical growth rates of SiO_2 as a function of temperature: (*a*) dry O_2; (*b*) oxygen plus water vapor (adapted from ref. 9.)

typical growth rates of SiO_2 as a function of temperature for the cases of dry or wet oxygen (i.e., oxygen plus water vapor).

Masking Properties of SiO_2

The diffusion coefficients of most dopants in SiO_2 are about two to four orders of magnitude smaller than silicon. Therefore, for these impurities, which include all of those listed in Table 1.1 with the exception of Ga and Al, an SiO_2 layer of proper thickness can serve as a diffusion barrier. The minimum oxide thickness necessary to mask against a given diffusion step depends to a large extent on the specifics of the diffusion process, such as type of dopants used, surface concentrations, and predeposition temperature and time. Figure 1.11(*a*) and (*b*) gives some typical curves showing the minimum oxide thickness needed to mask against the two most commonly used dopants, boron and phosphorous.

Impurity Redistribution During Oxidation

A significant amount of impurity redistribution can occur at the Si-SiO_2 interface during oxidation. Some of the dopants, such as boron, show an affinity for diffusing from the silicon surface into the oxide layer. This results in a depletion of the actual boron concentration at the silicon surface during the SiO_2 growth. This depletion effect is particularly noticeable

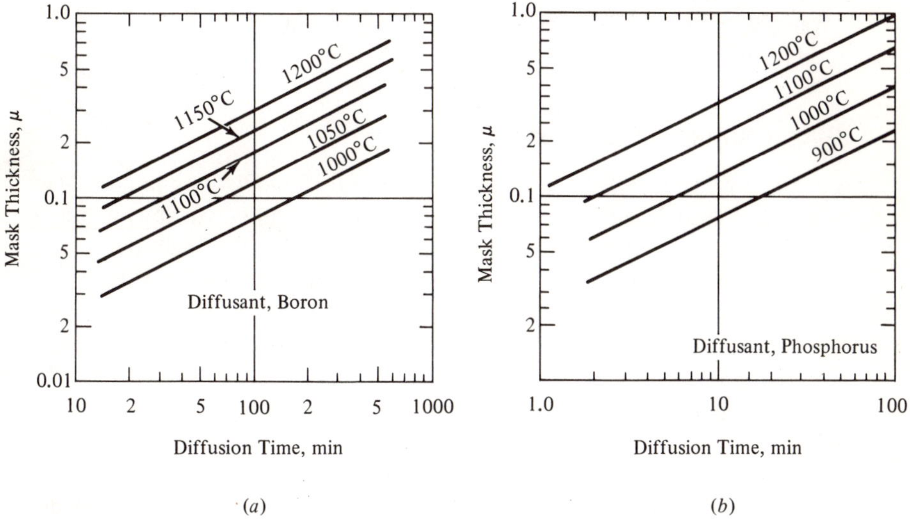

Figure 1.11 Masking oxide thickness for (*a*) boron, and (*b*) phosphorus.[10]

for rapid oxide growth rates where the surface concentration of boron can deplete as much as an order of magnitude. However, the penetration depth of this depletion region into the silicon substrate is quite small, typically of the order of 1000 to 1500 Å.

During oxidation, phosphorous behaves in an opposite manner to boron. It shows a negative affinity for the SiO_2 layer, and tends to stay in the silicon substrate. Therefore, during the oxide growth, the concentration of phosphorus tends to increase slightly (typically by 10 to 20 percent) at the Si-SiO_2 interface. Typical penetration depth of this increase into the substrate is normally less than 1000 Å.

Several positively charged ionic species, such as sodium (Na^+) or hydrogen (H^+) ions can diffuse through the SiO_2 layer with relative ease, for temperatures as low as 150°C. Therefore, oxide passivation is prone to ionic contamination. These ions tend to generate a positive space-charge within the SiO_2 side of the Si-SiO_2 interface, which in turn leads to an increased free electron concentration of the silicon side of the oxide-silicon interface. As a consequence, the surface layer of silicon directly against the SiO_2 layer tends to appear less *p*-type or more *n*-type than would be expected from the dopant impurity concentration. This effect, when coupled with the depletion of the *p*-type boron concentration during the oxide growth cycle may result in the formation of a parasitic *n*-type inversion layer at the Si-SiO_2 interface. This parasitic inversion layer, known as "channeling," is a dominant failure mechanism for integrated devices containing lightly doped *p*-type regions. It can be eliminated by maintaining a relatively high surface concentration of boron within the *p*-type regions (typically $\geq 10^{17}$ atoms/cm^3), and by avoiding ionic contamination of the SiO_2 layer.

Silicon Nitride Passivation

Silicon nitride (Si_3N_4) is far more resistant to ionic contamination than SiO_2. Therefore, it is frequently utilized as a passivating layer for integrated circuit structures whose performance can be easily degraded by surface contamination. This is paticularly true for analog integrated circuits involving MOS devices or operating at low current levels. An additional advantage of silicon nitride over the thermally grown oxide is its superior masking properties against the dopant impurities. Even such dopants as Ga or Al which readily diffuse through SiO_2 can be effectively masked by Si_3N_4.

The silicon nitride passivating layer is most conveniently formed by a pyrolytic deposition process, at a temperature range of 800 to 1000°C.

The deposition is obtained by the decomposition of silane (SiH_4) and ammonia (NH_3) in the presence of hydrogen gas, in accordance with the reaction,

$$3\ SiH_4 + 4\ NH_3 \rightarrow Si_3N_4 + 12\ H_2 \tag{1.17}$$

Silicon nitride is often used to complement the SiO_2 passivation process. In such an application, a layer of Si_3N_4 (typically 1000Å thick) is sandwiched between two SiO_2 layers on the wafer surface, to provide an added degree of surface passivation. The second layer of SiO_2 over the nitride layer is normally formed by pyrolitic deposition. This second oxide layer also serves as mask during the photomasking and etching the contact windows through the nitride layer. The problem of etching contact or diffusion windows through the Si_3N_4 layer will be described further in the next section in connection with photolithography and masking techniques.

Post-Metal Passivation

Sometimes it is advantageous to form an inert dielectric coating over the integrated circuit surface by pyrolytically depositing SiO_2 or silicon glass, commonly referred to as "S-glass," after the metal interconnection step is completed. This post-metal passiviation involves an added oxide deposition and masking step beyond the conventional planar process steps listed in Fig. 1.1. In return, it offers the following advantages:

1. It protects the aluminum interconnections and thin film resistors deposited on the chip surface from accidental scratches during subsequent testing and packaging steps.
2. It allows the use of multiple metal interconnection layers and simplifies the circuit layout.

During the deposition of silicon glass for post-metal passivation, the wafer is maintained at a low temperature (typically 400°C); therefore thin film resistors or the metal interconnections are not affected by the deposition. Normally some boron or phosphorus is introduced into the deposited oxide layer to increase its density.

After the post-metal passivation step, the external contacts to the circuit are obtained by etching windows on the portions of the deposited silicon glass over the metallized bonding pads. The structural diagram of a monolithic circuit chip utilizing the post-metal passivation step is shown in Fig. 1.12.

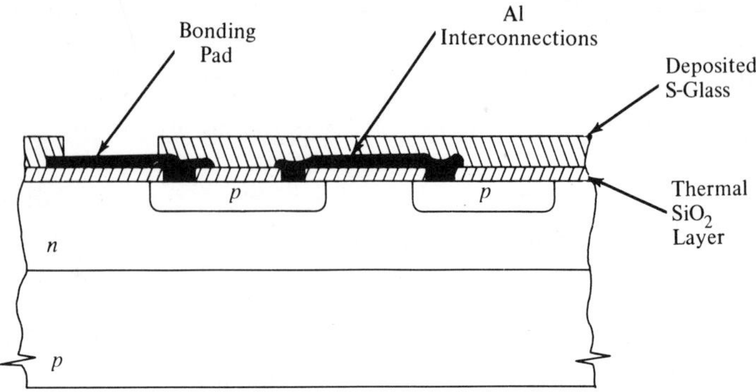

Figure 1.12 Post-metal passivation of wafer surface by pyrolytic deposition of silicon glass.

1.5 PHOTOLITHOGRAPHY

The initial layout of an integrated circuit is normally done at a scale several hundred times larger than the final dimensions of the finished monolithic chip. This initial layout is then decomposed into individual mask layers, each corresponding to a masking step during the fabrication process. The individual mask layers are then reduced photographically to the final dimensions of the integrated unit. The reduced form of each of these patterns is then contact-printed on a transparent glass slide to form a photographic "mask" of the patterns to be etched on to the SiO_2 surface. To facilitate batch-processing, a large number of such masks are contact-printed on the same glass slide, forming a "masking plate." The plate is sufficiently large to cover the entire surface of the silicon wafer to be masked. Thus, in a single masking operation, an array of a large number of identical masks can be applied simultaneously over the wafer surface.

During the masking operation, the mask is transferred from the masking plate to the wafer surface by photolithographic techniques. The wafer surface to be masked is initially coated with a photosensitive coating known as "photoresist" or "resist." The resist-coated wafer surface is then brought into intimate contact with the masking plate and exposed under an ultraviolet light. The portions of the photosensitive resist not covered by opaque portions of the mask polymerize and harden as a result of this exposure. Then, the unexposed parts of the resist can be washed away, leaving a "photoresist mask" on the wafer surface. As a consequence of the masking step, the pattern to be etched through the oxide is transferred to the wafer surface in the form of a hardened, etch-resistant, photoresist pattern.

The photomasking step is followed by an etching step during which the parts of the SiO_2 layer not protected by the exposed resist mask are etched away, forming the diffusion or the contact windows on the oxide. In this process, a buffered hydroflouric acid (HF) solution is used as the etchant. Following the etching step, the photoresist is washed away by a special cleaning solution, and the silicon wafer is ready for the next diffusion step. A similar photomasking step is also used in forming the metal interconnection patterns.

If silicon nitride (Si_3N_4) is used for surface passivation, the etching of the diffusion and the contact windows becomes a more complicated problem. The Si_3N_4 layer can only be etched in boiling phosphoric acid (H_3PO_4 at 160°C). Ordinary photoresist layers cannot form a protective coating for this type of an etch. On the other hand, the SiO_2 layer is resistant to phosphoric acid. Therefore, where silicon nitride is involved, a two-step etch process is utilized as follows: A thin layer of SiO_2 (typically 1500 Å) is deposited on the nitride layer. An initial window is cut over the SiO_2 layer by the conventional masking and etching steps. Then the SiO_2 layer is used as a mask for etching the contact window through the nitride layer, with phosphoric acid as the etchant.

Dimensional Tolerances

In most monolithic circuit structures, the lateral dimensions of the integrated components are determined by the limitations of the photolithographic reduction, masking, and etching processes. The two fundamental limitations on the photolithography process are the alignment and the resolution of the mask patterns.

Since the monolithic circuit fabrication steps require successive application of a number of masks, it is necessary that each new mask applied to the silicon surface should align with the previous set of masks, over the entire surface of the wafer. This requires a good degree of dimensional accuracy associated with the initial layout of the circuit. To ensure this dimensional accuracy, the initial layout is carried out at the largest possible magnification within the capabilities of the photoreduction system. Typically a 500X size is preferred for the initial layout for circuits having final reduced dimensions up to approximately 70 mils square. For larger overall chip dimensions a smaller initial layout scale, such as 400X, may be preferred to avoid optical distortion during the reduction process. Note that the drafting inaccuracies associated with the initial layout are also reduced at the same scale as the original layout. Thus, for example a 0.01 in. dimensional inaccuracy in the initial layout leads to a $\pm 0.5\ \mu$ error in the final dimension, as a 500X reduction.

A possible source of error in the masking step is the tolerance associated with the "step and repeat" process in contact-printing the mask array on the masking plate. The source of error in this case is the mechanical advance mechanism involved. An additional factor limiting the alignment tolerances of a mask set is the accuracy of positioning the mask on the wafer surface. This is done with the aid of a mechanical alignment jig, under a high powered microscope; however it is still subject to some operator error. To minimize the alignment errors at the stage of the masking operation, it is customary to use concentric alignment patterns on successive mask layers. The alignment accuracy for a typical mask set, under production conditions, is approximately $\pm 1\ \mu$, for concentric patterns.

The ability of the mask to define or reproduce fine details on the wafer surface is determined by the resolution of the photomasking step. A good measure of the resolution is the minimum line width needed to resolve, reproducibly two parallel lines spaced one line width apart. The main limitations to the resolving power of the photomasking techniques are the statistical fluctuations in the molecular structure of photographic emulsions, and the diffraction of the light at the mask edges. At the present, the minimum line width that can be resolved under production conditions is approximately $2\ \mu$.

The etching step also introduces random irregularities or errors which tend to reduce the overall mask resolution. The grainy structure of the exposed and polymerized photoresist does not define a true edge during the etching step, but can cause random irregularities of the order of $\pm 0.5\ \mu$ along straight edges. This effect, along with the nonuniform etching properties of the oxide layers also tend to round the sharp corners on the masks, to a typical radius of 2 to 3 μ.

The alignment and the resolution tolerances associated with the photographic reduction and masking processes set a limit on the lateral dimensions of the integrated circuit components. For a given photomasking step, the absolute values of the dimensional tolerances are constant. Therefore, the percentage tolerances associated with the device dimensions become poorer as the dimensions are made smaller. For example, the matching of two 5 μ wide resistors is considerably poorer than those having a width of 10 μ.

1.6 ISOLATION TECHNIQUES

Since all the integrated circuit components are fabricated simultaneously and on the same silicon substrate, it is necessary to employ some means of electrical isolation between them. This is achieved by fabricating the monolithic devices within electrically isolated regions of the substrate, known as isolation "tubs" or "pockets." The electrical separation between each isolation pocket and the rest of the circuit is generally achieved by reverse biased junctions (junction isolation) or by dielectric barrier layers (dielectric isolation). In this section, the fabrication techniques and the electrical properties associated with each of these isolation methods will be examined.

Junction Isolation

Junction isolation techniques are by far the most economical and commonly utilized fabrication method for monolithic circuits. In forming junction-isolated pockets, the nonconductive, or current-blocking properties of a reverse biased *pn* junction is used to obtain electrical isolation. Figure 1.13 shows the cross-section diagram of a junction-isolated pocket containing an *npn* bipolar transistor. In fabricating the electrically isolated device structure of Fig. 1.13, one uses the basic planar fabrication processes outlined in Fig. 1.1, with a few minor additions: Prior to the epitaxy step, a selective n^+ diffusion* is made into the *p*-type substrate. This subepitaxial n^+ diffusion, commonly called a "buried layer," provides a low resistivity ohmic current conduction path from the physical collector contact on the wafer surface to the active collector area directly below the base region (see Section 2.1).

Following the epitaxy and surface oxidation steps, an "isolation mask"

* In this context, (+) is used to imply heavy impurity concentration, not the electrical charge.

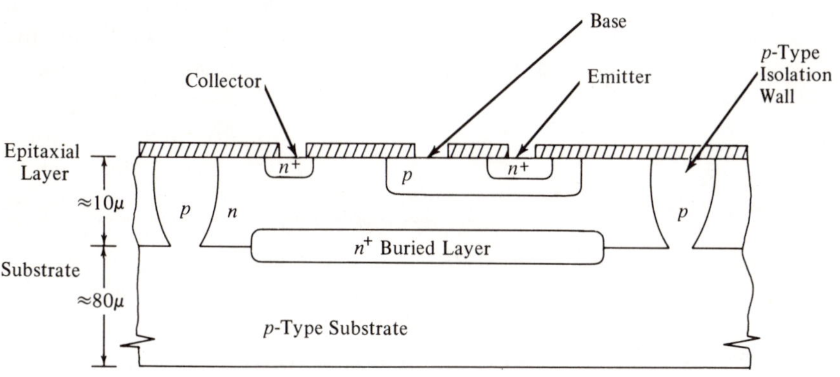

Figure 1.13 Structural diagram of a junction-isolated pocket containing *npn* transistor.

is applied to the wafer surface which opens the diffusion windows outlining the isolation grid. The *p*-type isolation walls are diffused from the wafer surface, through the *n*-type epitaxial layer into the *p*-type substrate. This isolation diffusion step forms a continuous *p*-type wall around a selected region of the *n*-type epitaxial layer. When the substrate and the isolation walls are at negative dc potential with respect to the *n*-type pocket, the reverse biased *pn* junction surrounding the pocket electrically isolates it from the rest of the wafer.

Isolation diffusion is a relatively noncritical diffusion step with the basic requirement that the final depth of the isolation wall be greater than the epitaxial layer thickness. After completion of the isolation diffusion, the transistor base and the emitter regions are formed by respective diffusions into the *n*-type pocket, which serves as the collector of the *npn* transistor.

In device structures which require a thick epitaxial layer, the time required for isolation diffusion can be excessively long. Such a long diffusion cycle may be detrimental to overall device characteristics, particularly due to out-diffusion of the n^+ buried region into the epitaxial layer. To avoid long diffusions at elevated temperatures, it is possible to diffuse the isolation walls simultaneously from both sides of the epitaxial layer. This is commonly referred to as the "double-diffused" isolation wall. It can be achieved by selectively predepositing a subepitaxial *p*-type impurity layer at the episubstrate interface, directly below the path of the topside isolation diffusion. Then, during the drive-in cycle of the isolation diffusion, the two *p* regions diffuse toward each other, and meet within the epitaxial layer to complete the isolation wall.

The double-diffused isolation method can reduce the isolation diffusion time by about 75 percent. However, it requires an additional masking and predeposition step beyond what is needed for the conventional diffused isolation structure. Therefore, its use and applications are generally limited to special device structures, such as simultaneously fabricated *npn* bipolar and *n*-channel junction-gate FETs. This will be discussed further in Chapter 2.

Dielectric Isolation

In certain applications, the parasitic junction capacitances or leakage currents associated with the junction isolation methods may not be acceptable. In such cases, a superior electrical isolation is obtained by insulating each pocket by a dielectric layer, as shown in Fig. 1.14. Normally, thermally grown SiO_2 is used as the dielectric material.

In forming the dielectrically isolated pockets on the wafer surface, a number of alternate fabrication techniques can be utilized. Figure 1.15

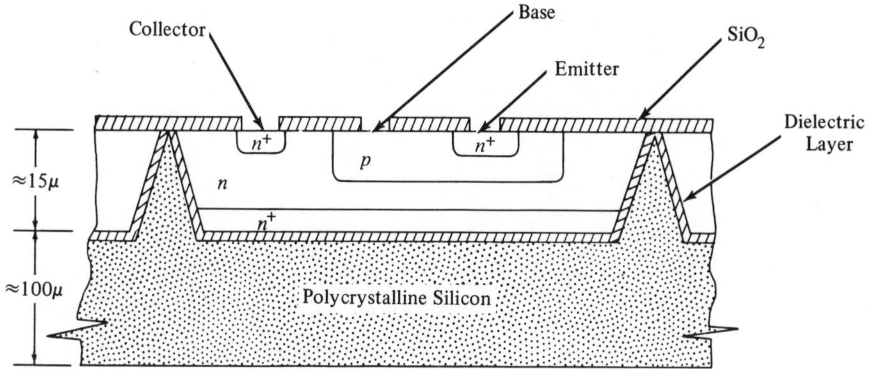

Figure 1.14 Structural diagram of a dielectrically isolated pocket containing *npn* transistor.

shows a typical sequence of fabrication steps in forming the dielectrically isolated single crystal silicon pockets or islands. Starting with an n-type substrate, a nonselective n^+ layer is diffused into the wafer surface. For reasons which will be explained shortly, a $<100>$ oriented crystal is utilized as the starting material, as opposed to the $<111>$ oriented crystal normally used for junction isolation.*

Following the initial n^+ diffusion, the wafer surface is oxidized, and a mirror image mask of the desired isolation grid pattern is applied to the wafer, to remove the oxide along the isolation grid. The exposed silicon surface is then etched by a potassium hydroxide (KOH) based etch. The etchant used in this step etches away the exposed silicon anisotropically, i.e., the etch rate is much faster along the [111] planes than along the [100] crystal planes. This preferential etching results in the formation of a "V" shaped isolation "groove" or "moat" on the wafer surface, as shown in Fig. 1.15(*b*).

Figure 1.16 shows a detailed close-up of the isolation groove obtained in $<100>$ oriented silicon by preferential etching along the [111] planes. The etching process stops when the two [111] planes outlining the isolation groove or the moat intersect at a point at a depth d, below the silicon surface. Since the [100] and the [111] planes in the lattice are oriented at an angle of 35.8° with respect to each other, the depth d of the isolation groove can be determined by the initial oxide cut width W as

$$d_1 = \frac{W}{2}\tan \quad (35.8°) = \frac{W}{\sqrt{2}} \tag{1.18}$$

* During the initial crystal growth, $<111>$ crystal is easier to grow defect-free than the $<100>$ orientation, and is therefore less expensive. This is the main reason for the choice of $<111>$ oriented substrate for most device structures.

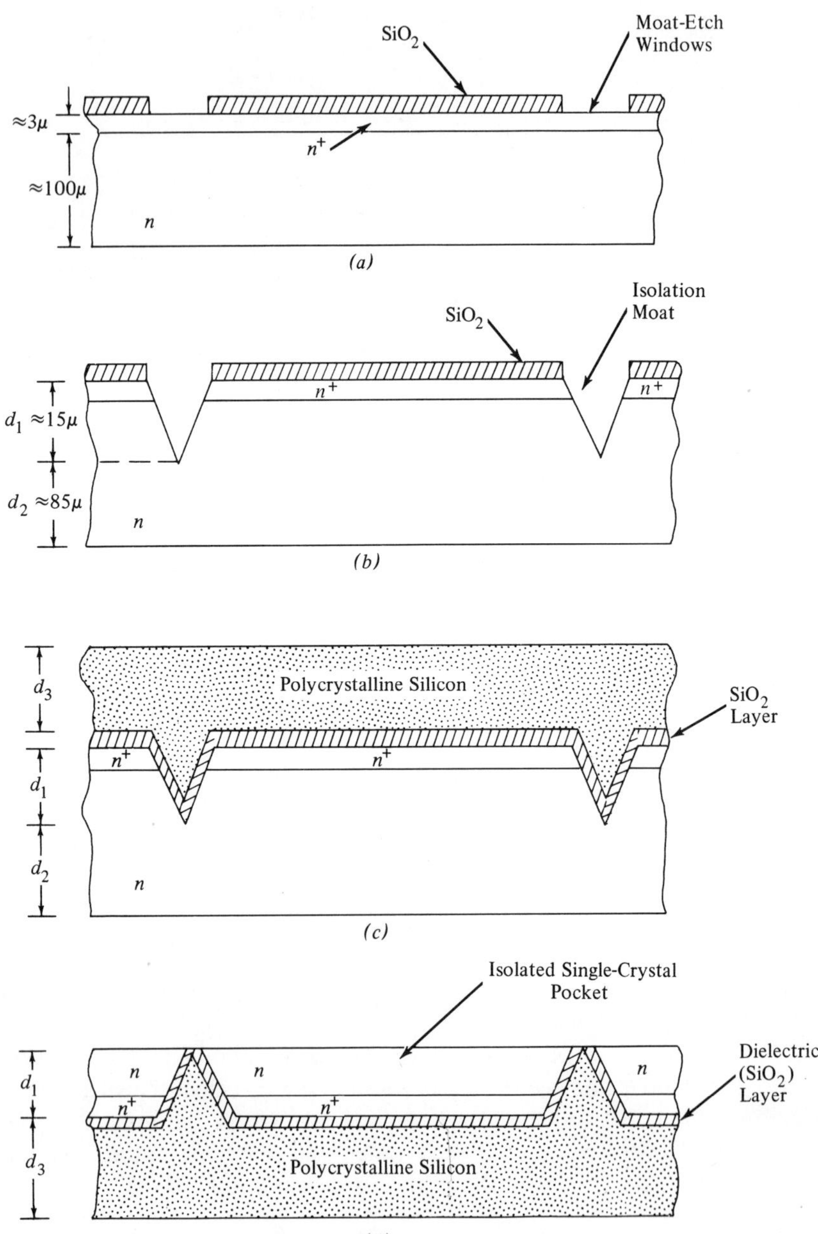

Figure 1.15 Sequence of processing steps in dielectric isolation.

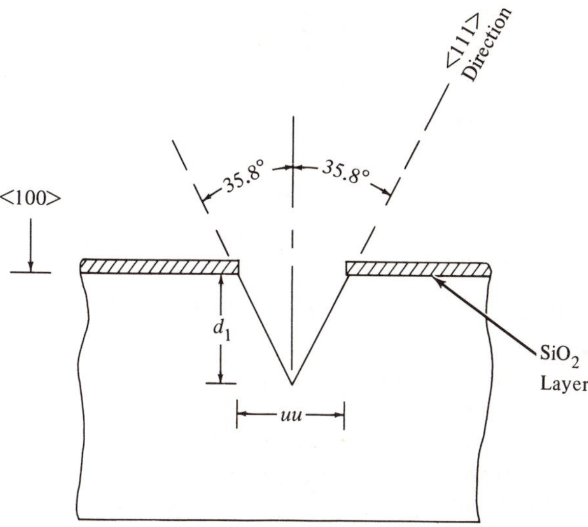

Figure 1.16 Isolation moat formed by preferential etching along [111] planes.

Referring back to Fig. 1.15(*b*), it should be noted that the vertical dimension of the drawings are not to scale, and the depth d_1 of the isolation moat is a small fraction of the total wafer thickness. After the preferential etching step, the exposed silicon is reoxidized, and a thick layer of polycrystalline silicon is deposited over the oxide layer, as shown in Fig. 1.15(*c*). The thickness and the electrical characteristics of this polycrystalline silicon layer is of no consequence since its main function is to serve as a mechanical support for the wafer. Then, the original wafer is flipped around, the bottom surface of Fig. 1.15(*c*) now corresponding to the top of the device structure. Then, the single crystal n layer of thickness d_2 is back-lapped until the isolation grid appears on the wafer surface, resulting in the isolated n-type single crystal pocket shown in Fig. 1.15(*d*). After the isolated pockets are formed, the fabrication of integrated devices within the pockets is completed by a sequence of conventional masking and diffusion steps, resulting in the isolated device structure of Fig. 1.14.

In some cases, it may be desirable to reverse the order of the initial n^+ layer which covers the sides, as well as the bottom of the isolated pocket. Such a structure offers lower ohmic resistance from the topside collector contact to the active collector region directly below the base. However, in this case a larger clearance is required between the p-type base and the dielectric sidewall, to prevent the n^+ layer from touching the base diffused regions.

The basic sequence of processing steps has been known for quite some time. However, the process has become practical only after the development of anisotropic etch techniques[11] which allow very accurate control of the etch depths by the width of the initial oxide window, as given by Eq. 1.18.

1.7 ION IMPLANTATION

Ion implantation is a relatively new semiconductor processing technique[12] which is rapidly gaining wide acceptance. With this technique, the desired impurities are introduced into the silicon lattice by bombarding the silicon surface with high energy impurity ions.

The impurity ions are generated from an ion source and accelerated to energy levels of 50 to 150 keV and are made to impinge on the silicon target placed in the path of the ion beam. The whole operation takes place in vacuum and a mass-spectrometer is used to deflect the undesired impurities off the beam. The remaining ion beam is then focused to a narrow area (typically no more than ¼ in. square) and is scanned across the silicon wafer which serves as the target.

The impurity ions impinging on the silicon surface penetrate into the silicon lattice. This depth of penetration is very shallow, typically of the order of 0.1 to 1 μ. The depth of penetration can be carefully controlled by the choice of target crystal orientation or the beam energy. The semiconductor surface can be masked against the implanted ions by using a metal layer (such as aluminum) or a thick oxide layer as a mask. Thus, ion implanted regions of the silicon surface can be readily patterned in the same manner as the diffused regions, using photomasking techniques.

Ion implantation is performed at or near room temperature; therefore, it does not interfere with the high temperature diffusion cycles. The implantation step is normally done after the diffusion cycles are completed and is followed by a low temperature anneal to remove the structural damage sustained by the silicon lattice during the implantation process. This anneal temperature is typically in the range of 450 to 900°C. In certain applications, ion implantation can also be used to introduce impurities into the silicon prior to a diffusion cycle. In this application it is functionally equivalent to the "predeposition" step described in Section 1.2.

Some specific applications of ion implantation are in the fabrication of shallow-emitter high frequency transistors; in shifting or adjusting the "threshold voltage" of insulated-gate field-effect transistors; and in fabricating high value resistors.[13] Some of these applications will be discussed further in later chapters.

1.8 THIN-FILM PROCESSES

The electrical interconnection of integrated components is achieved by evaporation of conductive thin films on the wafer surface. In addition, resistor and capacitor structures can be formed on the passivated silicon surface by deposition of resistive or dielectric thin-film layers. The term "thin-film" is used to imply approximate film thickness of 1 μ or less, as compared with the larger geometry and thicker films associated with hybrid integrated circuits. The latter are commonly referred to as "thick films," and will be descibed briefly in the following section.

The deposition of dielectric films has been described earlier, in connection with the surface passivation technology (see Section 1.4). In this section particular attention will be given to deposition of conductive and resistive films.

Thin-film resistors formed by deposition and patterning of resistive thin-film layers on the wafer surface have some distinct advantages over the conventional diffused resistors: thin-film resistors in general exhibit lower temperature coefficient, and offer a wide range of sheet resistivity values which can be chosen independently of active device design requirements. Furthermore, in some cases, they can be trimmed to a final value by post-deposition heat treatment or anodization techniques.

For interconnection purposes, aluminum (Al) is the most commonly used thin-film material, because of its high electrical conductivity and good adherence to the SiO_2 surface. In forming the resistor patterns, resistive thin films such as tantalum (Ta), nickel-chromium (Ni-Cr) alloys or tin oxide (SnO_2) are the most commonly used materials. Various deposition techniques can be utilized in forming thin films on the passivated silicon surface. Some of these are outlined below.

Deposition Techniques

Vacuum Evaporation. The passivated substrate, together with the source of the material to be evaporated is placed in a bell jar under high vacuum conditions (10^{-5} to 10^{-6} torr). The material to be evaporated is heated by an electrical element until it vaporizes. Under the high vacuum conditions used, the mean-free path of the vaporized molecules is comparable to the dimensions of the bell jar, therefore the vaporized material radiates in all directions within the bell jar. Some of the vaporized material then deposits on the substrate, which is placed some distance from the source to ensure uniformity of deposition. The substrate is also maintained at an elevated temperature to provide a good adhesion of the deposited film.

Both conductive and resistive films can be deposited by vacuum evaporation. Aluminum, gold, and silver are among the conductive films formed in this manner. Nickel-chromium resistors can also be deposited by vacuum evaporation techniques, except in this case, due to high power densities required to vaporize the source, electron beam bombardment, rather than thermal heating of the source material, is used.

The films deposited by vacuum evaporation exhibit a fine-grained structure. The gain structure of the film becomes finer as the evaporation rate is increased and the angle of incidence of the radiating vapor on the wafer surface is made steeper. For uniformity and repeatability of film properties, a fine-grained structure is desirable.

Cathode Sputtering The sputtering process takes place in a low pressure gas atmosphere. A glow discharge is formed by applying a high voltage (typically 5000 V) between the cathode and the anode sections of the sputtering apparatus. The cathode is coated with the material to be evaporated, and the substrate is attached to the anode or placed within the glow-discharge region. Normally an inert gas, such as argon (A) is used as the sputtering medium. The A^+ ions generated by the glow discharge accelerate toward the cathode, due to the negative cathode potential. When these high energy ions impinge on the cathode, they cause the atoms or the molecules of the cathode to break away, or sputter, from the surface. Then, some of these cathode particles which float away are intercepted by the substrate, and deposit in the form of a thin layer. Under the low vacuum conditions used in sputtering, the mean-free path of the source atoms is much shorter than the source to substrate spacing. Therefore, the deposition rates in the sputtering process are much slower than vacuum evaporation.

By adding small amounts of reactive gases, such as oxygen or nitrogen, to the inert argon atmosphere, the chemical composition of the deposited layer can be modified. This is known as "reactive sputtering" and is particularly useful for tantalum deposition. In this case, nitrogen is used as the reactive gas, resulting in a nitride of tantalum rather than pure Ta in the deposited film. By controlling the amount of reactive sputtering, the sheet resistance of the Ta film can be increased over a 5 to 1 range, with a corresponding temperature coefficient change from +1000 ppm/°C to −60 ppm/°C.

Vapor-Phase Deposition. In vapor or gas phase deposition, halide compounds of the material to be deposited are chemically reduced, and the resulting metal atoms are deposited on the substrate. This basic deposition process very closely resembles the epitaxial growth step of the planar process. Vapor deposition is particularly useful for obtaining thick layers of deposited films (up to 20 μ). It is commonly utilized for forming aluminum oxide ($Al_2O_3 \cdot SiO_2$) dielectric layers or tin oxide (SnO_2) resistive films. The sheet resistance of SnO_2 films can be controlled by introducing Group III or Group V ions (such as In or Sb) to increase or reduce the sheet resistance. In this manner, sheet resistances of 100 to 5000 Ω per square can be obtained.

Plating Techniques. There are two basic kinds of plating techniques used in forming metallic films on semiconductor substrate: electroplating and electroless plating. In electroplating, the substrate to be plated is placed at the cathode terminal of the plating apparatus and is immersed in an electrolytic solutions. An electrode made up of the metal to be plated serves as the anode. When a direct current is passed through the solution, the positively charged metal ions which dissolve into the solution from the anode migrate and plate at the cathode. This method of plating is often used for forming conductive films of gold or copper.

In electroless plating, simultaneous reduction and oxidation of a chemical agent is used in forming a free metal atom or molecule. Since this method does not require electrical conduction during the plating process, it can be used with insulating substrates. Nickel, copper, and gold are among the most common metals which can be deposited in this manner.

Patterning and Etching of Thin Films

With minor modifications, the basic photomasking and etching techniques described in Section 1.5 can be utilized in patterning the thin-film components. One significant exception is the case of multiple thin film layers (such as aluminum interconnections over thin film resistors) where additional care should be taken in the choice of the etchant to ensure that the bottom film is not damaged by the patterning of the top layer. In some cases, this may require the use of a multilayer device structure similar to that shown in Fig. 1.12.

In the case of Al, SnO_2, Ta, or $Al_2O_3 \cdot SiO_2$ layers, patterning and etching can be achieved by direct photoresist techniques. In the case of very thin (300 to 500 Å) nickel-chromium films, an "inverse metal-masking" technique can be used. In this process, a thin layer of metal film (typically copper) is deposited and etched into an inverse, or a negative, of the desired final metal pattern. Then, the desired thin film layer is deposited on this inverse metal pattern. In the final etching step, the inverse metal pattern of the initial metal film is etched away, taking with it the layer of desired metal deposited on it, and only the portions of the desired metal layer which directly adhere to the substrate are left behind.

In a simpler version of the inverse metal-masking technique, a photoresist layer may be used in place of the first metal layer, and on which the desired metal can be deposited. Then, the photoresist can be etched and cleared away to leave behind the desired metal pattern, which adheres directly to the substrate. Since photoresist is an organic polymer, it cannot withstand exposure to high temperatures. Therefore, the application of this inverse photoresist technique is limited to thin-film processes where the substrate is maintained at a low temperature (typically ($\leq$250°C).

Ohmic Contacts

The basic prerequisite for the conductive films used for interconnections is that they should make good ohmic contact with the diffused components or other metallic films deposited on the device surface. A good "ohmic contact" is defined as one that exhibits a linear voltage-current relationship which passes through the origin of the I-V characteristic.

The exposure of contact areas to the ambient atmosphere often results in the formation of parasitic oxide layers over the chip areas to be interconnected. Therefore, to provide good ohmic contact, the interconnecting metal must be chemically active so that it can alloyed or sintered through these parasitic oxide layers. The most commonly used interconnection metal is aluminum, it can be readily alloyed into the silicon substrate to form ohmic contacts. Since Al is a *p*-type dopant (see Table 1.1), to avoid the formation of a nonohmic, rectifying contact the contact areas on the lightly doped *n*-type semiconductor regions are n^+ doped prior to metallization (see for example the collector contacts of the *npn* bipolar transistors of Figs. 1.13 and 1.14). The heavy n^+ doping causes a high degree of damage to the Si lattice at the surface; therefore the parasitic *pn* junction formed by the alloying of *p*-type Al interconnection into *n*-type silicon is very leaky, and nearly ohmic in its conduction properties.

In conventional monolithic circuit fabrication, the alloying of the Al interconnections into silicon is the last step of the planar process. It is normally accomplished by a short heat treatment in an inert atmosphere, typically about 10 min at 500°C.

A troublesome metal interconnection problem can occur in devices which employ two active dissimilar metals in their interconnection scheme. At the interface of two dissimilar metals, parasitic intermetallic compounds and oxides can form. A typical example of this is the intermetallic gold-aluminum compounds forming between the aluminum bonding pads and the gold wires which may be used to connect the chip to the package terminals. These compounds, which are brittle and nonconductive, are commonly referred to as the "purple plague" because of their dark color.

They are a serious detriment to the reliability of integrated circuit connections using gold wire bonds and aluminum pads.

1.9 THICK-FILMS

No clear-cut definition exists for separating the "thick" and the "thin" films. Generally, film layers of greater than 10 μ thickness are referred to as thick-films. The structure and the dimensions of most thick films are not compatible with monolithic circuits. Therefore, their use is mostly limited to hybrid structures.

The resistive thick-films are normally deposited and patterned on a ceramic substrate, using silk-screening techniques. These resistors are composed of a suspension of conductive metal particles in a ceramic matrix, with an organic resin as the filler. Following the silk-screening operation, the substrate is fired at 1400 to 1600°C to ensure a permanent adhesion of the resistive film. Because of their ceramic and metal base, these resistors are known as "cermets." Typical cermet structures are composed of Cr, Ag, or PbO in a SiO or SiO_2 matrix. By proper control of the original resistor composition, sheet resistivities as high as 10,000 Ω/square can be obtained. However, this process in general does not lend itself to tight control of sheet resistance; therefore, where resistor accuracy is desired, the deposited resistor values need to be adjusted by trimming techniques.

The electrical contacts to the resistive thick-films are normally made by gold or nickel interconnections which are deposited by electroless plating techniques, and patterned by photographic etching.

1.10 ASSEMBLY AND PACKAGING

The fabrication of the monolithic circuit on the surface of a silicon wafer represents only one part of the total manufacturing process. Additional assembly and packaging steps are required to make the circuit electrically functional.

Electrical Sorting

The assembly and the packaging steps are not as readily adaptable to batch-processing as the planar process steps described earlier. Therefore, at the end of the fabrication cycle, each finished unit has to be individually mounted and bonded to a separate package and electrically tested to ensure it meets desired performance requirements. Since these final steps of chip assembly, packaging, and testing are rather costly, it is necessary to screen the finished monolithic circuits for defects and electrical failure, while the units are still on the wafer. For this purpose, an automatic probing station is utilized. During the electrical probing operation, the bonding pads on each circuit chip are contacted by needlelike metallic probes, and various current and voltage levels at the terminals of the circuit are measured. Normally, these electrical tests are performed by programmed automatic testers which can perform up to 100 tests/sec on a single chip. If the electrical characteristics of the circuit are not acceptable, an automatic marking pen at the probing station is activated to mark the unit as a reject. When the test sequence for a given circuit is completed, the wafer is automatically indexed; and the probes advance to the next chip. In this manner, the entire wafer is automatically sorted.

Die Separation and Attachments

Following the electrical sorting step, the wafer is scribed by a diamond tool along the rectangular grid separating the individual chips or dice. Then, the wafer is fractured along these scribe lines; the individual chips are physically separated, and the rejects from the electrical sorting step are dis-

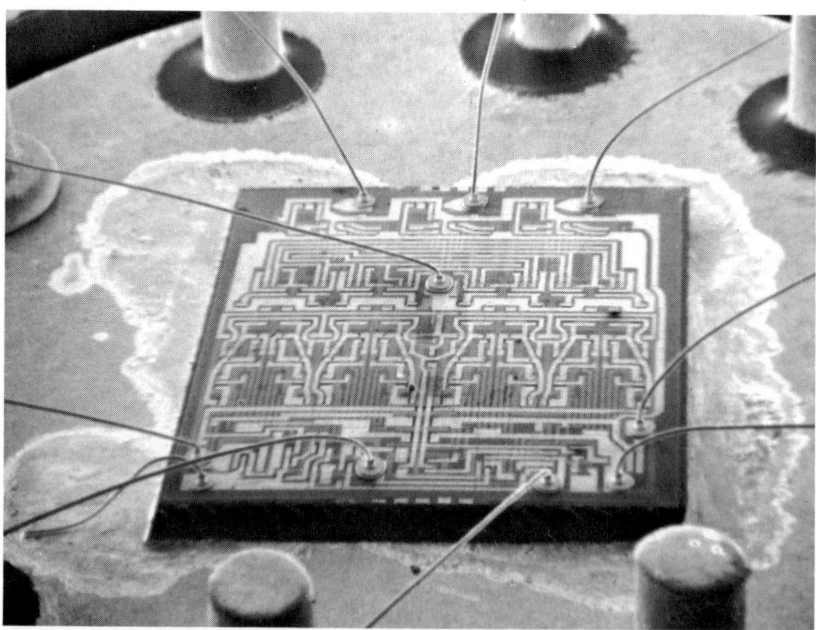

Figure 1.17 Photomicrograph of a monolithic circuit mounted on a TO-5 header. (*Photo: Signetics.*)

carded. At this point, the circuit chip, or die, is ready to be mounted into the final circuit package.

During the die-attach step, the die is normally bonded to a gold plated header by means of a gold-germanium eutectic preform which forms good thermal and electrical contact with the package. After the die is permanently secured in the package, gold or aluminum wire leads are used to connect the bonding pads in the circuit chip to the package leads, or posts. In bonding the wire leads to the chip, any one of several techniques such a "ball-bonding," "stitch-bonding" or "ultrasonic-bonding" can be used. Figure 1.17 is a photograph of an integrated circuit chip mounted on a To-5 header, with ball-bonded gold leads. After the die-attach and lead-bonding steps, the package is sealed and the fabrication process is completed, except for a final cycle of electrical testing and classification

Integrated Circuit Packages

An integrated circuit package is expected to satisfy a large number of partly conflicting requirements: low cost, mechanical strength, high packing density, hermeticity, low parasitic reactances, low thermal resistance, and ease of handling and testing. No single circuit package exists which ideally fulfills these characteristics. For a majority of monolithic analog circuits, the package choice is narrowed down to three most commonly used package types. These are the TO-5, the Flatpack, and the Dual-In-Line (DIP) packages.

The TO-5 package is a modification of the well established transistor package, and is available with four to twelve leads. It can normally handle a circuit chip up to 100 mils square, and has a typical junction to ambient thermal resistance of 75°C/W, for a 50 × 50 mil chip. The typical outer dimensions of the package is 0.18 in. (height) and 0.335 in. diameter.

The flatpack offers smaller overall dimensions (approximately) 0.05 in. high and 0.25 × 0.25 in. square, and a higher packing density than the TO-5. Its thermal resistance is approximately 60 percent higher than the TO-5.

The DIP package is normally available in 8-, 14- or 16-lead versions. It is somewhat larger (0.25 × 0.75 in) and easier to handle than the flatpack. The in-line bent structure makes it convenient for plugging into circuit sockets or automatic assembly onto printed wiring boards. Its thermal resistance is comparable to that of a TO-5 package.

Packaging and assembly are two of the major considerations in the low cost, high volume manufacture of monolithic circuits. As such, an extensive amount of engineering and development effort has gone into these areas within the recent years. The cost element, in particular, has brought about a rash of new assembly and packaging developments, such as "flip-chip," "beam-lead," or "spider-bond" techniques. A detailed analysis and comparison of these technique is a highly specialized subject, and is well covered in the literature.[14]

REFERENCES

1. J. A. Hoerni, U.S. Patent No. 3,025,589, assigned to Fairchild Camera and Instrument Corp., New York.
2. C. S. Fuller and J. A. Ditzenberger, "Diffusion of Donor and Acceptor Elements in Silicon," *J. Appl. Phys.,* **27** (1956): 544–553.
3. F. A. Trumbore, "Solid Solubilities of Impurity Elements in Germanium and Silicon," *Bell System Tech. J.,* **39** (1960): 205–234.
4. D. P. Kennedy and R. R. O'Brien, "Analysis of the Impurity Atom Distribution Near the Diffusion Mask for a Planar P-N Junction," *IBM J. Res. Develop.,* **9**(3) (1965): 179–186.
5. J. C. Irvin, "Resistivity of Bulk Silicon and of Diffused Layers in Silicon," *Bell System Tech. J.,* **41** (1962): 387–410.
6. A. S. Grove, A. Roder, and C. T. Sah, "Impurity Distribution in Epitaxial Growth," *J. Appl. Phys.,* **36** (1965): 802.
7. B. E. Deal, "The Oxidation of Silicon in Dry Oxygen, Wet Oxygen and Steam," *J. Electrochem. Soc.,* **110** (1963): 527.
8. A. S. Grove, *Physics and Technology of Semiconductor Devices,* Wiley, New York, 1967, Chap. 2.
9. B. E. Deal and A. S. Grove, "General Relationships for the Thermal Oxidation of Silicon," *J. Appl. Phys.,* **36** (1965): 3370–3378.
10. S. K. Ghandhi, *The Theory and Practice of Microelectronics,* Wiley, New York, 1968, Chap. 6.
11. H. A. Waggener, R. C. Kragness, A. L. Tyler, "Anisotropic Etching for Forming Isolation Slots in Silicon Beam-Leaded Integrated Circuits," *IEEE Int. Electron Dev. Conf.,* Washington, D.C., Oct., 1967.
12. J. T. Burrill, W. J. King, S. Harrison, and P. McNally, "Ion Implantation as a Production Technique," *IEEE Trans. Electron. Devices,* **ED–14** No. 1 (Jan., 1967): 10–17.
13. J. D. Macdougall and K. E. Manchester, "Implanted Components in Microcircuits," *Proc. Nat. Electron. Conf.,* **25** (Dec., 1969): 140–145.
14. G. Fehr, "A Survey of Today's Microcircuit Packaging," *1970 NEP/CON Proc.* (Feb., 1970), pp. 5. 12–5.20.

2

Integrated Devices: Transistors and Diodes

In monolithic circuits, the device and the design functions cannot be clearly separated from each other. The limitations imposed by planar fabrication techniques, and the strong interdependence of device parameters and parasitics, often require significant design and layout compromises on the part of the circuit designer. Therefore, it is necessary for the designer of an analog integrated circuit to be familiar with the characteristics and the limitations of integrated devices. The purpose of this chapter is to provide a comprehensive overview of the active components in monolithic integrated circuits. It will be assumed that the reader is already familiar with the fundamentals of semiconductor device theory, as applied to discrete devices. In surveying the integrated device and component structures, special attention will be given to relating the electrical parameters to the physical properties of integrated structures, as pertinent to analog circuits.

The bipolar transistor is by far the most significant single component in monolithic circuits. The planar process described in the preceding chapter was initially developed for fabrication of discrete bipolar transistors, and later extended to integrated circuits. Therefore, in going from a discrete to an integrated device, the bipolar transistor structure usually involves the least amount of design compromise. This is particularly true for the *npn* bipolar which is the workhorse of almost all analog integrated circuits.

2.1 *npn* TRANSISTORS

The *npn* bipolar is the key circuit component in the design. Once chosen, it serves as the starting point for the rest of the design and the layout of the circuit. Then, the rest of the components are chosen or designed to be compatible with the basic *npn* structure. Figure 2.1 shows a comparison of the discrete and the integrated planar *npn* bipolar structures. In the case of the discrete device, a heavily doped (low resistivity) substrate is used as the starting material, on which a higher resistivity *n*-type epitaxial layer is grown to serve as the active collector region. The basic difference between

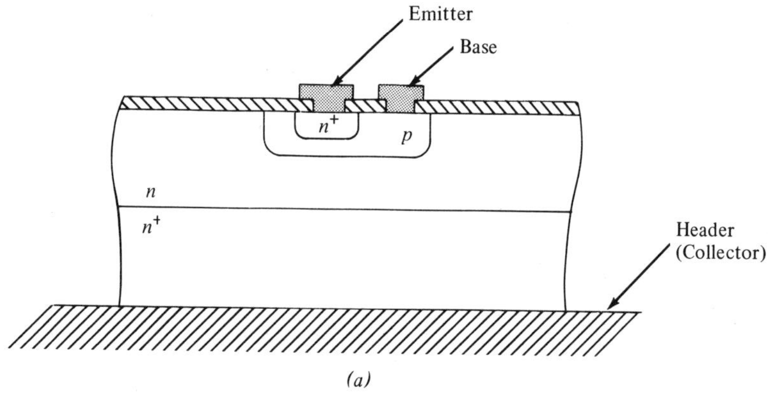

(*a*)

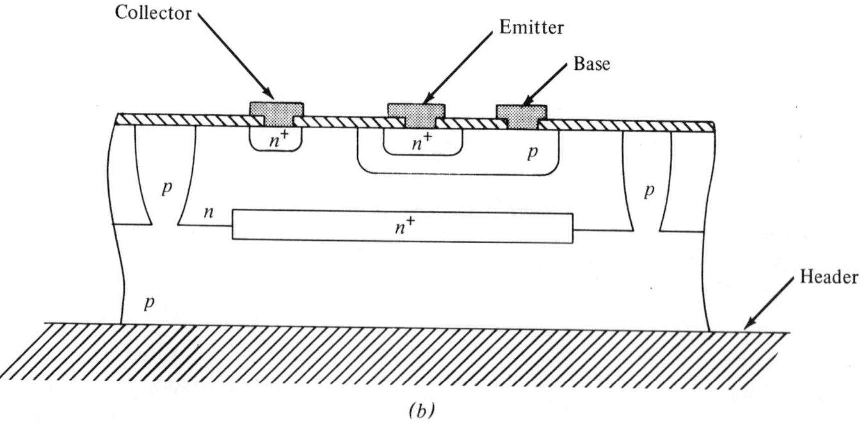

(*b*)

Figure 2.1 A comparison of discrete (*a*) and integrated (*b*) *npn* bipolar transistor structures.

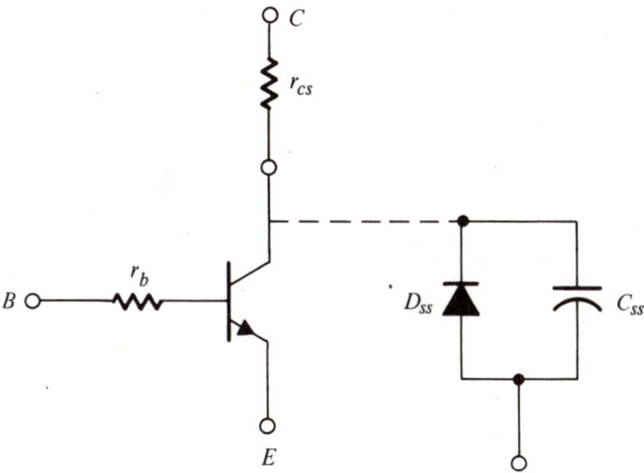

Figure 2.2 Inherent parasitics associated with junction isolated transistors.

the two device structures shown in Fig. 2.1 is that, in the case of an isolated integrated device structure, the collector region is only accessible from the top surface of the wafer. This in turn results in a considerably higher collector series resistance r_{cs}, for the integrated device. The presence of a junction isolation pocket around the collector of the integrated device also introduces two additional parasitics into the structure. These are the parasitic collector-substrate diode, D_{ss}, and the junction capacitance C_{ss} associated with as shown in Fig. 2.2. In the figure, the intrinsic transistor, between the terminals *E*, *B*, and *C* is electrically equivalent to the discrete planar device of Fig. 2.1(*a*). The actual values of r_{cs} and C_{ss} depend on the choice of the lateral geometry and the layout of the device. The subepitaxial n^+ layer in the integrated structure serves a dual purpose: it provides a low resistivity current path from the active collector region to the physical collector contact; and it reduces the possibility of a parasitic *pnp* action between the substrate and the *p*-type base, if the collector-base region of the *npn* were to become forward-biased.

Figure 2.3 shows the typical impurity diffusion profile for an analog integrated circuit transistor, with the impurity concentration (log-scale) plotted as a function of distance, in microns, into the device. The numbers associated with each of the impurity profiles correspond to the order of processing steps. The impurity profile (1) is formed by the out-diffusion of the subepitaxial n^+ layer during the subsequent epitaxy and diffusion steps. Considering a typical junction-isolated device structure with a 10 to 12 μ epitaxial layer thickness, the out-diffusion of the n^+ layer is of the order of 3 to 4 μ, as measured from the nominal epi-substrate interface. Typical sheet resistivity of this subepitaxial layer is approximately 12 to

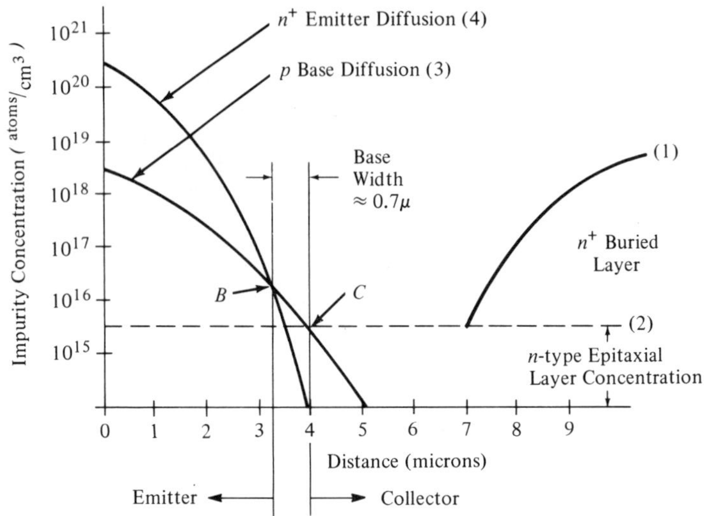

Figure 2.3 Typical impurity profile for an analog integrated circuit transistor.

18 Ω per square. To minimize excessive out-diffusion of this profile, a slow diffusant such as As or Sb is used as a dopant.

The impurity profile (2) is formed by the epitaxial layer, and forms a uniform background into which the subsequent base and emitter diffusions can be made. For analog circuit applications the resistivity of the n-type epitaxial layer is chosen to be in the range of 0.5 Ω-cm to 5 Ω-cm ($N_0 \approx 10^{16}$ at ohms/cm³ to $N_0 \approx 10^{15}$ at ohms/cm³). The particular choice of epitaxial resistivity within this range is normally determined from the base-collector breakdown considerations. The third impurity profile is the p-type base diffusion. This step normally represents an approximate Gaussian distribution, with a typical sheet resistivity range of 120 to 200 Ω per square. As will be described in the following sections, the diffused resistor and the lateral pnp transistors are also formed by this diffusion cycle. The fourth profile is the emitter diffusion. This diffusion normally uses phosphorus as the dopant, and closely approximates a complementary error-function distribution with the surface concentration maintained at or near the solid solubility level. The base-emitter and the base-collector junctions are formed when each corresponding diffusion profile intersects the other one since at these points the diffusing impurity levels are equal to the existing background concentrations. The locations of these junctions are shown as points (B) and (C) in the Figure. The base width W, of the transistor is normally of the order of 0.8 to 0.6 μ, with a typical control tolerance of ± 0.1 μ.

Figure 2.4(a) shows the lateral geometry of a typical integrated small

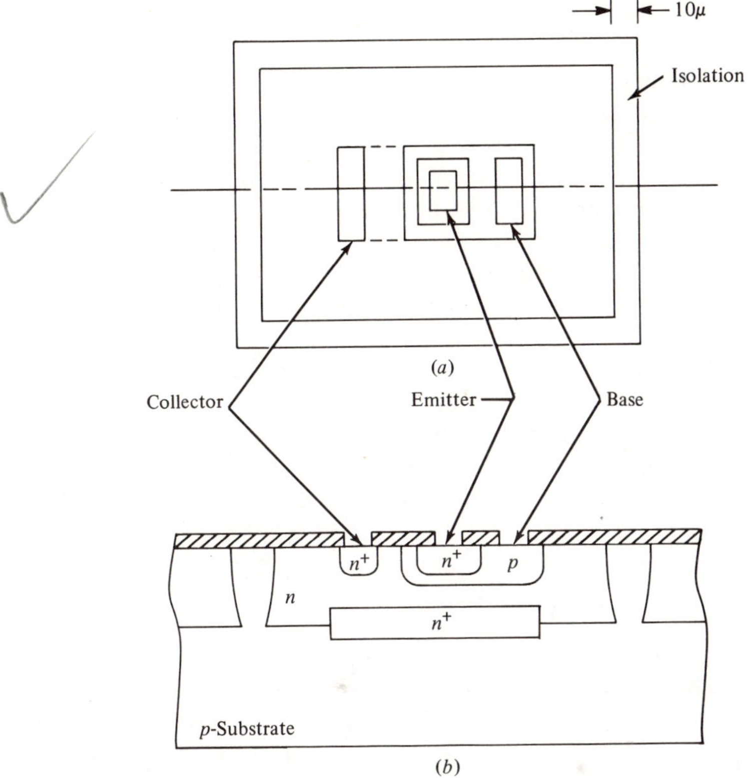

Figure 2.4 Lateral geometry and structural diagram of a small-signal *npn* transistor.

signal *npn* transistor. To give an idea of the lateral device dimensions and tolerances, the device geometry is shown drawn at 500X scale, with each small division of the background grid corresponding to 5 μ. The structural cross-section of the same device, sectional normal to the wafer surface, is given in Fig. 2.4(*b*). Note that, the vertical dimensions in this latter view are not drawn to scale.

The minimum size for the lateral dimensions of the device is limited by two significant factors: masking and mask alignment tolerances, and the side-diffusion effects. Normally, a clearance of approximately 5 μ is left around an oxide contact cut, and the edge of the corresponding diffusion, as is the case for the base and the emitter contacts. This tolerance is left to account for any possible misalignment of the mask during the masking operation, or the over-etching of the oxide window during the subsequent etching step.

The transistor action takes place directly below the emitter region, as indicated in Fig. 2.4(*b*). Therefore, to be able to supply the collector current with a minimum amount of series voltage drop, the collector contact is located as close to the emitter as possible. The distance between the base region and the collector contact is chosen to be significantly more than the respective side-diffusions of the *p*-base and the n^+ collector contact areas (see Fig. 1.7). If this precaution is not taken, the n^+ collector contact region may touch the base, thus resulting in a low collector-base voltage breakdown. For the typical junction depths given in Fig. 2.3, this spacing would be of the order of 15 μ. The subepitaxial n^+ layer is directly located below the base region, and extends to the area directly below the collector contact. The distance between the *p*-type isolation wall and the inner transistor structure is again set by the side-diffusion effects. Since the isolation diffusion is a deep one, it also tends to side-diffuse significantly more than the base. Therefore, it is customary to leave a typical clearance of approximately ($2X_e$) between the isolation wall and the inner transistor structure, where X_e is the epitaxial layer thickness.

Electrical Characteristics

When biased in its active region, the bipolar transistor functions as a nonideal current-controlled current amplifier. A given amount of base current, I_B, injected into the base terminal causes a much larger collector current, I_C, to flow. Figure 2.5 shows a simplified equivalent circuit for an *npn* bipolar transistor which describes its direct current-voltage characteristics. When the transistor is in its active region, the base-emitter junction is forward-biased, and the base-collector junction is under reverse bias. Therefore, the diode D_{BE} in Fig. 2.5 represents the physical base-emitter junction

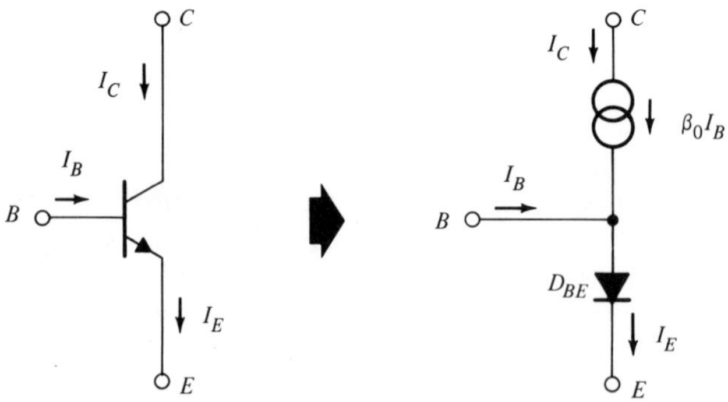

Figure 2.5 Simplified equivalent circuit for transistor dc characteristics.

diode under forward-bias condition. The current I_E through this diode, as a function of the voltage V_{BE} across it, can be expressed by the well-known diode equation:

$$I_E = I_{EO}(e^{qV_{BE}/kT} - 1) \approx I_{EO}\, e^{qV_{BE}/kT} \tag{2.1}$$

where I_{EO} is the reverse saturation current associated with the base-emitter diode
q = electronic charge
k = Boltzmann's constant
T = temperature in °K

Similarly, the base voltage to sustain a given emitter current I_E can be written as

$$V_{BE} = \frac{kT}{q} \ln (I_E/I_{EO}) \tag{2.2}$$

In the simplified equivalent circuit of Fig. 2.5 β_0 is the dc gain in the common-emitter (c-e) configuration and can be expressed as

$$\beta_0 = \frac{I_C}{I_B} = \frac{\alpha_0}{1 - \alpha_0} \tag{2.3}$$

where α_0 is the common-base (c-b) current gain factor defined as

$$\alpha_0 = \frac{I_C}{I_E} = \frac{I_C}{I_C + I_B} \tag{2.4}$$

In terms of the physical current conduction mechanism within the transistor, α_0 can be expressed as the product of a number of device parameters:

$$\alpha_0 = \gamma\beta^* M \tag{2.5}$$

where γ = emitter efficiency
β^* = base transport factor
M = collector avalanche multiplication factor

The emitter efficiency γ is defined as the ratio of the electron current (for the case of an *npn*) injected into the base from the emitter, to the total hole and electron current crossing the base-emitter junction. It can be closely approximated by an expression of the form:[1]

$$\gamma = \frac{1}{1 + (\rho_E/\rho_B)(W/L_E)} \tag{2.6}$$

where I_E is the diffusion length of the minority carriers into the emitter, and ρ_E and ρ_B are the resistivities of the emitter and the base regions within a diffusion length of the junction. As implied by Eq. (2.6), γ approaches unity as the emitter is more heavily doped with respect to the base, and as the base width, W, is narrowed. For an integrated device structure similar to that shown in Fig. 2.3 and Fig. 2.4, the emitter efficiency is of the order of 0.99 to 0.995, for I_E in the low milliampere range.

The base transport factor β^*, is the fraction of the minority carriers injected from the emitter that reach the collector. The average length an excess minority carrier can travel before it recombines is known as the diffusion length, L, and it is given as

$$L = \sqrt{D\tau} \tag{2.7}$$

where D is the carrier diffusion coefficient, and τ is the excess minority carrier lifetime in the base region. If the physical dimensions of the base width W are much shorter than L, the base transport factor can be approximated as

$$\beta^* \approx 1 - \frac{W^2}{2\,D\tau} \tag{2.8}$$

For typical integrated devices, β^* is of the order of 0.995.

The collector avalanche multiplication factor M is the ratio of the carriers entering the collector to the number of minority carriers arriving at the base side of the base-collector depletion layer. Because of the secondary ionization effects within the collector depletion region, M is usually slightly larger than unity. It can be related to the base-collector reverse bias, V_{CB} by the empirical relationship,[1]

$$M = \left[1 - \left(\frac{V_{CB}}{BV_{CB}}\right)^m\right]^{-1} \tag{2.9}$$

where BV_{CB} is the breakdown voltage of the base-collector junction. The exponent ^{m} has the approximate value of 4 and 2 for the *npn* and the *pnp* transistors, respectively.

Typical current-voltage characteristics of an integrated *npn* transistor (similar to that of Fig. 2.3 and Fig. 2.4.) are shown in Fig. 2.6 for common-emitter configuration. The finite slope of the collector current-voltage characteristics is due to the modulation of the effective base width by the widening of the collector-base depletion layer as the collector voltage is increased. This base width modulation effect becomes more significant as W is made smaller, or as the resistivity of the base region is increased with respect to the collector. The rapid increase of the current gain just before the breakdown is reached is the result of the avalanche multiplication within the collector depletion layer, as predicted by Eq. (2.9).

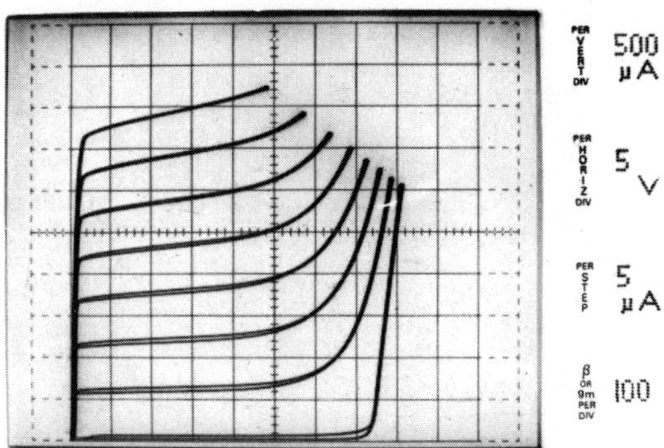

Figure 2.6 Current-voltage characteristics of an integrated *npn* transistor (Impurity profile and geometry similar to those shown in Figures 2.3 and 2.4). (*Photo: Exar Integrated Systems.*)

The transistor dc gain, β_0, depends somewhat on the value of the collector current. The optimum current range at which β_0 reaches a maximum is a function of the lateral dimensions of the device, particularly the emitter area. Figure 2.7 shows the typical β_0 vs. I_C characteristics for an

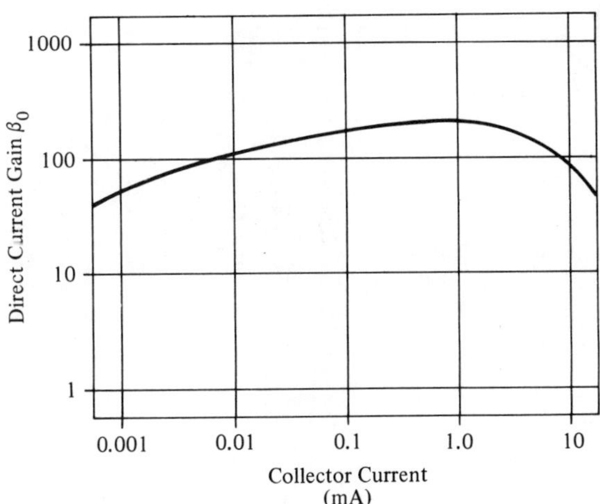

Figure 2.7 Dependence of β_o on I_C for a typical integrated *npn* transistor (see Figures 2.3 and 2.4).

integrated circuit transistor similar to that given in Fig. 2.4. At low current levels, parasitic surface recombination effects are responsible for the β_0 falloff. Therefore, to improve the low current β_0, it is necessary to minimize the surface recombination effects. This can be done through additional surface passivation steps such as silicon nitride deposition, or by reducing the emitter periphery and the emitter-base junction area.

The decrease of β_0 at high currents is due to two dominant factors: decrease of emitter efficiency and emitter-crowding effects. The decrease of emitter efficiency at high currents results from the presence of a large number of excess minority carriers in the base, reducing the effective base resistivity, ρ_B, near the base emitter junction. The emitter-crowding effect is caused by the ohmic drop within the active base region, due to the flow of base current. As a consequence of this, a voltage gradient is created within the active base region, and the edge of the emitter becomes more forward-biased than the bottom of the emitter area. Thus, the emitter region injects carriers preferentially along its periphery, and only the edge of the emitter is electrically active. To reduce the β_0 falloff at high current levels, it is necessary to maximize emitter periphery-to-area ratio, and to minimize the base-spreading resistance. This in turn leads to an interdigitated transistor structure as shown in Fig. 2.8 for high current applications.

β_0 is also a function of temperature, and decreases as the temperature is reduced. This temperature dependence of β_0 is mostly due to the decrease

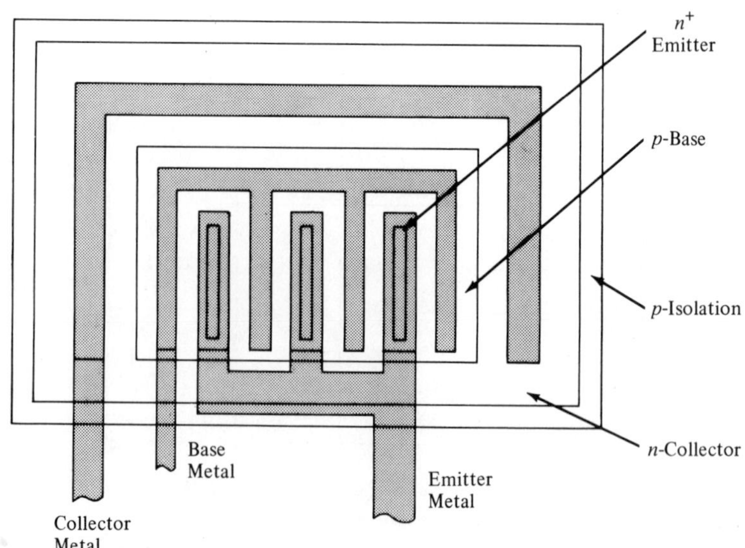

Figure 2.8 Lateral geometry of a high current transistor.

of the minority carrier limetime τ, which in turn decreases the base transport factor (see Eq. 2.8). The typical temperature coefficient of β_0 is of the order of +5000 ppm/°C.

Voltage Breakdown

As the reverse bias across a *pn* junction is increased beyond a critical value, the current through the junction increases rapidly. This critical voltage is known as the junction breakdown voltage *BV*. In silicon, two separate breakdown mechanisms exist. These are the avalanche and Zener breakdowns.

If the impurity concentration on *either* side of the junction is less than $\approx 10^{18}$ atoms/cm^3 the breakdown voltage is determined by the onset of avalanche multiplication. It occurs when the electric field within the depletion layer provides sufficient energy for the free carriers to knock off additional valence electrons from the lattice atoms. These secondary electrons in turn generate additional free carriers, leading to an avalanche multiplication of the free carriers within the depletion layer. This phenomenon is similar to the ionization breakdown in gases. Avalanche breakdown voltage is normally determined by the impurity concentration on the lighter doped side of the junction. Figure 2.9 shows the avalanche breakdown voltage versus concentration for a *pn* junction, where the lighter doped side is assumed to have a uniform impurity distribution. It can be shown that the avalanche breakdown can be approximated by an expression of the form[1]

$$BV \approx (2.7)(10^{12})\, N^{-2/3} \text{ volts} \tag{2.10}$$

where N is the impurity concentration in atoms/cm^3 on the lighter doped side. The avalanche breakdown voltage shows a strong temperature dependence, with a typical temperature coefficient of the order of +300 ppm/°C to +1300 ppm/°C, the dependence being stronger for higher breakdown values.

The second breakdown mechanism is the Zener type, which happens if both sides of the junction are very heavily doped. Zener breakdown is a quantum-mechanical phenomenon where the electrons from the valence band on one side of the junction can "tunnel" to an available energy state in the conduction band on the other side. For junctions which break down at 5 V or less, Zener breakdown is the main conduction mechanism. In analog integrated circuit structures, impurity concentration levels which can give way to Zener breakdown are not normally encountered.

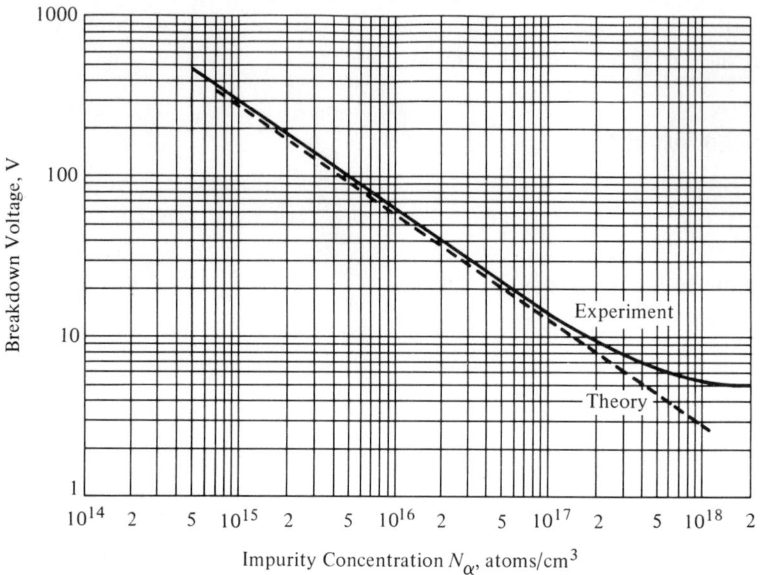

Figure 2.9 Avalanche breakdown voltage vs. concentration for a step junction (After Ref. 1).

The basic breakdown voltages associated with the integrated transistor structures are the base-collector (BV_{CBO}), base-emitter (BV_{EBO}) breakdowns. In designating the transistor breakdown voltage, a triple subscript designation is used where the last subscript O designates that the remaining terminal is open-circuited. Both the BV_{EBO} and the BV_{CBO} values can be closely approximated from the data of Fig. 2.9 once the base and the collector impurity levels are known. The typical values of BV_{EBO} are in the range of 6 to 8 V for integrated transistors, which make the base-emitter breakdown diode a convenient device for dc voltage reference and level shifting within the circuit.

Due to carrier multiplication effects within the base region of the transistor, the collector emitter breakdown, BV_{CEO}, is smaller than BV_{CBO}. It can be approximated as[1]

$$BV_{CEO} \approx \frac{BV_{CBO}}{\sqrt[m]{\beta_0 + 1}} \tag{2.11}$$

where the exponent m has the values discussed in connection with Eq. (2.9).

An additional voltage breakdown mechanism in a transistor is the "punch-through" breakdown. In a narrow and lightly doped base structure, the collector depletion layer can extend through the entire base region into the emitter, thus forming a current path between the collector and the emitter, thus forming a current path between the collector and the emitter. Punch-through is the dominant breakdown mechanism in the punch-through or "super-β" transistors—which will be covered later in this section.

Matching of Device Characteristics

The absolute value tolerances in integrated devices are in general poorer than their discrete counterparts. However, the monolithic components are far superior to discrete devices in the matching and the tracking of device parameters. Table 2.1 shows the typical range of values of β_0, V_{BE} and the base emitter reverse breakdown, V_{EB}, for integrated *npn* transistors. Typical values of the absolute value and the matching tolerances, as well as the temperature coefficients are also listed in the Table. In all cases, the tolerance and the temperature coefficients shown refer to a parameter design value falling halfway between the range limits i.e., $\beta_0 \approx 100$, $V_{BE} \approx 0.65$ V and $V_{EBO} \approx 7$ V).

Frequency Response

For small signal applications, the frequency response of the bipolar transistor can be closely approximated by the "hybrid-π" model shown in Fig.

TABLE 2.1 TYPICAL PARAMETER VALUES AND TOLERANCES FOR MONOLITHIC *npn* TRANSISTORS

Device Parameter	Typical Range of Values	Absolute Value Tolerance	Matching Tolerance	Temperature Coefficient	Thermal Tracking
β_0	50 to 200	±20%	±5%	+5000ppm/°C	±500ppm/°C
V_{BE}	0.6 V to 0.7 V	±20 mV	±1 mV	−2 mV/°C	±10μV/°C
V_{EBO}	6 V to 9 V	±200 mV	±25 mV	+2 mV/°C to +6 mV/°C	±200 μV/°C

2.10. In the equivalent circuit, the parameters r_b, C_c, and r_o are the parasitics inherent to the basic bipolar transistor structure. The base-spreading resistance r_b is the resistance of the current path which the base current must traverse from the physical base contact to the active base region below the emitter. C_c is the collector-base junction capacitance which depends on the area A_B of the collector base junction, as well as the reverse bias V_{CB} across it. Assuming a uniformly doped collector region it can be approximated by an expression of the form,

$$C_c \approx A_B \sqrt{\frac{q\epsilon N_c}{V_{CB}}} \tag{2.12}$$

where ϵ = dielectric constant of silicon
N_c = collector doping concentration
A_B = collector-base junction area

The output resistance, r_o, of the transistor is in common-emitter configuration, and comes about as a result of base width modulation. The collector series resistance r_{cs} and the collector-substrate capacitance C_{ss} are the two additional parasitics particularly associated with the integrated transistor structures (see Fig. 2.2).

The device transconducture, g_m, can be expressed as:

$$g_m = \frac{\partial I_C}{\partial V_{BE}} \approx \frac{\partial I_E}{\partial V_{BE}} = \frac{qI_E}{kT} \tag{2.13}$$

The dynamic resistance of the base-emitter junction, r_π, *is* also related to the direct bias current I_E as

$$r_\pi = \beta_0 \frac{kT}{qI_E} = \frac{\beta_0}{g_m} \tag{2.14}$$

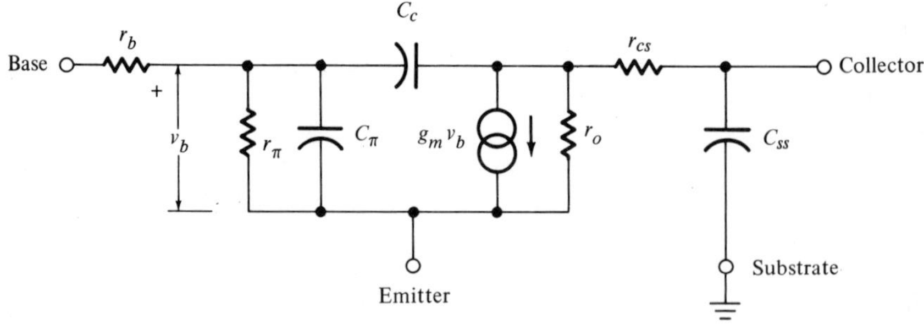

Figure 2.10 Small signal equivalent circuit for an integrated transistor.

C_π represents the total effective capacitance associated with the base-emitter junction and can be expressed as

$$C_\pi = C_t + C_d \tag{2.15}$$

where C_d is the emitter diffusion capacitance corresponding to the charge in transit through the base region and C_t is the physical junction capacitance associated with the emitter-base junction. The latter is analogous to C_c of the collector-base junction, and can be expressed as

$$C_t = A_E \sqrt{\frac{q\epsilon N_B}{V_j}} \tag{2.16}$$

where A_E = area of emitter-base junction
N_B = base impurity concentration at emitter-base junction
V_j = net voltage across junction (i.e., built-in potential minus V_{BE})

The frequency dependence of the common-base current gain α can be closely approximated as a one-pole gain roll-off function

$$\alpha(\omega) = \frac{1}{1 + j(\omega/\omega_\alpha)} \tag{2.17}$$

where $\omega_\alpha = 2\,\pi f_\alpha$ is known as the "alpha cutoff" frequency. Since the common-emitter current gain is related to α by Eq. (2.3), the frequency dependence of β can be, to a first order, approximated as

$$\beta(\omega) = \frac{1}{1 - \alpha} = \frac{\beta_0}{1 + j(\omega/\beta_0\omega_\alpha)} \tag{2.18}$$

However, in the case of graded base transistor structures it has been experimentally observed that an additional phase delay is introduced into the c-e current gain expression, above and beyond that predicted by Eq. (2.18). This "excess phase" effect can be incorporated into $\beta(\omega)$ by means of an emperical phase-delay factor, by rewriting (2.18) as

$$\beta(\omega) = \frac{\beta_0\, e^{-j(m\omega/\omega_\alpha)}}{1 + j(\omega/\beta_0\omega_\alpha)} \tag{2.19}$$

where the excess phase factor m is approximately 0.4 for integrated transistors.

Figure 2.11 shows a plot of the magnitude of $\beta(\omega)$ as a function of

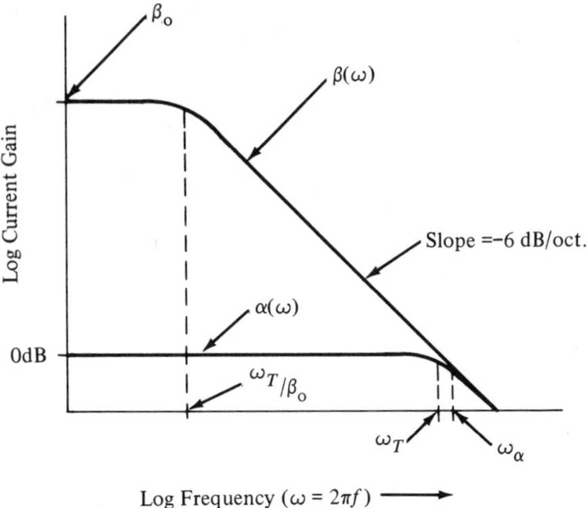

Figure 2.11 Frequency dependence of transistor current gain.

frequency, in a logarithmic scale. The frequency at which the c-e current gain reaches unity is known as the "unity current-gain-bandwidth product," ω_T, for the transistor. It can be shown that ω_T is related to ω_α[2]

$$\omega_T = \frac{\omega_\alpha}{1 + \alpha_0 m} \simeq \frac{\omega_\alpha}{1 + m} \tag{2.20}$$

Due to -6 db/octave roll-off of $\beta(\omega)$ at high frequencies, the beta cut-off frequency, ω_β, can be related to ω_T as

$$\omega_\beta = (\omega_T/\beta_0) \tag{2.21}$$

Normally ω_T is measured with a current-source input into the base terminal; and with the collector ac grounded. Then, referring to the equivalent circuit of Fig. 2.10, and neglecting r_{cs}, ω_T can be expressed as

$$\omega_T \approx \frac{\beta_0}{(C_\pi + C_c)r_\pi} = \frac{g_m}{(C_\pi + C_c)} \tag{2.22}$$

Thus, once the ω_T of the transistor is known at a given bias setting, the base-emitter capacitance, C_π, of Fig. 2.10 can be calculated as

$$C_\pi \approx \left(\frac{qI_E}{kT}\right)\left(\frac{1}{\omega_T}\right) - C_c \tag{2.23}$$

Figure 2.12 shows the typical f_T vs. current characteristics for a small signal integrated *npn* transistor. The reduction of f_T at low current levels is mostly due to the emitter-base transition capacitance C_t. The drop at

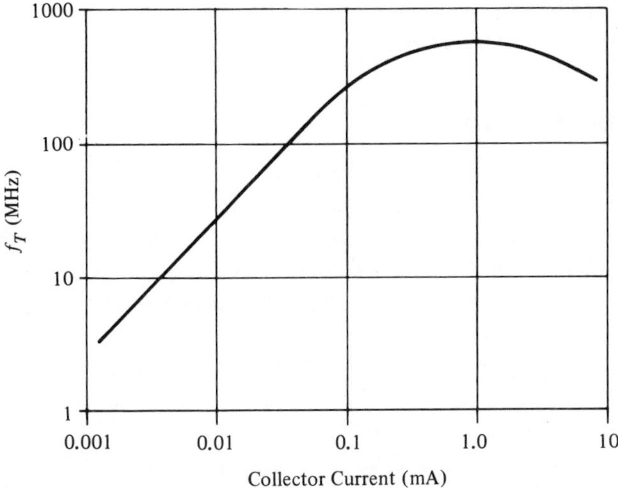

Figure 2.12 Typical f_T vs. current characteristics for an integrated *npn* transistor.

high currents is a result of current-crowding and the conductivity modulation effects in the base region. Table 2.2 shows the typical parameter values for the small signal analog transistor structure of Fig. 2.3 and 2.4.

An alternate figure-of-merit for the high-frequency performance is the frequency at which the unilateral power gain of the transistor drops to unity. This frequency is also known as the maximum frequency of oscillation, $f_{\max}$, for the device. In the case of a bipolar transistor, it can be related to f_T of the transistor as

$$f_{\max} \approx \sqrt{\frac{f_T}{8\ \pi r_b C_c}} \tag{2.24}$$

For most transistors $f_{\max}$ is greater than f_T. For a typical integrated

TABLE 2.2 TYPICAL ELEMENT VALUES FOR ALTERNATING EQUIVALENT CIRCUIT FOR INTEGRATED TRANSISTOR OF FIGURE 2.3 AND 2.4

$r_b = 50$
$r_{cs} \leq 100$
$r_o \approx 100$ K (at $I_C = 1$ mA, $V_{CE} = 5$ V)
$C_c = 0.6$ pF*
$C_t = 1.2$ pF
$C_{ss} = 2$ pF*

*At 3 V reverse bias.

transistor (see Fig. 2.3), the values of f_T and f_{max} are of the order of 400 MHz and 900 MHz respectively.

Punch-Through Transistors

In certain analog circuits such as the input stages of operational amplifiers, it is necessary to have very high input impedance and low input bias currents. For such an application, β_0 of a typical integrated *npn* transistor is not high enough since the actual device design requires a compromise between the current gain and the voltage breakdown requirements (see Eq. 2.11).

It is possible to increase the value of β_0 by improving the base transport efficiency β^* (see Eq. 2.5 and 2.8). This can be done by using a very narrow base structure ($W_B \approx 2500$ Å). However, the collector-emitter voltage breakdown of such a device structure is limited to 2 to 3 V range because of the collector-base depletion layer punching through the active base region into the emitter. The punch-through or "super-β" transistors show a typical β_0 value of 2000 to 5000 at a collector current level of 20 μA, with $V_{CE} = 0.5$ V. The output resistance, r_o, of the punch-through transistor is lower than that of a conventional *npn,* due to excessive base width modulation effects.

In circuit design, punch-through transistors are often used together with a conventional integrated circuit transistor which can provide the necessary voltage protection for it. Figure 2.13 shows a composite connection of a punch-through transistor with a lateral *pnp* transistor where the base-emitter diode of the *pnp* clamps the voltage swing across the punch-through transistor to a level below its breakdown voltage. Thus, the resulting transistor structure has the effective β_0 of the punch-through *npn*, and the breakdown voltage of the lateral *pnp*.

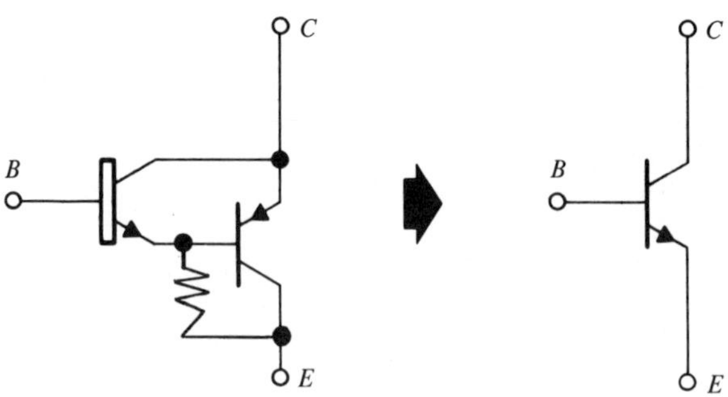

Figure 2.13 Composite connection of a punch-through *npn* with lateral *pnp*.

Punch-through transistors can be fabricated simultaneously with conventional *npn* bipolar transistors, with the addition of one extra masking and diffusion step. After the base diffusion step for conventional *npn* transistors, a special mask is applied which opens the diffusion windows for the emitters of the punch-through devices. Then, an n^+ diffusion is made to form the emitter regions of the punch-through transistors. Normally, the emitter of the punch-through device is only partially diffused during this special diffusion cycle. This is then followed by the masking and the n^+ diffusion steps for the conventional *npn* emitters. Due to the special diffusion step, the emitter of the punch-through transistor is diffused slightly deeper than that of the normal *npn*, resulting in a significantly reduced base width.

2.2 *pnp* TRANSISTORS

Some analog circuit functions may require the use of complementary bipolar transistors. In such cases, it is necessary to fabricate functional *pnp* transistors on the same substrate with the *npn* devices. For this purpose, a number of monolithic *pnp* transistor structures have been developed[3-5] which are totally or partially compatible with the standard *npn* bipolar process technology. Some of these will be reviewed in this section.

Lateral *pnp*

The simplest *pnp* transistor structure which can be fabricated simultaneously with the *npn* bipolars is the lateral *pnp* transistor,[3] which requires no additional masking or diffusion steps. Figure 2.14 shows the plan view and the structural diagram of a lateral *pnp* transistor. The base region of the device is formed by the *n*-type epitaxial layer which serves as the collector of the *npn* transistors. The *p*-type base diffusion of the *npn* is used to form the emitter and the collector regions of the lateral *pnp*; the n^+ emitter diffusion of the *npn* is used to form the n^+ contact region for the *pnp* base. In such a device structure, the transistor action takes place in the lateral direction, i.e., parallel to the device surface. The minority carriers injected into the base diffuse laterally toward the collector region. The carrier transport across the base region is most efficient at or near the surface of the device, where the separation between the collector and the emitter is minimal. This minimum spacing is the effective base width, W_B, for the device. Due to masking tolerances and voltage breakdown requirements W_B is constrained to be of the order of 6 to 12 μ. This value for the effective base width is much larger than that of a vertical *npn* transistor ($W_B \approx 0.8\ \mu$; see Fig. 2.3). Thus, the current gain and the frequency

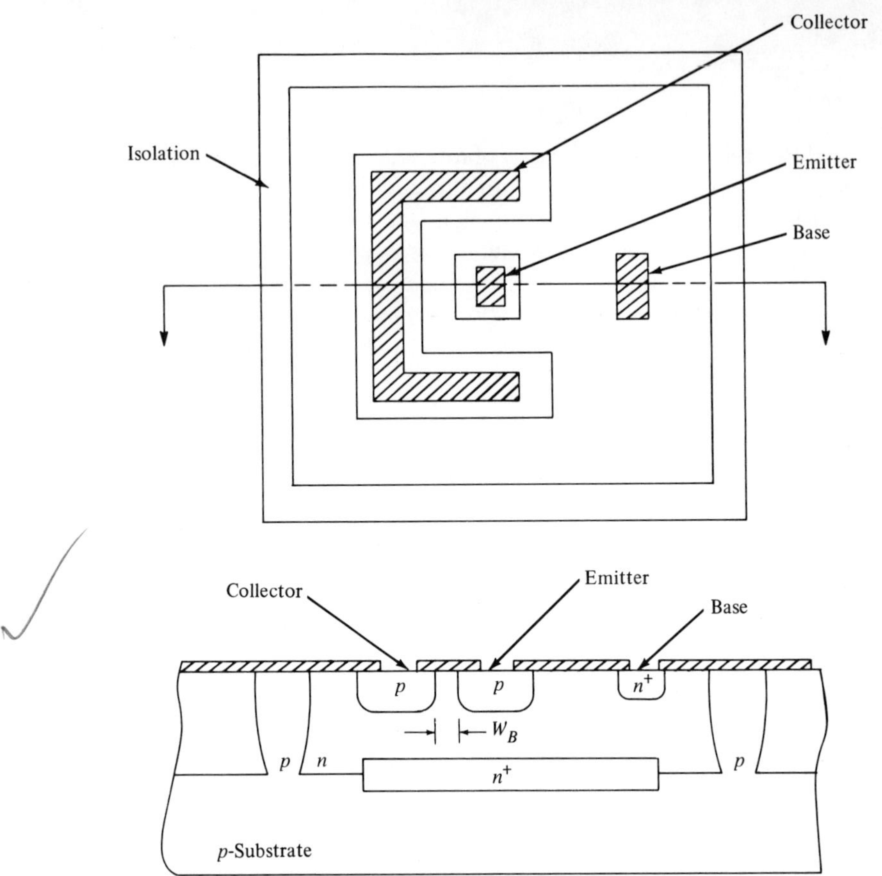

Figure 2.14 Plane view and structural diagram of a lateral *pnp* transistor.

performance characteristics of the lateral *pnp*s are inferior to the *npn* devices. In the fabrication of the lateral *pnp*, an n^+ buried layer is used under the device, along the epi-substrate interface. This is the same $n+$ buried layer used in the *npn* devices. For the lateral *pnp,* this n^+ layer is used to avoid parasitic *pnp* action between the lateral *pnp* emitter and the *p*-type substrate, and to provide a low resistivity path for the *pnp* base current.

The direct current gain terms, α_0 and β_0 are limited by the relatively poor emitter efficiency, wide base-width and the surface-recombination effects. The emitter efficiency of the lateral *pnp* is degraded due to two factors: (1) low impurity concentration in the *p*-type emitter region, and (2) small effective area of the emitter. This latter effect is due to the fact that only the lateral emitter edge facing the collector is active; the holes injected from the rest of the emitter have a much lower probability of reaching the collector. Since most of the current flows near or along the device surface, the lateral *pnp* current gain is also very strongly effected by

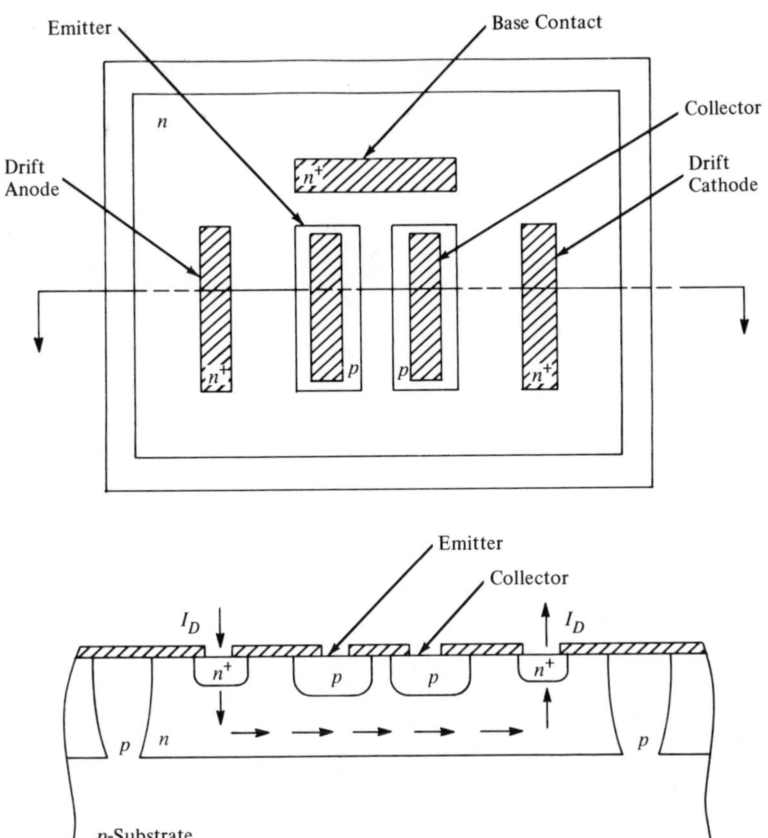

Figure 2.15 Drift-aided lateral *pnp* transistor.

surface recombination effects. Under production environment, the typical values of α_0 for the lateral *pnp* are of the order of 0.6 to 0.95, corresponding to typical β_0 values of 1.5 to 20.

The frequency response characteristics of the lateral *pnp* are dominated by the base transit time. Its cutoff frequency can be approximated as[6]

$$f_T \approx f_\alpha \approx (2.43)\,\frac{D_p}{W_B{}^2} \tag{2.25}$$

where D_p is the diffusion constant of the holes in the base region. For typical lateral *pnp* structures, the values of f_T and f_α are of the order of 2 to 5 MHz.

The lateral *pnp* current gain and the frequency response can be significantly improved by creating a voltage gradient, or a drift field within the base region.[7] This can be achieved using a "drift-aided" *pnp* structure, shown in Fig. 2.15. A bias current is forced through the two drift electrodes at opposite ends of the base, setting up an electric field within the

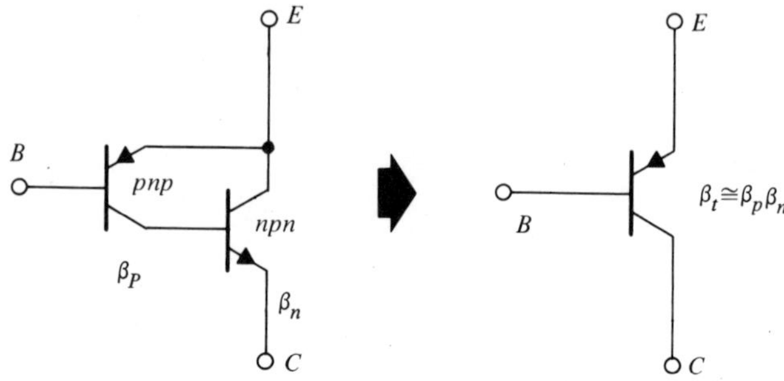

Figure 2.16 Composite connection of complementary transistors.

base region. This bias current has a twofold effect on the electrical characteristics of the device: (1) it preferentially biases the emitter such that it emits only along the edge facing the collector and (2) it sets up an aiding field within the base region to reduce the base transit time for the emitted holes. Using such a device structure, it is possible to extend the f_T of the lateral *pnp* transistor to in excess of 50 MHz.[7]

The low β_0 of the lateral *pnp* can be improved by combining it with an integrated *npn* transistor, forming a composite transistor shown in Fig. 2.16(*a*). The polarity and the electrical characteristics of such a composite device are equivalent to that of a single *pnp* transistor having a higher current gain, shown in Fig. 2.16(*b*). The current gain β_t of the composite device is related to β_p and β_n of the *pnp* and the *npn* transistors as

$$\beta_t = (\beta_p)(\beta_n) \tag{2.26}$$

However, the frequency response of the composite device is determined by the lateral *pnp*.

Substrate *pnp*

A functional *pnp* transistor can also be obtained by using the base region of the *npn* as the emitter, the *n*-type epitaxial layer as the base, and the *p*-type substrate as the collector. Such a device is known as the "substrate *pnp*" and has the structure shown in Fig. 2.17. The substrate *pnp* can be fabricated simultaneously with the *npn* bipolar transistors, without requiring additional masking or diffusion steps. However, since the epitaxial layer thickness is now directly related to the effective base width, W_B, of the *pnp*, a tighter control of the epitaxial layer thickness is necessary (typically to $\pm 1\ \mu$).

The collector of the substrate *pnp* is formed by the *p*-type substrate which is common to the rest of the circuit, and is at all times ac grounded. Therefore, the substrate *pnp* is only available in the grounded collector

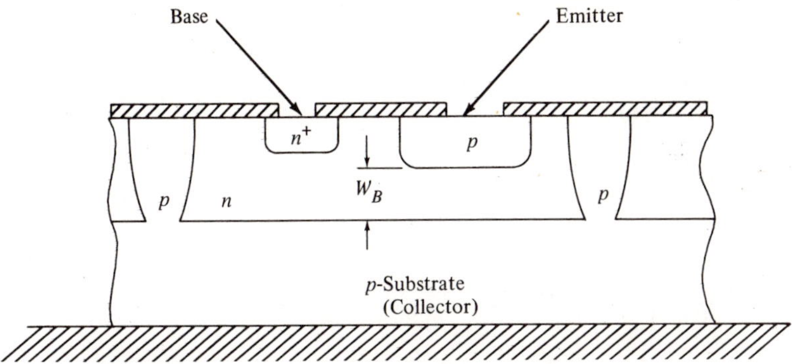

Figure 2.17 Structural diagram of a substrate *pnp*.

configuration, and cannot be used for level shifting or voltage amplification. However, it still provides current amplification, and can be used as a low-impedance output device in class-B complementary stages.

The current gain and the frequency response of the substrate *pnp* is also limited by its relatively large base width ($W_B \approx$ 6 to 10 μ), and relatively poor emitter efficiency. The typical values of β_0 for the device are in the range of 5 to 30. Since the entire bottom surface of the emitter is electrically active, the substrate *pnp* can handle a higher amount of current than the lateral *pnp* of comparable geometry.

Similar to the case of the lateral *pnp*, the frequency response of the substrate *pnp* is also dominated by the base transit time. The f_T of the device is inversely related to the base transit time, and can be approximated as

$$f_T \approx \frac{D_n}{W_B{}^2} \tag{2.27}$$

For typical values of device parameters associated with integrated structures (i.e., $D_p = 20\ \text{cm}^2/\text{sec}$ and $W_B \approx$ 6 to 8 μ), typical values of f_T for a substrate *pnp* are in the 10 to 30 MHz range.

High Performance *pnp* Transistors

In the design of high performance analog circuits, particularly for operation within a radiation environment, the performance characteristics of the lateral or the substrate *pnp* transistors are not acceptable. A number of high performance *pnp* transistor structures have been developed for these applications. However, each of these device structures requires additional processing steps above and beyond what is required for the basic *npn* transistor and are, therefore, limited to special design applications where the added fabrication cost or complexity can be justified.

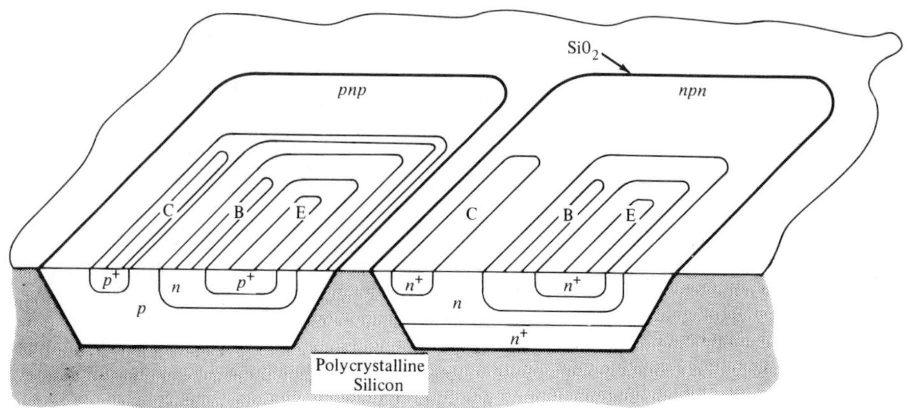

Figure 2.18 Dielectrically isolated complementary bipolar transistors.

The use of dielectric isolation techniques provides an added degree of freedom in fabricating high performance complementary devices by providing access to the backside of the device structure. Therefore, using dielectric isolation methods, it is possible to fabricate high performance *pnp* transistors, simultaneously with the *npn* bipolars, using a device structure as shown in Fig. 2.18. The process steps for such a device structure follow the basic sequence of steps shown in Fig. 1.15, except that prior to an n^+ deposition on the wafer, a selective *p* diffusion is made from the backside. This is a deep ($\approx 20\ \mu$) noncritical diffusion step, forming the *p*-type collector region for the *pnp* transistors. This step is then followed by the selective n^+ predeposition for the *npn* collectors. After moat-etching, polycrystalline silicon growth, and backlapping operations, one ends up with dielectrically isolated *p*- and *n*-type pockets on the silicon surface. The *p*-type pockets do not have a buried layer similar to the n^+ layer below the *npn* collector. However, due to the backside diffusion step in forming the *p*-type islands, the *pnp* collector region has a reverse impurity gradient, being more heavily doped near the bottom of the pocket. This reverse impurity profile results in a low collector series resistance for the *pnp*, and eliminates the need for a separate p^+ buried layer. Once the dielectrically isolated *p*- and *n*-type islands are formed, the device structure of Fig. 2.18 is completed by the following sequence of diffusion steps: (1) *pnp* base, (2) *npn* base, (3) *pnp* emitter, (4) *npn* emitter.

In this process, the *pnp* diffusion steps are interleaved with the standard *npn* diffusion cycles without changing the *npn* impurity profile. Therefore, the electrical characteristics of the resulting *npn* devices remain unchanged. However, the *pnp* devices show a significant improvement over the lateral or the substrate type *pnp* transistors, with the following typical performance characteristics:

$$\begin{aligned} \beta_0 &\approx 50 \text{ to } 100 \text{ (at } I_C = 1mA) \\ LV_{CEO} &\approx 60\ V \\ BV_{EBO} &\approx 9\ V \\ BV_{CBO} &\approx 80\ V \\ f_T &= 150\ MHz \end{aligned}$$

Typical absolute value tolerances, matching and thermal tracking properties of the high performance *pnp* transistors are comparable to those listed in Table 2.1 for the *npn* devices.

2.3 INTEGRATED DIODES

Any one of the semiconductor junctions forming the monolithic circuit structure can be used as a diode. Figure 2.19 shows the basic diodes associated with an integrated *npn* transistor. D_{BE} and D_{BC} represent the diodes formed by the base-emitter and the base-collector junctions; D_{CS} is the collector-substrate diode in junction isolated circuits. The resistors R_b and R_{cs} represent the parasitic bulk resistances between the device terminals and the actual diode junctions. It is worth noting that a parasitic *pnp* transistor is formed when the diode D_{BC} is forward-biased and D_{CS} is reverse biased. This corresponds to the substrate *pnp* transistor described in the previous section, and is shown with dotted lines in Fig. 2.19. This parasitic *pnp* action can be significantly decreased by the presence of the n^+ buried layer which reduces the current gain of the parasitic *pnp*.

The diode current I_D is exponentially related to the voltage V_D applied across the diode as

$$I_D = I_0(e^{qV_D/kT} - 1) \tag{2.28}$$

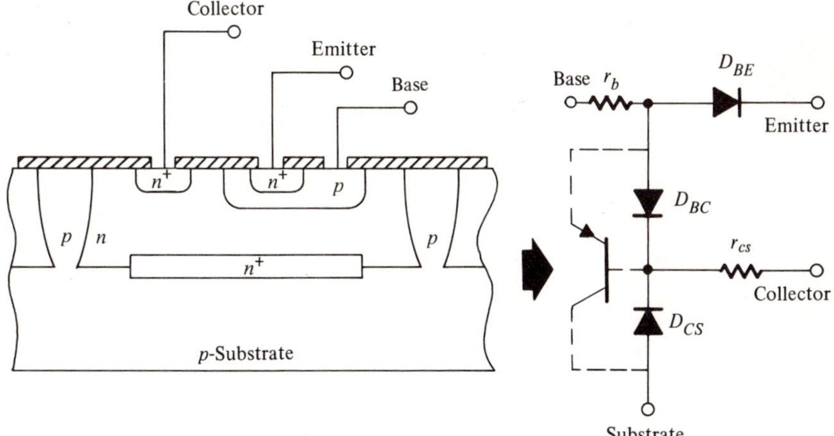

Figure 2.19 Possible diodes available in an isolated *npn* transistor structure.

where I_0 is the reverse saturation current, and is proportional to the junction area. Note that this equation is the same as that described earlier, in connection with the transistor base-emitter characteristics (see Eq. 2.1). For a typical integrated diode the forward (I_D, V_D) characteristics predicted by Eq. (2.28) are valid over six orders of magnitude in current. For a 1 mil-square base-emitter junction area, this range covers current values of 10 nA to 10 mA. At high current densities, due to high level injection effects, the diode forward characteristics can be approximated as

$$I_D \approx I_0 e^{qV_D/2kT} \tag{2.29}$$

The dynamic diode forward conductance g_d can be evaluated by differentiating Eq. (2.28) with respect to V_D, i.e.,

$$g_d = \frac{1}{r_d} = \frac{\partial I_D}{\partial V_D} = \frac{qI_D}{kT} \tag{2.30}$$

Similarly, the diode voltage drop V_D at any given current level can be written as

$$V_D = \frac{kT}{q} \ln\left(\frac{I_D}{I_0}\right) \tag{2.31}$$

For forward current levels of greater than 10 μA, typical values of V_D are in the range of 0.5 to 0.7 V, neglecting the bulk resistances.

The temperature dependence of V_D at a given current level can be expressed as

$$\frac{\partial V_D}{\partial T} = -k/qI_0 \left(\frac{\partial I_0}{\partial_T}\right) \tag{2.32}$$

Note that the temperature dependence of V_D is independent of the current level, I_D. The thermal drift given by (2.32) is a highly predictable and repeatable effect. For the case of the base-emitter diode, which is the most commonly utilized diode connection, the thermal drift of V_D falls within the narrow range of -1.9 to -2.1 mV/°C.

Only two of the diodes, D_{BE} and D_{BC}, associated with the *npn* structure of Fig. 2.19 are readily suitable for circuit applications. The collector-substrate diode D_{CS} is not as useful since its cathode, the substrate, is common to the rest of the circuit. The semiconductor junctions which make up the *npn* bipolar transistor can be interconnected as a diode in any one of the five possible configurations listed below:

1. Base-emitter junction with the collector left open.
2. Base-emitter junction with the collector shorted to base.
3. Base-collector junction with the emitter open.
4. Base-collector junction with the emitter shorted to base.
5. Base-emitter and the base-collector junctions in parallel.

	Diode Connection	Series Resistance	Reverse Breakdown	Parasitic *pnp*
(1)	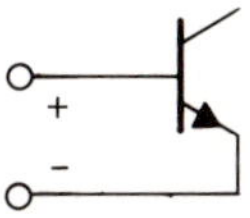	Low ($\approx r_b$)	Low ($\approx$ 7 V)	No
(2)	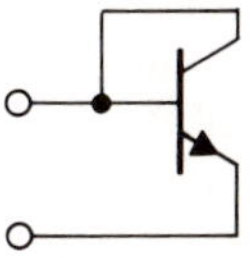	Low ($\approx r_{cs} + r_b/\beta_0$)	Low ($\approx$ 7 V)	No
(3)	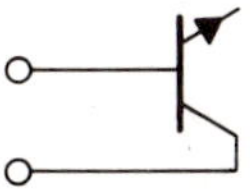	High ($\approx r_b + r_{cs}$)	High ($>$ 40 V)	Yes
(4)	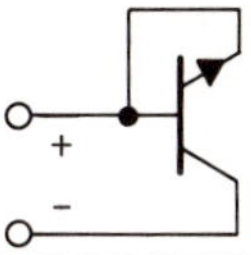	High ($\approx r_b + r_{cs}$)	High ($>$ 40 V)	Yes
(5)	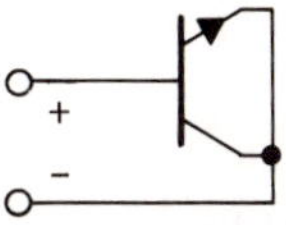	High ($\approx r_b + r_{cs}$)	Low ($\approx$ 7 V)	Yes

Figure 2.20 A comparison of practical diode connections for an *npn* transistor.

Figure 2.20 gives a relative comparison of each of these diode connections with respect to the parasitics and the breakdown characteristics associated with each configuration. The series bulk resistances associated with a given diode structure are in general the most significant parasitics. With respect to the bulk resistances, the diode connection (2) has a distinct advantage over the other configurations since in this connection any parasitic resistance in the base or terminal of the device appears as divided by β_0 of the transistor.

All the diode-connected transistor configurations which include the base-emitter junction have low reverse breakdown voltages (typically 6 to 9 V). The diode configurations which utilize the collector-base junction under forward bias can have parasitic *pnp* action between the base, collector, and the substrate regions (see Fig. 2.19).

The diode connection (2) of Fig. 2.20 combines two desirable electrical properties: low series resistance and no parasitic *pnp* action to the substrate. Therefore, it is the most commonly used configuration for integrated circuits, as long as its low breakdown voltage does not present a problem.

Avalanche Diodes. The avalanche breakdown characteristics of the junctions can be used for voltage reference or the dc level-shift purposes. Typical breakdown voltages associated with each of the five diode connections are listed in Fig. 2.20. The base-emitter breakdown which falls within the 6 to 9 V range is the most commonly used avalanche diode since its breakdown voltage is compatible with the voltage levels available in analog circuits. For this purpose, either one of the diode configurations (1) or (2) can be utilized. Each of these diodes has a reverse breakdown resistance approximately equal to r_b of Fig. 2.19.

The avalanche breakdown voltage BV_{EB} associated with the base-emitter junction shows a positive temperature coefficient, typically in the range of +2 mV/°C to +5 mV/°C. Since the thermal drifts of the diode forward voltage V_D and BV_{EB} are in opposite directions, it is possible to partially compensate the thermal drift of an avalanche breakdown diode by connecting a forward-biased diode in series with it. The resultant composite diode has a breakdown voltage of $(V_D + BV_{EB})$, with a significantly reduced temperature coefficient. In a monolithic circuit, this partial compensation can be obtained with a minimum increase of chip area by connecting two transistors back-to-back in the diode connection (2) as shown in Fig. 2.21. Since both transistors now have their collector and base regions in common, they can be designed as a single transistor with two separate emitters, as shown in the Figure.

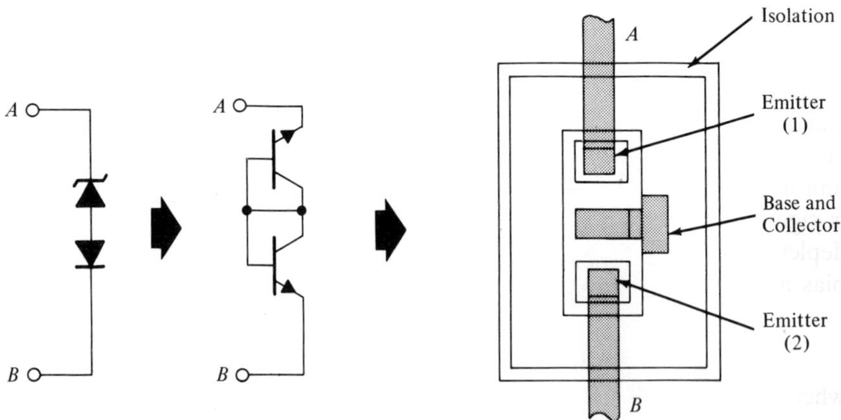

Figure 2.21 Temperature-compensated avalanche diode and its circuit layout.

Field-Effect Transistors

The principle of operation of the field-effect transistor (FET) differs significantly from that of the bipolar type. In the case of the FET, the current transport or conduction mechanism within the device relies solely on majority carriers. Therefore, the name "unipolar transistor" is also used interchangeably with the name FET, to distinguish it from the bipolar transistor, where both majority and minority carriers actively participate in the current transport process.

A detailed coverage of the FET device theory is readily available in the literature,[8-11] and will not be repeated here. Instead, this section is aimed at covering the salient features and properties of integrated FET structures. Since the *npn* bipolar transistor is by far the most important integrated device for analog circuits, the main emphasis will be limited to those FET structures which are readily compatible with the *npn* bipolar technology. The FET is a voltage-controlled device where the current conduction between source and drain regions is controlled or modulated by means of a control voltage applied to the gate terminal. Depending on the physical structure of the gate region, the FETs can be classified into two categories: (1) junction-gate (JFET) and (2) insulated gate (IGFET) devices. In the following sections, the electrical properties of each of these will be examined separately.

2.4 JUNCTION GATE FET

Figure 2.22 shows a cross-section diagram of an integrated JFET with *n*-type channel region. In the structure of such a device, the reverse bias is applied to the gate-channel depletion layer so as to extend it into the channel region and modulate the effective width of the conductive path between the source and the drain. If we assume that the *n*-type channel is uniformly doped and that the impurity concentration of the gate regions is significantly higher than the channel, the thickness of the gate-channel depletion layer d, extending into the channel, can be related to the reverse bias across the junction as

$$d(cm) = \sqrt{\frac{2\epsilon V_{GC}}{qN_D}} \tag{2.33}$$

where N_D = channel impurity concentration (carriers/cm^3)
ϵ = dielectric constant of silicon (farads/cm)
V_{GC} = total dc bias across gate-channel (volts) junction

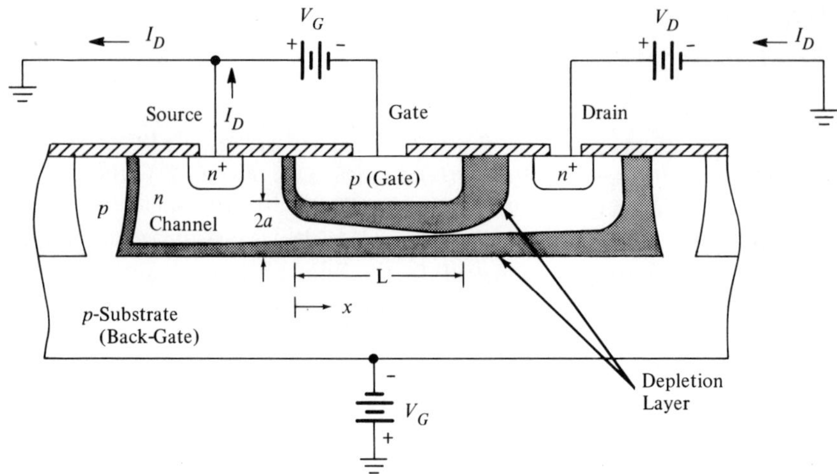

Figure 2.22 Structural diagram of an *n*-channel JFET (cross-hatched regions denote gate-channel depletion layer).

The reverse bias across the junction which would cause d to extend into the channel from both gates far enough to totally deplete or "pinch" the channel is known as the pinch-off voltage V_p. In other words, for $V_G = V_p$, the depletion layer associated with each gate is exactly equal to the channel half-width a. From Eq. (2.33), V_p can be written as

$$V_p = \frac{qN_D a^2}{2\epsilon} \tag{2.34}$$

Referring to the dc voltage levels defined in Fig. 2.22, for low values of the drain potential, the ohmic drop along the channel due to the flow of the drain current I_D is negligible, and the JFET operates as a voltage-controlled resistor. In this mode of operation, the effective source-drain resistance R can be given as

$$R = R_0 \frac{1}{\left(1 - \sqrt{\frac{V_G}{V_p}}\right)} \tag{2.35}$$

where R_0 is the bulk resistance of the channel, with zero depletion layer width. In terms of channel dimensions and resistivity, it can be expressed as

$$R_0 = \frac{\rho L}{2\ az} \tag{2.36}$$

where ρ = resistivity of the channel region
z = depth of the channel, measured normal to dimensions L and a

At any given gate bias $V_G < V_p$, as the drain voltage is increased, the drain current also increases, thus increasing the ohmic drop along the channel. At any point along the channel, this voltage drop adds to the net bias across the gate-channel interface, and thus causes the depletion region to extend further into the channel, in the vicinity of the drain. Consequently for drain voltages in excess of the pinch-off voltage, a space-charge region is formed near the drain end of the channel. This space-charge layer then causes I_D to reach a saturation level and be relatively insensitive to the further increase of the drain potential. This is known as the "pinched" operation of the FET, where the device functions as a voltage-controlled current source. For operation below the drain current saturation (i.e., $V_D \leq (V_p - V_G)$), the drain current-voltage characteristics can be approximated as

$$I_D = \frac{2\,azV_p}{3L}\left[3(V_D/V_p) - 2\left(\frac{V_D + V_G}{V_p}\right)^{3/2} + 2\,(V_G/V_p)^{3/2}\right] \qquad (2.37)$$

Figure 2.23 shows the typical I_D vs. V_D characteristics of a JFET, with gate bias V_G as a parameter. The region of validity of Eq. (2.37) corresponds to the area to the left of the dotted line where the net voltage

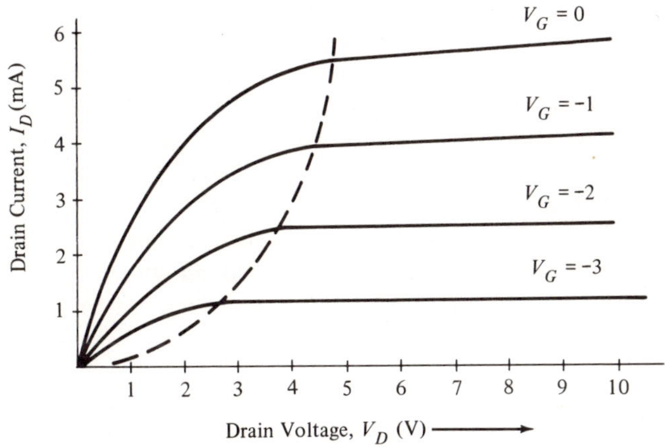

Figure 2.23 Current-voltage characteristics of a typical JFET.

across the gate-drain junction is less than the pinch-off voltage. For higher values of the drain voltage, the first order theory predicts a total saturation of I_D. However, in practical devices, I_D still exhibits a slight increase with increasing V_D, thus leading to a nonzero value of dynamic drain conductance, g_d. This effect comes about due to the modulation of the effective channel length L by the space-charge region near the drain, in a manner analogous to the base-width modulation effect in bipolar transistors. It can be shown that this finite conductance beyond pinch-off can be approximated as[9]

$$g_d = \frac{\partial I_D}{\partial V_D} \simeq \frac{2\, I_{DO}}{\pi L(V_{DG} - V_p)} \tag{2.38}$$

where I_{DO} = saturation value of I_D predicted by Eq. (2.37) for a given V_G

V_{DG} = Net dc bias across gate-drain junction $= V_D + V_G$

The saturation value of the drain current can be expressed from Eq. (2.37) by setting $(V_D + V_G = V_p)$ as

$$I_{DO} = \frac{V_p}{3\, R_0}\left[1 - \frac{V_G}{V_p} + 2\left(\frac{V_G}{V_p}\right)^{3/2}\right] \tag{2.39}$$

The device transconductance, g_m, for pinched operation can be derived from the above equation:

$$g_m = \frac{\partial I_D}{\partial V_G} = \frac{1}{R} = \frac{1 - \sqrt{\dfrac{V_G}{V_p}}}{R_0} \tag{2.40}$$

The parasitic bulk resistances R_S and R_D at the source and the drain end of the channel, between the active channel region* and the external source and drain contacts, introduce an additional ohmic drop within the FET. The exact values of R_S and R_D depend on the channel resistivity and the geometrical layout of the device. For practical FET structures compatible with monolithic circuits, the values of R_S and R_D are in the range of 30 to 80 Ω. When the FET is operated in its pinched region, R_D is in series with the large dynamic output resistance r_d (where $r_d = 1/g_d$) and has negligible effect on device performance. However, parasitic source resistance R_S provides a degenerative feedback between the source and the gate terminal of the device and reduces the available transconductance of the device as

$$(g_m)_{\text{actual}} \approx \frac{g_m}{1 + R_S g_m} \tag{2.41}$$

where g_m is the ideal transconductance predicted by Eq. (2.40).

* The active channel region is that section of the channel directly between the two gate regions.

Breakdown Characteristics

The high voltage capability of the JFET is determined by the avalanche breakdown of the gate-channel junction, BV_{GC}. The drain-source breakdown voltage BV_{DS} is slightly less than the gate-channel breakdown due to the additional reverse bias provided by the gate voltage (see Fig. 2.22). It can be expressed as

$$BV_{DS} = BV_{GC} - V_G \tag{2.42}$$

For typical integrated FET structures, using the *n*-type collector region of the epitaxial *npn* transistor, the gate-channel breakdown is the same as the collector-base breakdown BV_{CBO}, associated with the integrated *npn* transistors.

High Frequency Characteristics

For operation of the JFET in its pinched region (i.e., $V_D + V_G > V_p$), the frequency performance of the device can be closely approximated by the ac equivalent circuit of Fig. 2.24. In the Figure, R_S represents the parasitic bulk resistance in series with the source contact. For typical integrated JFET structures R_S is of the order of 30 to 80 Ω. The capacitances C_{gs} and C_{gd} are the gate-source and the gate-drain capacitances, and g_d is the dynamic drain conductance defined in Eq. (2.38). The values of C_{gs} and C_{gd} can be closely approximated from the junction capacitance expressions of Section 2.6 and from the knowledge of the FET gate geometry. In the actual device layout, the drain area is made as small as possible

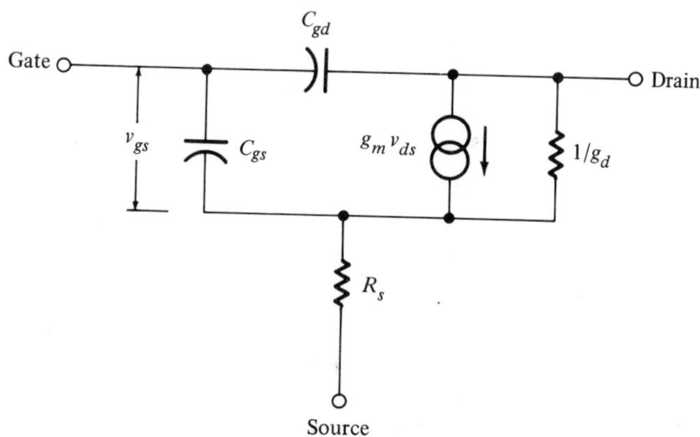

Figure 2.24 An ac small-signal equivalent circuit for a FET.

to minimize the C_{gd} since this capacitance provides parasitic coupling between the drain and the gate terminals and reduces the frequency capability of the FET in a manner similar to the C_c of the bipolar transistors.

A useful figure-of-merit for the high frequency capability of an FET is the transconductance cutoff frequency, f_c. This frequency is reciprocally related to the time t_c needed for the drain current to make up the change in the total charge in the gate-channel depletion region. The transconductance cutoff frequency can be written as

$$f_c = \frac{1}{2\pi t_c} \approx \frac{g_m}{2\pi C_g} \tag{2.43}$$

where C_g is the total capacitance seen looking into the gate terminal. For a uniform channel, doping f_c can be related to the intrinsic device parameters as

$$f_c = \frac{\rho}{2\pi\epsilon}\,(a/L)^2 \tag{2.44}$$

In a typical integrated JFET structure with a 5 Ω-cm channel resistivity and having a channel width of 3 μ and a length of 10 μ, f_c is of the order of 150 to 200 MHz.

Integrated JFET Structures

For monolithic integrated circuits, the most useful JFET structures are those which are compatible with the *npn* bipolar technology, and can be fabricated simultaneously with *npn* transistors. Figure 2.25 shows the layout and the cross-section of an *n*-channel JFET structure which is readily compatible with the bipolar diffusion schedules. This device uses the *n*-type epitaxial collector region of the *npn* transistor as the channel of the FET. Similarly the *p*-type base diffusion of the *npn* transistor is used to form the control gate with the source and the drain contacts formed by the n^+ emitter diffusions for the *npn* transistor.

To obtain a narrow channel width without degrading the *npn* bipolar breakdown characteristics, a p^+ subepitaxial layer is diffused under the FET gate region, as well as under the isolation walls. Then, during the isolation diffusion, this subepitaxial p^+ layer out-diffuses into the epitaxial layer and reduces the effective channel width of the *n*-channel FET. In such a structure, the *p*-type substrate is also a part of the FET gate. However, since the substrate is common to the rest of the circuit, it cannot be used as a control terminal. Therefore, only the top gate functions as the control electrode, and the bottom one is at all times connected to a fixed negative potential, with respect to the rest of the device.

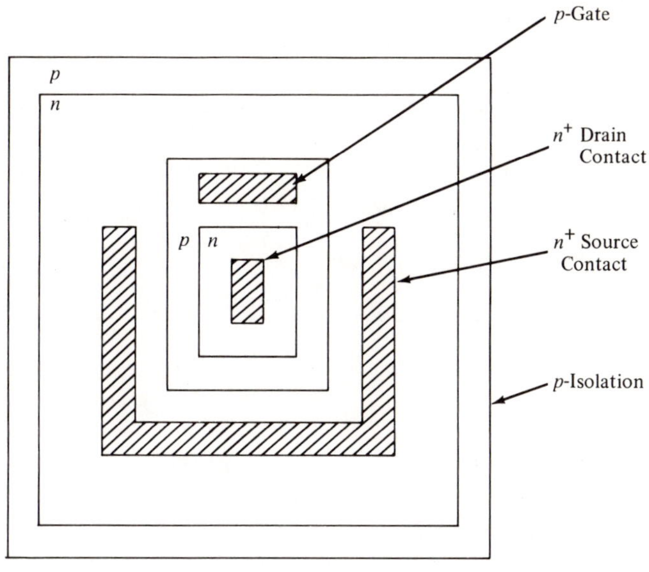

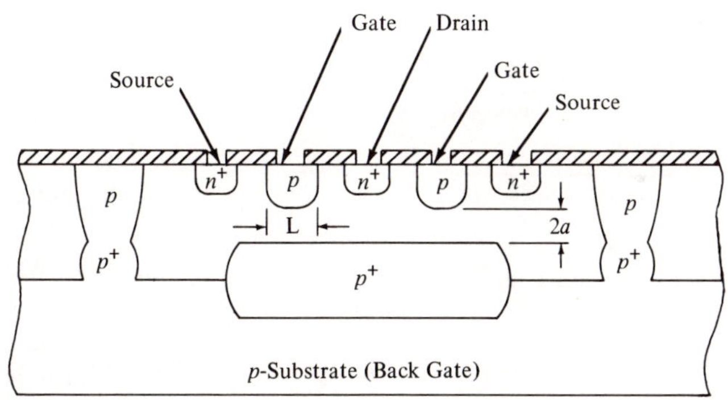

Figure 2.25 Layout and cross-section of *n*-channel JFET compatible with *npn* bipolar transistors.

As shown in Fig. 2.25, the control gate has to be electrically disconnected from the substrate. This can be achieved using the continuous gate structure shown in (*a*), which totally surrounds the drain. The dimensions of the device, and particularly the channel (width/length) ratio are determined by the saturation current (I_{DO}) and the transconductance requirements.

Figure 2.26 shows two additional JFET structures which are also compatible with the *npn* bipolar technology. The structure of Fig. 2.26(*a*)

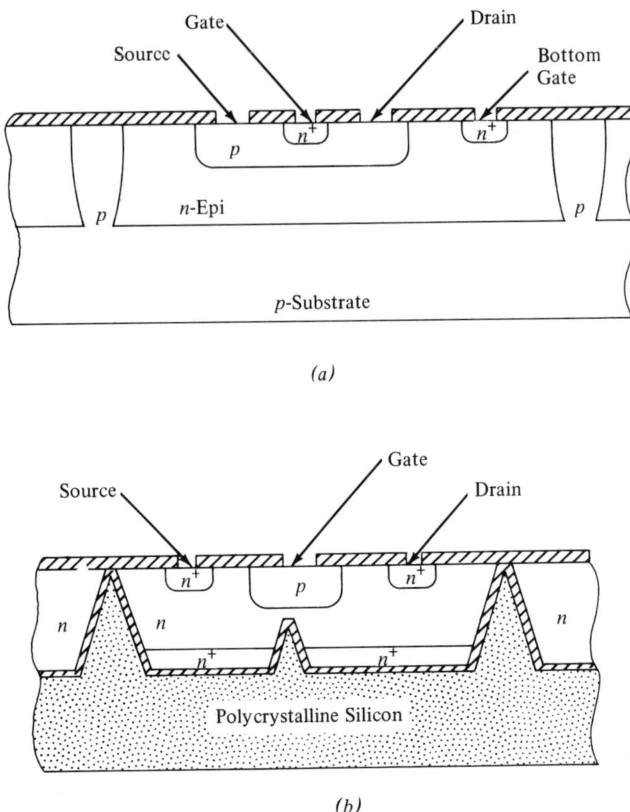

Figure 2.26 Other JFET structures compatible with integrated circuits.

is a *p*-channel device, with a diffused channel region. The *p*-type channel region can be either formed by *npn* base diffusion or by an additional *p*-diffusion step resulting in a deeper junction structure. If the ordinary base and emitter diffusions of *npn* transistors is used to form the *p*-channel FET, the resulting device is referred to as a "pinched-resistor." The structure of such a device exhibits a very narrow channel width, equal to the base width, W_B, of the *npn* transistor. Since the transistor base-emitter junction is used as the gate-channel junction of the FET, such a device has a very low voltage breakdown, with $BV_{GD} = BV_{EBO} \approx 7$ V.

In some circuit applications, a separate, lightly doped *p*-type diffusion is used to form the FET channel to avoid the voltage breakdown limitations.[12] Since the channel width of the FET structure in Fig. 2.26(*a*) is determined by the diffused junction depths and not by the epitaxial layer thickness, the channel width can be controlled to a much tighter tolerance than that of Fig. 2.25. Such a structure also offers a higher transconductance than that of Fig. 2.25 since both of its gate regions can function as control electrodes.

The structure of the device in Fig. 2.26(*b*) is particularly suitable for high frequency applications, and uses dielectric isolation and anisotropic etch techniques. The gate and the source and drain contacts are again formed by the *npn* base and emitter diffusions. The channel is formed by a wedge-shaped groove etched under the gate region. To obtain such a sharp "V" shaped grove anisotropic etching methods are used (see Section 1.7). Using such a device, it is possible to reduce the (width/length) ratio of the channel to approximately unity, and thus greatly increase the cutoff frequency (see Eq. 2.44). However, since the channel width of such a device structure is determined by the etch depths and the back-lapping step associated with the dielectric isolation process, the exact value of the width is difficult to control accurately. Since the structure in Fig. 2.26(*a*) has only one control gate, the gate structure has to be continuously wrapped around the drain region as shown in the lateral geometry of Fig. 2.25.

One of the major drawbacks of JFET structures is the strong dependence of device parameters on the channel geometry, and particularly on the channel half-width, *a*. Table 2.3 lists the dependence of some of the significant JFET parameters on the length and width of the channel. For most analog design applications, the most significant device characteristic that needs to be controlled is the pinch-off voltage V_p, which shows a strong dependence on the channel width. The change ΔV_p introduced by a corresponding change of (a) can be written as

$$\frac{\Delta V_p}{V_p} = 2(\Delta a/a) \tag{2.45}$$

Due to the tolerances associated with the epitaxial growth or diffusion steps, the absolute value tolerance associated with the channel width is of the order of ± 10 percent. For a nominal 5 V pinch-off voltage, this results in an absolute value tolerance of ± 1 V. As a consequence of these large absolute value tolerances, the matching of adjacent FET characteristics is also much poorer than those of monolithic bipolar transistors. Both the diffusion depths and the epitaxial layer thickness can show an rms variation of the order of ± 0.5 percent over a distance along the wafer surface comparable to the device dimensions. Therefore, the matching of V_p in adjacent FET structures having identical lateral geometry is no better than ± 1 percent. To equalize the drain current levels in two identical FET structures, it is necessary to apply a gate bias offset equal to the V_p mismatch, ΔV_p. From Eq. (2.45), this implies that for a typical pinch-off

TABLE 2.3 DEPENDENCE OF JUNCTION-GATE *FET* PARAMETERS ON CHANNEL DIMENSIONS (Uniform Channel Doping Assumed)

Device Parameter	Dependence On Channel Dimensions	
	Channel half-width, a	Channel length L
Saturation current: I_{do}	$(a)^3$	$(1/L)$
Transconductance: g_m	(a)	$(1/L)$
Cut-off frequency: f_c	$(a)^2$	$(1/L)^2$
Drain conductance beyond pinch-off: g_d	$(a)^3$	$(1/L)^2$
Pinch-off voltage: V_p	$(a)^2$	—
"On" resistance: R_o	$(1/a)$	(L)
Input capacitance: C_{in}	—	(L)

voltage of 5 V, the expected offset voltage ΔV_p is of the order of 50 mV, which is at least an order of magnitude worse than the V_{BE} matching of adjacent bipolar transistors.

2.5 INSULATED-GATE FET

In an insulated-gate FET structure, a thin dielectric barrier is used to isolate the gate and the channel. The control voltage applied to the gate terminal induces an electric field across the dielectric barrier and modulates the free carrier concentration in the channel region. The IGFET structures can be classified as *p*-channel and *n*-channel devices, depending on the conductivity type of the channel region. In addition, these devices can also be classified according to their mode of operation as "enhancement" or "depletion" type devices.

In a depletion-mode FET, a conducting channel exists under the gate, with no applied gate voltage. The applied gate voltage controls the current flow between the source and the drain by depleting a part of this channel. This is very similar to the operation of the JFET described previously.

In the case of the enhancement-mode IGFET, no conductive channel

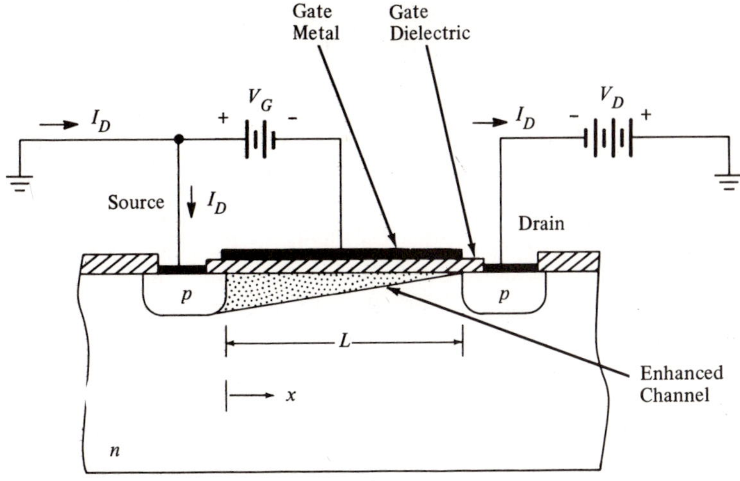

Figure 2.27 Cross-section of a *P*-channel enhancement-mode IGFET.

exists between the source and the drain at zero applied gate voltage. As a gate bias of proper polarity is applied and increased beyond a threshold value V_T, then a localized inversion layer is formed directly below the gate; this serves as a conducting channel between the source and the drain electrodes. If the gate bias is increased further, the resistivity of the induced channel is reduced, and the current conduction from the source to the drain is enhanced. Figure 2.27 shows a cross-section diagram of a *p*-channel enhancement-mode IGFET. The particular polarities of the gate and the drain bias for the proper operation of the device are also identified on the figure. Normally an SiO_2 layer is used as the gate dielectric. The thickness of this oxide layer is normally much less than the oxide layers commonly used for masking or surface passivation.

The enhancement-mode IGFET is preferred over its depletion-mode counterpart because it is a "self-isolating" device, does not require tight control of diffusion cycles, and can be fabricated by a single diffusion step forming the source and the drain pockets. Since all the active regions of the IGFET are reverse-biased with respect to the substrate, adjacent devices fabricated on the same substrate are electrically isolated, without requiring a separate isolation diffusion. Because of this "self-isolation" advantage, IGFET devices offer a much higher packing density per unit area of silicon surface than the bipolar transistors.

When fabricating the *p*-channel enhancement IGFET simultaneously with the *npn* bipolar transistor, the *n*-type background material shown in Fig. 2.27 can be formed by the epitaxial collector region of the *npn* bipolar transistor. For proper operation and low threshold voltages, the background resistivity for the IGFET channel is in the range of 3 to 8 Ω-cm. The lower end of this resistivity range is compatible with the *npn* collector resistivity requirements for most analog integrated circuits. Therefore, this type of FET can be readily fabricated with the *npn* bipolar process steps, where the *p*-type base diffusion can be used to form the source and the drain pockets.

Referring to Fig. 2.27, with the drain voltage $V_D = 0$, a uniform, conducting channel exists under the gate region, for values of the gate voltage V_G in excess of V_T. It can be shown that this uniform channel has a sheet conductance, g_c, given as

$$g_c = \mu C_0(V_G - V_T) \tag{2.46}$$

where C_0 = capacitance per unit area of the gate electrode
μ = majority carrier mobility in the induced channel.

C_0 can be related to the thickness T_x and the dielectric coefficient ϵ_x of the gate dielectric layer as

$$C_0 = \frac{\epsilon_x}{T_x} \tag{2.47}$$

If the drain voltage is increased in the polarity shown in Fig. 2.27, a finite drain current, I_D, flows through the induced channel. This current also causes an ohmic drop along the channel which subtracts from the net gate voltage $(V_G - V_T)$. Since at any point along the channel, the apparent sheet conductance is proportional to this net gate voltage, the voltage gradient due to I_D causes the channel to deplete along its length; and the sheet conductance becomes a function of the distance x along the channel:

$$g_c(x) = \mu C_0[V_G - (V(x) - V_T)] \tag{2.48}$$

where $V(x)$ is the channel potential at a distance (x) from the source. Thus, an increase of V_D causes the drain current to increase, which in turn causes the channel to deplete or pinch off near the drain. For values of the drain voltage in excess of the net gate bias, i.e., for

$$V_D \geqq (V_G - V_T) \tag{2.49}$$

a space-charge layer is formed at the drain end of the chanel, since the net gate bias (i.e., the applied gate voltage—the voltage drop along the channel) at this point is no longer sufficient to maintain an induced channel. This leads to a saturation of drain current similar to the case of JFET; and results in a set of drain current vs. voltage characteristics shown in Fig. 2.28.

For values of V_D less than $(V_G - V_T)$, the drain current can be related

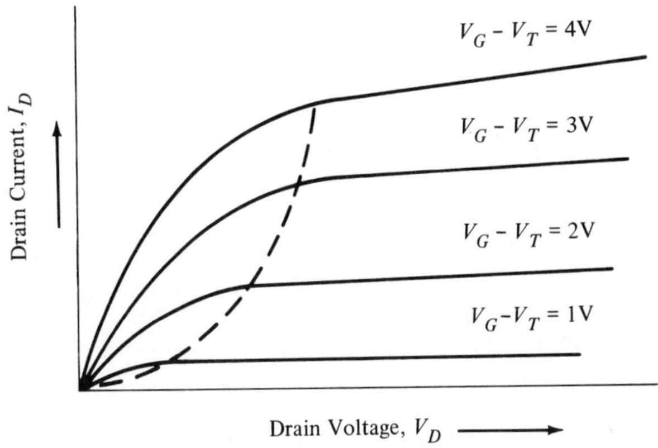

Figure 2.28 Current-voltage characteristics of an enhancement-mode IGFET.

to V_D as

$$I_D = \frac{\mu C_0 z}{L} [(V_G - V_T) - V_D/2]V_D \tag{2.50}$$

where $V_G > V_T$. The dimension z is the depth of the channel measured normal to the cross section of Fig. 2.27. The device transconductance can be derived by differentiating Eq. (2.50) as

$$g_m = \frac{\partial I_D}{\partial V_G} = \frac{\mu C_0}{L} [V_G - V_T] \tag{2.51}$$

As shown by Eq. (2.51), g_m increases linearly with the gate voltage, V_G. A device exhibiting this characteristic is known as a "square-law" device, and is particularly useful for multiplication of ac signals (i.e., mixer applications) or automatic gain control (AGC) applications.

High Frequency Characteristics

For small signal operation, the frequency performance of an IGFET can be closely approximated by the equivalent circuit of Fig. 2.29, for $V_D > (V_G - V_T)$. It should be noted that with the exception of the source-substrate and the drain-substrate capacitances, C_{ss} and C_{ds}, this equivalent circuit is identical to that of Fig. 2.24. Normally, the IGFET is used in the grounded-source configuration, with the substrate either ac grounded or shorted to the source. In this latter case, C_{ss} is shorted out, and C_{gs} appears between the drain and the ground terminal. As in the JFET case, C_{gd} and C_{gs} refer to the components of the gate capacitance associated

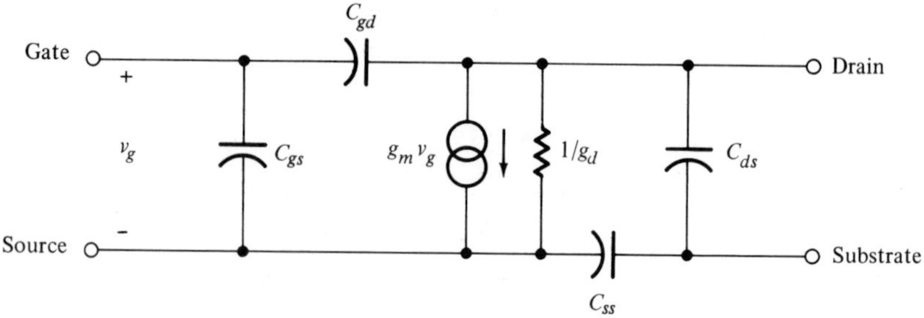

Figure 2.29 Ac equivalent circuit for an IGFET.

with the source and the drain electrodes. The drain output conductance, g_d, is again due to the modulation of the effective channel length L by the drain space-charge layer width, for $V_D > (V_G - V_T)$; and it can be approximated as[13]

$$g_d = \frac{\partial I_D}{\partial V_D} \approx \frac{2\, I_{DO}}{\pi L(V_D + V_T - V_G)} \tag{2.52}$$

for $V_D > 2(V_G - V_T)$, where I_{DO} is the saturation value of the drain current. It can be obtained from Eq. (2.50) by setting $V_D = (V_G - V_T)$, i.e.,

$$I_{DO} = \frac{\mu C_0 z}{2L}(V_G - V_T)^2 \tag{2.53}$$

Although not shown explicitly in Fig. 2.29, the parasitic bulk resistances R_S and R_D are still present in the IGFET devices, and have the same degradation effects on the device performance as discussed earlier in connection with junction-gate transistors.

Since the basic nature of operation of the IGFET is similar to its junction-gate counterpart, a useful figure of merit for the high frequency capability of the device is again the transconductance cutoff frequency, f_c, given as

$$f_c = \frac{g_m}{2\pi z L C_0} \tag{2.54}$$

Since g_m and C_0 can be related by Eq. (2.51), the maximum value of f_c can be written as

$$f_c = \frac{\mu V_D}{2\pi L^2} \tag{2.65}$$

for $V_D \leq (V_G - V_T)$. Note that for optimizing the high frequency performance of the device, a very short channel length is required.

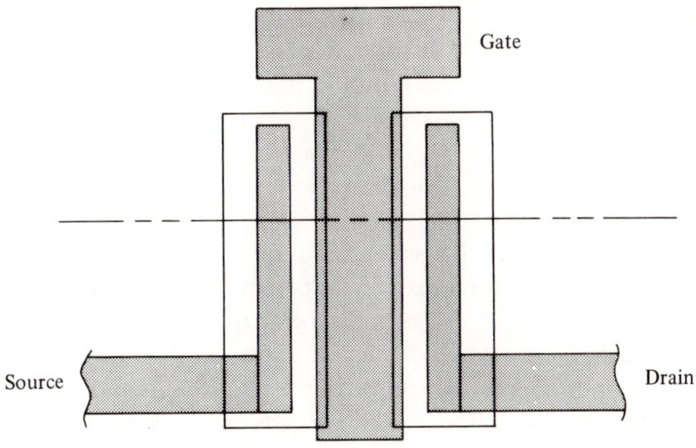

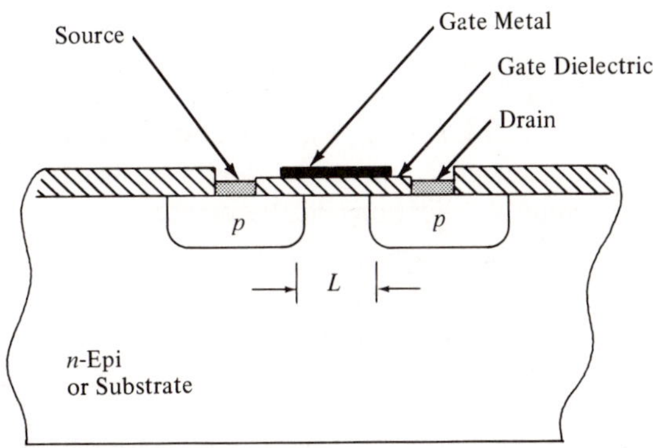

Figure 2.30 Typical device layout for a *p*-channel IGFET.

Integrated IGFET Structures

The *p*-channel enhancement device is the simplest insulated-gate FET structure which can be fabricated simultaneously with *npn* bipolar transistors, without requiring additional diffusion cycles. In such a device, the *n*-type collector and the *p*-type base regions of the *npn* are used as the "substrate" and the source and drain pockets for the IGFET. Figure 2.30 shows the layout of a typical *p*-channel enhancement device which can be fabricated in this manner. The channel length L is determined by the final separation between the source and the drain pockets; and it is normally chosen to be $\geq 5\ \mu$ to avoid punch-through breakdown between the source and the drain. In considering the final dimensions of the channel, the side diffusion of the p-type source and drain islands must be taken into consideration. Thus, if the source and the drain diffusion windows are 12 μ apart, and if the pockets are diffused to a depth of 4 μ, the resulting channel length is approximately 6 μ (see Fig. 1.7). Furthermore, in an enhancement-mode FET, the gate metal layer must overlap the source and the drain pockets to form a continuous channel.

The threshold voltage, V_T, is a strong function of the cleanliness of the device surface. To obtain repeatable and stable values of V_T also requires that the gate dielectric be free of ionic contamination.[14] With the present-day process technology, typical values of V_T are in the range of 2.5 to 3 V, with a gate oxide thickness of approximately 1500 Å and a substrate resistivity of 4 to 6 Ω-cm. The minimum gate oxide thickness is set by the voltage breakdown requirements of the gate electrode. The gate breakdown voltage, V_{GB}, is determined by the dielectric strength γ_x, and the thickness T_x of the gate dielectric:

$$V_{GB} = \gamma_x T_x \tag{2.57}$$

If SiO_2 is used as the gate dielectric, ($\gamma_x \approx (5)(10^6)$ V/cm), the typical values of V_{GB} are of the order of 40 to 60 V for the gate oxide thickness of $\approx$1000 Å. In an IGFET, the gate breakdown results in a permanent damage to the gate dielectric; therefore it is a destructive failure mechanism. Hence, in circuit applications of the device, proper precaution must be taken to protect the gate electrode from transient voltage spikes.

2.6 OTHER INTEGRATED DEVICES

In addition to the bipolar and field-effect transistors, several other active device structures can be fabricated using the monolithic integrated circuit processes. The most significant among these are the unijunction transistor, the four-layer diode, and the thyristor or the controlled rectifier structures. In the following sections some of these devices will be reviewed briefly.

Unijunction Transistor (UJT)

The unijunction transistor or the UJT is a single junction device with two ohmic base contacts. Fig. 2.31 shows a simplified diagram of a UJT, along with its integrated realization. The device exhibits a voltage-controlled negative resistance characteristic as a result of the conductivity modulation within the high resistivity base region. For a given base bias V_B, negligible anode current I_j flows until the anode terminal is raised to a sufficiently positive voltage level V_q to cause the junction to be forward-biased. The

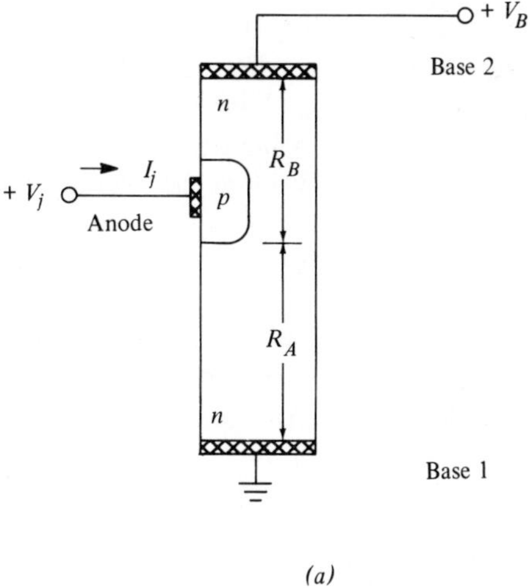

(a)

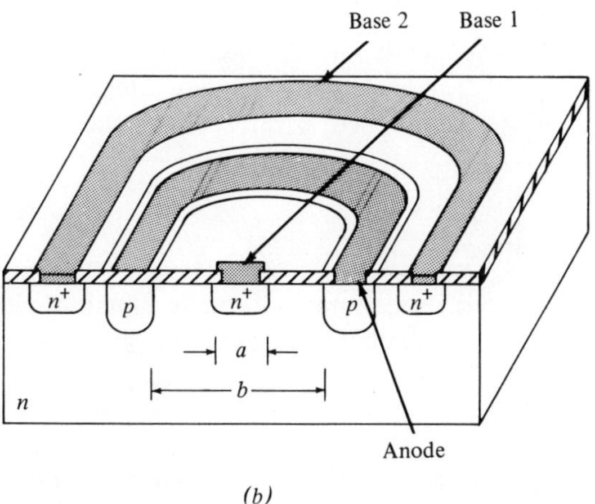

(b)

Figure 2.31 The unijunction transistor: (*a*) its schematic representation; (*b*) a practical integrated UJT structure.

minority carriers injected into the base region cause the effective value of R_A to decrease, due to conductivity modulation, thus further forward-biasing the junction. This results in a negative resistance characteristic where I_j increases rapidly, without requiring V_j to raise, and the device switches to its "on" state. Similarly, it can be returned to its stable "off" state by reducing the anode voltage to below V_q. The voltage-controlled negative resistance characteristics of the UJT make it useful for relaxation oscillator application. However, due to excessive charge storage effects in the high resistivity base region, the switching speed of the UJT is several orders of magnitude lower than the bipolar transistor.

The key design parameter for the UJT is the "stand-off ratio," η, defined as

$$\eta = \frac{V_q}{V_B} \approx \frac{R_A}{R_A + R_B} \tag{2.58}$$

For the circular integrated UJT structure of Fig. 2.31(*b*), the stand-off ratio can be related to the device dimensions as

$$\eta = \left(1 - \frac{a}{b}\right) \tag{2.59}$$

It should be noted that in a conventional junction isolated structure, if one uses the *n*-type epitaxial layer as the base region of the UJT in Fig. 2.31(*b*), this may lead to a parasitic *pnp* action between the *p*-type substrate and the UJT anode. This can be eliminated using dielectric isolation techniques.

Since the UJT depends on conductivity modulation effects for its operation, its I-V characteristics are strongly temperature dependent. This effect, along with its slow speed, severely limits the use of UJT in analog integrated circuits.

Four-Layer Devices

The four-layer (*pnpn*) diode and its three-terminal version, the thyristor, each contain four semiconductor regions. As shown in Fig. 2.32(*a*), for analysis purposes a *pnpn* diode can be decomposed into a set of cross-coupled *pnp* and *npn* transistors. It can be shown that if one gradually raises the anode voltage, the device initially remains in a nonconductive state, until a turn-on voltage, V_{BO}, is reached. Then, the device suddenly switches to its "on" state where it can carry a large amount of current with a very little voltage drop across it. The device can be turned off by reducing the current to a level below the critical current level, known as the "holding current." This is shown as current level I_H in Fig. 2.32(*b*).

By examining the two-transistor equivalent circuit of a four-layer diode, one can show that the device switches from its nonconductive to conductive state at a point where the product of the current gains ($\beta_1\beta_2$) reaches unity. At low values of the anode voltage with the device in its off state, the two center junctions are reverse-biased; therefore very little current flows through the device. But, as V_{BO} is reached, the leakage currents and

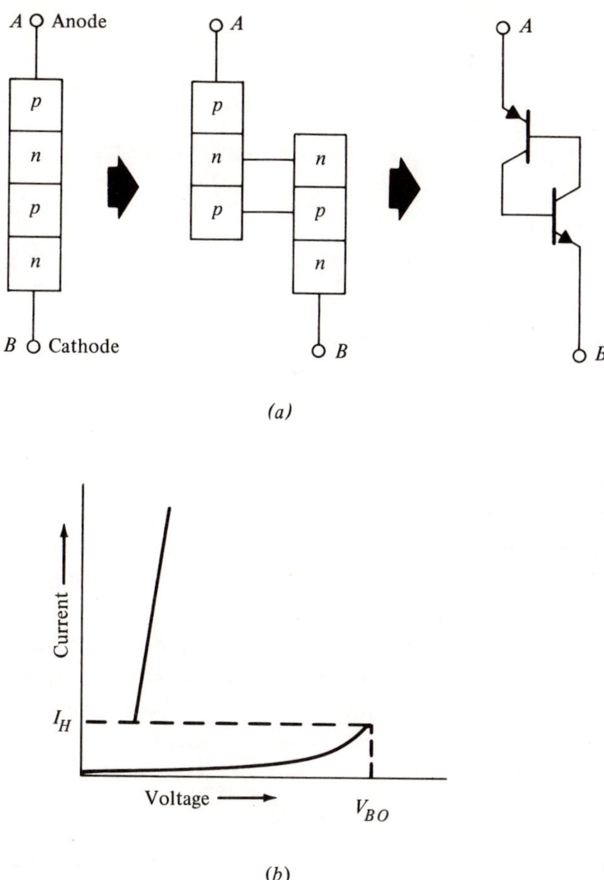

Figure 2.32 The four-layer diode: (*a*) transistor equivalent circuit; (*b*) I-V characteristics.

the avalanche multiplication effects are sufficient to raise the current gains of the intrinsic transistors to a level where $(\beta_1) \cdot (\beta_2) \geq 1$, and the diode switches to its conducting state.

An additional control of the turn-on characteristics can be obtained by connecting a "gate" electrode to either of the two center regions. By injecting a small amount of current into this gate, the turn-on voltage can be kept at a value below V_{BO}, giving the device an additional degree of control. Such a three-terminal *pnpn* device is known as a thyristor, or controlled rectifier.

The four-layer diode is readily available in a junction isolated inte-

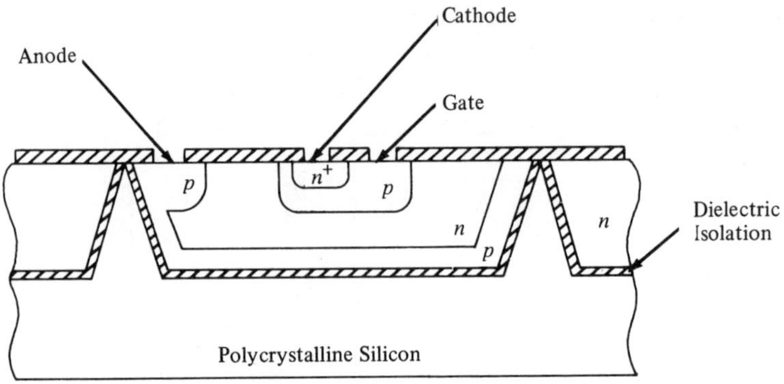

Figure 2.33 A four-layer diode structure using dielectric isolation.

grated circuit structure, using the p-substrate, n-epitaxial layer, p-base and the n-type emitter regions of a conventional *npn* transistor. However, such a structure is not directly useful, since its anode is formed by the substrate, which is also constrained to be at the most negative potential in the circuit to ensure that the isolation junctions are reverse-biased.

Figure 2.33 shows an alternate four-layer diode structure which overcomes this difficulty, using dielectric isolation. Such a structure is obtained by a p^+ predeposition step after the isolation moat etching cycle, and prior to oxidation and polycrystalline silicon growth.

References

1. S. K. Ghandhi, *The Theory and Practice of Microelectronics,* Wiley, New York, 1968, Chaps. 13 and 14.
2. Motorola Semiconductor Products Div., Engineering Staff, *Integrated Circuits: Design Principles and Fabrication,* McGraw-Hill, New York, 1965, Chap. 4.
3. H. C. Lin, T. B. Tan, G. Y. Chang and B. van der Leest, "Lateral Complementary Transistor Structure for Simultaneous Fabrication of Functional Blocks," *Proc. IEEE,* **49** (1964): 1491–1495.
4. A. B. Grebene, "Monolithic Structure with Three-Region Complementary Transistors," U.S. Patent No. 3,412,195, assigned to Sprague Electric Co., North Adams, Mass.
5. B. Polata, "Compatible, High Performance Complementary Bipolar Transistors for Integrated Circuits," IEEE Int. Electron Devices, Mtg, Washington, D.C., 1969.
6. H. C. Lin, *Integrated Electronics,* Holden Day, New York, 1967, Chap. 8.

7. E. L. Long, "A High-Gain 15-W Monolithic Power Amplifier with Internal Fault Protection," Int. Solid State Ckts. Conf., *Digest Tech. Papers,* **13** (1970): 108–109.

8. W. Shockley, "A Unipolar Field-Effect Transistor," *Proc. IRE,* **40** (1952): 1365.

9. A. B. Grebene and S. K. Ghandhi, "General Theory for Pinched Operation of the Junction-Gate FET," *Solid State Electronics,* **12** (1969): 573–589.

10. R. Cobbold, *Theory and Applications of Field-Effect Transistor,* Wiley, New York, 1970.

11. A. S. Grove, *Physics and Technology of Semiconductor Devices,* Wiley, New York, 1967, Chaps. 8 and 9.

12. G. R. Wilson, "A Monolithic Junction FET-NPN Operational Amplifier," Int. Solid State Ckts. Conf., *Digest Tech. Papers,* **11** (1968): 20–21.

13. S. R. Hofstein and G. Warfield, "Carrier Mobility and Current Saturation in the MOS Transistor," *IEEE Trans. on Electron Dev.,* **ED–12** (1965): 129–138.

14. A. S. Grove, *Physics and Technology of Semiconductor Devices,* Wiley, New York, 1967, Chap. 9.

3

Passive Components: Capacitors and Resistors

The choice and variety of passive components available in integrated circuits are severely limited. The heart of the integrated circuit technology is the planar process. This process is designed and developed around active devices; the passive components in general take second priority. In order to fabricate a number of active and passive devices simultaneously, with the same set of processes, a number of design and performance compromises have to be made. This, in turn, results in limited sizes and kinds of passive components available in monolithic form. Certain passive components, such as inductors, are not compatible with microminiaturization. The remaining ones, i.e., capacitors and resistors are available only with poor absolute value or temperature tolerances, and only over a relatively narrow range of values.

In going from a discrete circuit design to its integrated counterpart, the economics of design often become "inverted." In the discrete or nonintegrated design circuit cost and complexity can be related to the number of active devices needed. In the monolithic case it is usually the requirements on the absolute values or the sizes of passive components which can make the design a difficult or uneconomical one.

In spite of the limitations on their size or absolute value tolerances, passive components in integrated circuits still enjoy some of the basic advantages of monolithic structures, such as close matching and close thermal tracking. In many cases by proper choice of available components, and imaginative circuit design approaches, it is possible to overcome the limitations of monolithic components. The circuit techniques used for this purpose are covered in the following chapters of this book. However, before starting an actual design problem, it is imperative that the integrated circuit designer should know the basic choices of passive component types and values available to him.

The purpose of this chapter is to familiarize the designer with the important features of passive components, namely capacitors and resistors. Whenever applicable, the comparison between different classes of capacitors and resistors are presented in a tabular form for quick reference.

PART I Integrated Capacitors

The most fundamental limitation on integated capacitors is size. A general expression for the capacitance of a parallel-plate capacitor can be written as:

$$C = C_0A \tag{3.1}$$

where C_0 is the capacitance per unit area and A is the area of one of the plates. The value of C_0 is usually restricted to a narrow range (typically of the order of 0.05 pF/mil^2 to 0.5 pF/mil^2) due to the type of dielectric materials available in monolithic circuits and their voltage breakdown properties. Thus, the area requirement increases quite rapidly with the required capacitor value. Since the practical chip size of a monolithic circuit is dictated by yield considerations, this puts an upper limit on the value of capacitance which can be provided on the monolithic circuit.

The basic classes of capacitor structures are available in monolithic circuits: junction and thin film capacitors. Fundamental properties of each of these types of capacitors will be outlined below.

3.1 JUNCTION CAPACITORS

Application of a reverse bias across a semiconductor junction forces the mobile carriers to move away from the immediate vicinity of the junction. As shown in Fig. 3.1, this forms a depletion layer of finite width about the junction. This structure closely simulates a parallel-plate capacitor, with the plates separated by the total depletion layer width x. The depletion layer width can be readily related to the total voltage across the junction

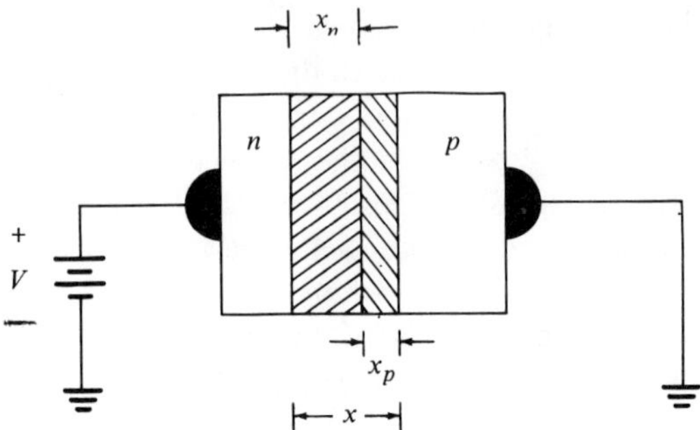

Figure 3.1 Schematic representation of junction capacitance.

(i.e., the applied voltage plus the built-in potential) by the solution of the one-dimensional Poisson's equation.[1] It can be shown that the overall charge neutrality conditions require that an equal number of positive and negative charges be depleted on either side of the junction. This implies that the depletion layer tends to spread more into the more lightly doped side of the junction. For a step-junction structure, with uniform impurity concentrations of N_A and N_D atoms/cm^3 on the P and N sides of the junction, the total depletion layer width can be related to the total applied voltage V as

$$V = \frac{q}{2\epsilon}\left(\frac{N_A N_D}{N_A + N_D}\right)(x)^2 \tag{3.2}$$

where $x = x_p + x_n$ = total depletion layer width
$\epsilon = \epsilon_r \epsilon_0$ = dielectric constant of silicon

For a step junction, the charge neutrality condition also requires that the total charge on each side of the junction be equal, i.e.,

$$N_A x_p = N_D x_n \tag{3.3}$$

Since the capacitance per unit area of a parallel-plate capacitor with a plate separation x is given as

$$C_0 = \frac{\epsilon}{x} \tag{3.4}$$

the capacitance per unit area of the junction can be related to the total reverse bias V across the junction, from Eqs. (3.3) and (3.4), as

$$C_0 = \sqrt{\frac{q\epsilon}{2V}\left(\frac{N_A N_D}{N_A + N_D}\right)} \tag{3.5}$$

With most junctions the impurity level on one side of the junction is much higher than the other. This is particularly true if one side of the junction, say the p side, is formed by diffusion into the uniform n background, such as the case of the base-collector junction of the npn transistor. In such a case, one can still invoke the step-junction approximation, with $N_A \gg N_D$. Then, Eq. (3.3) reduces to

$$C_0 = \sqrt{\frac{q\epsilon N_D}{2V}} \qquad \text{for } N_A \gg N_D \tag{2.64}$$

Figure 3.2 shows a plot of C_0 as a function of the total junction voltage, for different values of impurity concentration on the lighter doped side.

In general, the voltage dependence of most junction capacitances that can be fabricated with the integrated circuit processes can be described by an expression of the form

$$C_0 \approx K_1(1/V)^n \tag{3.7}$$

where the exponent n is within the narrow range of $1/3 < n < 1/2$, and K_1 is a constant of proportionality depending on the impurity concentration levels in the immediate vicinity of the junction. The $N = 1/2$ case corresponds to the step junction structure and $n = 1/3$ corresponds to a linearly graded junction. Both the Gaussian and the complementary error function diffusion profiles fall somewhere between these two limits. A detailed family of capacitance vs. voltage curves has been calculated by various workers in this field, and is available in the literature.[2]

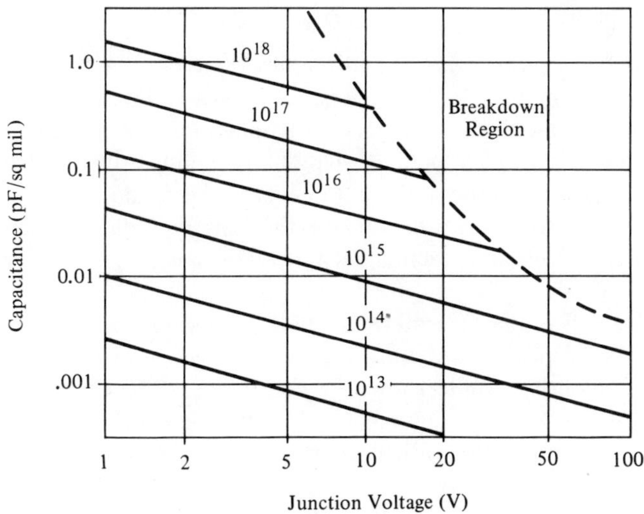

Figure 3.2 Capacitance per unit area vs. junction voltage for a step junction.

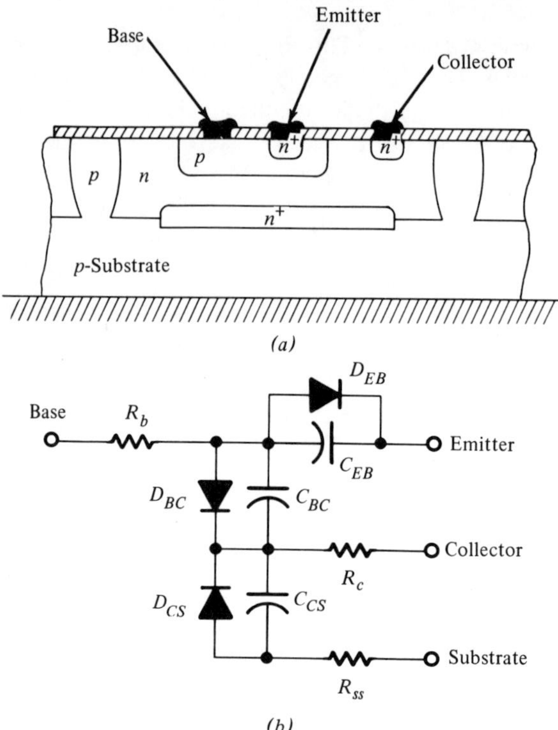

Figure 3.3 Junction capacitances in bipolar integrated circuits: (*a*) physical device structure; (*b*) equivalent circuit of junction capacitances.

In a planar epitaxial integrated circuit, there are three separate junctions which can be used as capacitors. As shown in Fig. 3.3, these are the base-emitter, base-collector, and collector-substrate capacitances associated with the integrated *npn* bipolar structure. Also shown in the figure is the resultant interconnection of these three capacitors, along with the ideal diodes in shunt with them to indicate the bias polarity requirements for each capacitor. Note that each capacitance also has a finite bulk resistance in series with it.

Table 3.1 lists the typical values of C_0 associated with each of these junctions, as a function of the reverse-bias voltage for an integrated device having the impurity profile shown in Fig. 2.3, i.e., 10 Ω-cm *p*-type substrate and 5 Ω-cm *n*-type epitaxial layer, with the Gaussian and the complementary error function distribution for the base and the emitter diffusions.

With reference to Fig. 3.3, the collector-substrate capacitance C_{CS} has only a very limited application since one of its terminals, the substrate, is

TABLE 3.1 TYPICAL VALUES OF CAPACITANCE PER UNIT AREA ASSOCIATED WITH INTEGRATED TRANSISTOR JUNCTIONS

Applied Voltage	Typical Junction Capacitance (pF/mil^2)			
	C_{EB}	C_{BC}	C_{CS} w/o n+ Layer	C_{CS} with n+ Layer
0 V	0.9	0.19	0.12	0.17
5 V	0.65	0.08	0.04	0.06
10 V	—	0.06	0.025	0.035

common to the rest of the circuit, and represents an ac ground point. However, C_{CS} is an inherent part of the device structure, and is always present in any junction isolated structure. The remaining capacitances, C_{EB} and C_{BC} can be eliminated when not needed by omitting the emitter or the base diffusion.

The emitter-base junction offers the highest capacitance per unit area; however, its low reverse breakdown voltage ($\approx$7 V) limits its use in some applications. The base-collector capacitance finds a wider range of applications than C_{EB} because of its high breakdown voltage (typically $\approx$50 V). However, its performance is hampered by the collector series resistance R_c, and the shunt C_{CS} to ground. A n^+ buried layer can be used to reduce R_c; however, this causes approximately a 30 percent increase in the value of C_{CS}, by increasing the impurity concentration at the epi-substrate interface.

In order to utilize C_{BC} efficiently for ac coupling purposes, it is necessary that the (C_{BC}/C_{CS}) ratio be kept as high as possible. This can be done by choosing the reverse-bias across the base-collector junction to be as low as possible, and by maintaining a large reverse-bias across the collector-substrate junction. In this manner, the (C_{BC}/C_{CS}) ratios in the range of 3 to 10 are feasible. In the case of dielectrically isolated integrated circuits (see Fig. 1.14), C_{CS} is much less than 0.01 pF/mil^2, and such bias precautions are not necessary.

3.2 THIN-FILM CAPACITORS

The thin-film capacitor is a direct miniaturization of the conventional parallel-plate capacitor. It is comprised of two conductive layers separated by a dielectric. In integrated circuits, it can be fabricated in either one of

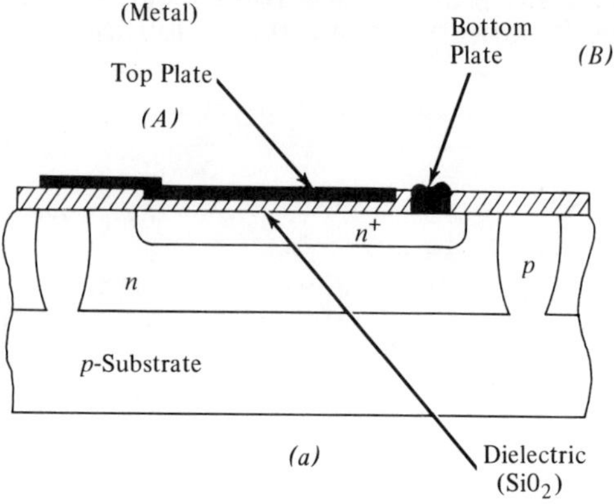

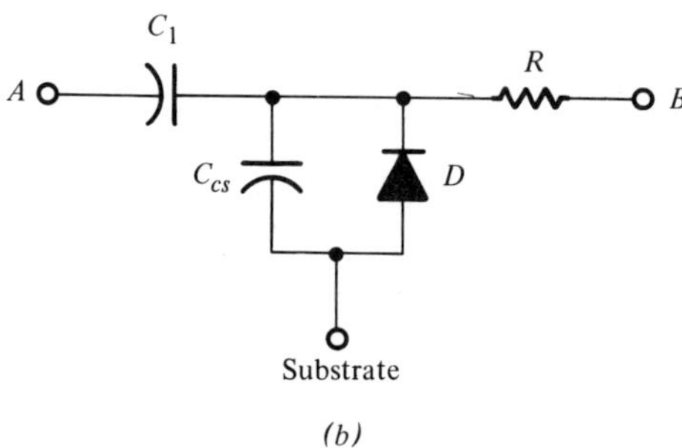

Figure 3.4 Metal Insulator semiconductor (MIS) capacitor: (*a*) device structure; (*b*) equivalent circuit.

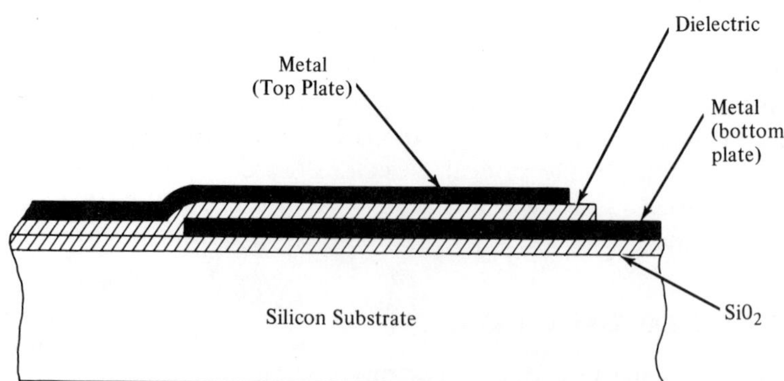

Figure 3.5 Thin-film capacitor using multiple metal layers.

two forms: (1) using a metal insulator semiconductor (MIS) structure as shown in Fig. 3.4; or (2) using a thin dielectric layer between two metal layers (see Fig. 3.5).

The MIS structure of Fig. 3.4 is the most commonly employed thin-film capacitor in monolithic circuits, since it is readily compatible with the conventional processing technology, and does not require multiple metallization layers.

In a thin-film capacitor, the capacitance per unit area is directly related to the dielectric constant and the thickness of the insulator layer T_x as

$$C_0 = \frac{\epsilon}{T_x} = \frac{\epsilon_r \epsilon_0}{T_x} \tag{2.66}$$

where ϵ_r = relative dielectric constant of the insulator
ϵ_0 = dielectric constant of free space = $(8.85)(10^{-14})$ F/cm.

In metal insulator semiconductor structures, either SiO_2 or silicon nitride (Si_3N_4) can be used as the dielectric layer. Although SiO_2 is more readily available, Si_3N_4 is preferred whenever the extra processing step can be justified, since it offers a higher value of C_0 because of its high relative dielectric constant. The minimum thickness of the dielectric layer is set by the yield and process control requirements, and is typically about 500 Å. In most applications, a thickness of 800 to 1200 Å is preferred.

As shown in Fig. 3.4, a low resistivity film is diffused below the dielectric to reduce the parasitic series resistance R in the equivalent circuit. For this purpose, normally an n^+ emitter diffusion is used, resulting in a value of $R \approx 3$ to 5 Ω. The values of C_0 obtainable by a dielectric thickness of $\approx$1000 Å are of the order of 0.2 to 0.4 pF/mil^2; therefore the parasitic effect of the substrate capacitance, C_{CS} is not as detrimental in the equivalent circuit of Fig. 3.4 as it was in the case of the junction capacitances.

An alternate thin-film capacitor structure, which uses a thin dielectric layer between two metal layers, is shown in Fig. 3.5. Although such a capacitor structure is virtually free from substrate parasitics, it requires a number of additional masking and deposition steps beyond the basic MIS structure. In such a structure one generally uses either aluminum or tantalum as the capacitor plates, with aluminum trioxide (Al_2O_3) or tantalum pentoxide (Ta_2O_5) as the dielectric material. Ta_2O_5 is particularly preferred for large value capacitors, since its dielectric coefficient is about an order of magnitude better than most other dielectrics used in thin-film capacitors.

Unlike their junction counterparts, thin-film capacitors are not a function of the magnitude or the polarity of the voltage applied across them, and they can offer higher capacitance per unit area with lesser parasitics.

TABLE 3.2 THIN-FILM CAPACITOR CHARACTERISTICS

	Dielectric Material			
	SiO_2	Si_3N_4	Al_2O_3	Ta_2O_5
Capacitance (pF/mil^2)	0.25–0.4	0.5–1.0	0.3–0.5	2–3.5
Relative dielectric constant (ϵ_r)	2.7–4.2	3.5–9	4–8.5	24–28
Breakdown voltage (V)	50	50	20–40	20
Absolute tolerance (%)	±20	±20	±20	±20
Matching tolerance (%)	±3	±3	±5	±5
Temperature coef. (ppm/°C)	+15	+4–+10	+300	+200–+500
Q (at 10 MHz)	25–80	20–100	10–100	10–100

However, there are two basic disadvantages associated with monolithic thin-film capacitors: (1) they require additional processing steps beyond the standard diffusion cycles; (2) they fail due to the breakdown of the gate dielectric when their voltage rating is exceeded. This is a destructive, irreversible failure mechanism; therefore, additional care should be taken in its circuit application to provide over-voltage protection.

Table 3.2 gives a summary of the electrical characteristics of thin-film capacitors available in integrated circuits. Note that the first two on the list are the MIS-type structures, and the last two columns correspond to the multilayer metal structure of Fig. 3.5. The maximum size of a monoltihic capacitance is limited by the yield and chip area considerations. With present-day technology, this upper limit is approximately 1000 square mils for high volume production circuits.

PART II Integrated Resistors

Two general classes of resistors are available in monolithic circuits: semiconductor and thin-film. The semiconductor resistors in turn can be categorized into any one of the following four groups, based on their physical structure:

1. Diffused resistors
2. Bulk resistors
3. Pinched resistors
4. Ion-implanted resistors

Semiconductor resistors are by far the most commonly used monolithic resistor structures. With the exception of ion-implanted resistors, they can be fabricated simultaneously with the rest of the circuit elements, without requiring extra processing steps. However, they are in general nonideal circuit components, with rather loose tolerances; and, by discrete-component standards, have poor temperature and frequency characteristics. The deposited thin-film resistors on the other hand offer superior electrical characteristics at the expense of additional processing steps.

In the following sections of this chapter, the physical structure and the electrical characteristics of each of these resistors will be examined briefly.

3.3 DIFFUSED RESISTORS

A diffused resistor structure is formed by the bulk resistance of a diffused semiconductor region. In forming a monolithic resistor, the two basic diffusion cycles—i.e., the base or the emitter diffusions—can be utilized. Figure 3.6 shows a typical geometry and the cross-section of a *p*-type diffused resistor obtained using the *npn* base diffusion step. In circuit applications of such a resistor structure, it is necessary that the *pn* junction outlining the resistor should at all times be reverse-biased, to confine the current flow to the volume of silicon defined by the resistor geometry. The total resistor value R can be expressed as:

$$R = \frac{\rho L}{A} = \frac{\rho L}{x_j W} \tag{3.9}$$

where ρ is the average resistivity; L is the effective length of the resistor; and A is the cross-sectional area. The latter area is equal to the resistor width W times the junction depth x_j.

In a diffused resistor, due to nonuniform impurity distribution, the average resistivity ρ is difficult to calculate analytically. In the literature, detailed tables of ρ are available for various impurity profiles and background concentrations. Figure 3.7 shows a family of such curves for the *p*-type diffused resistors, for a Gaussian impurity distribution.[3] Since the base diffusion of most integrated *npn* devices closely approximate a

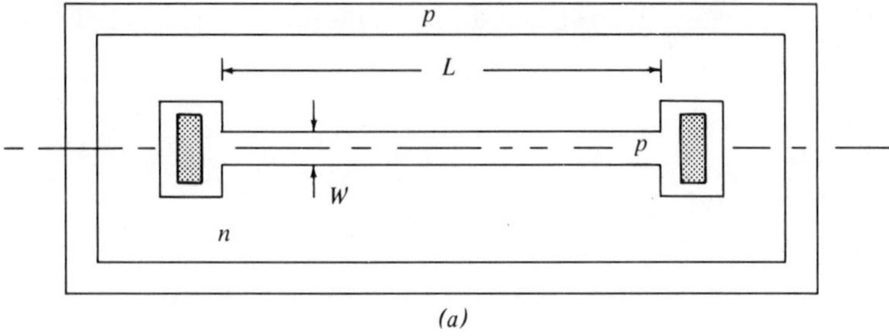

(a)

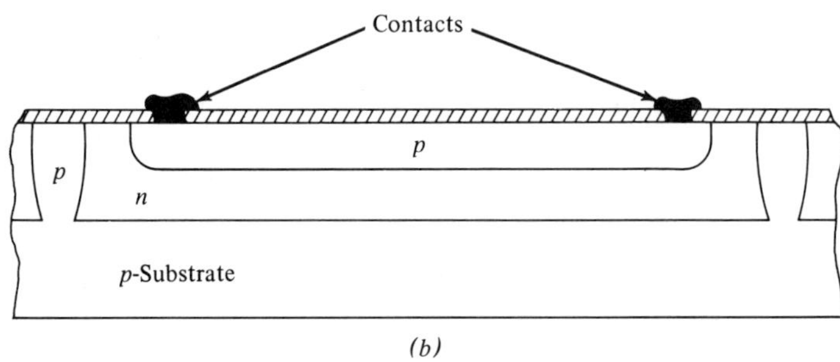

(b)

Figure 3.6 Basic diffused resistor structure: (*a*) lateral geometry; (*b*) cross-section.

Gaussian profile, these curves are suitable for most *p*-type resistor calculations.

In practical resistor calculations, a more convenient design parameter is the sheet resistance R_s, which can be defined as:

$$R_s = \frac{\rho}{x_j} \tag{3.10}$$

The sheet resistance has dimensions of ohms, and is numerically equal to the resistance of one unit square of the resistive material. Therefore, its value is commonly quoted as "ohms per square."

Since ρ and x_j are fixed by the transistor diffusion profiles, R_s automatically becomes a resistor design parameter associated with the given diffusion schedule. For a given value of R_s the total value of R can be readily expressed as:

$$R = R_s(L/W) \tag{3.11}$$

The sheet resistance, R_s has a positive temperature coefficient, and is typically in the range of 60 to 250 Ω. Similarly, for the case of n^+ emitter

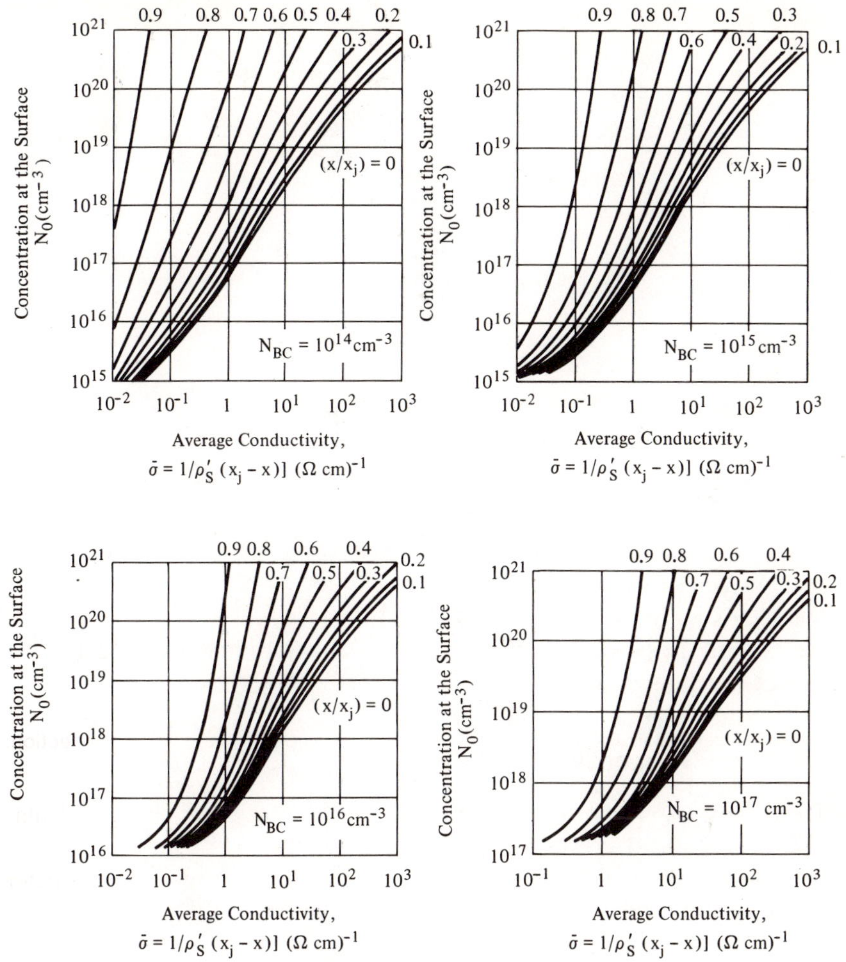

Figure 3.7 Average conductivity of *p*-type Gaussian layers in silicon.[3] (N_{BC} = background concentration; x/x_j = ratio of depth of two surfaces bounding the layer.)

diffusion, R_s is in the 2–5 Ω range. For the typical analog transistor structure shown in Fig. 2.3, R_s is approximately 150 Ω and 2.5 Ω respectively for the base and the emitter diffusions.

The sheet resistance, R_s has a positive temperature coefficient, due to the decrease of carrier mobility with temperature. Figure 3.8 shows the typical temperature dependence of R_s for a *p*-type diffused resistor. As will be described in later chapters, this strong temperature dependence of R_s

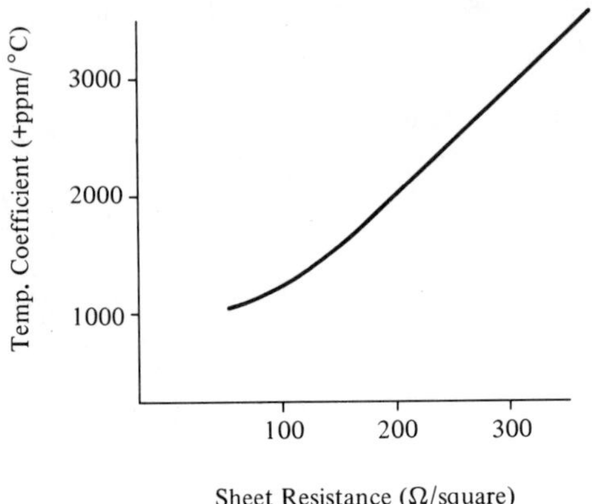

Figure 3.8 Typical temperature coefficient of *p*-type diffused resistors.

is one of the dominant thermal drift sources in a monolithic circuit. At high current levels, localized self-heating of the resistor can cause the I-V characteristics to be nonlinear. To avoid this effect, it is good design practice to limit the power dissipation of a diffused resistor to less than 3 mW/mil^2 of junction area.

Frequency Response

The diffused resistor structure also has a distributed capacitance associated with it, due to the reverse-biased junction which outlines the resistor. For a diffused resistor of total value R_1, the high frequency equivalent circuit looks as shown in Fig. 3.9, where C_1 is the total distributed capacitance

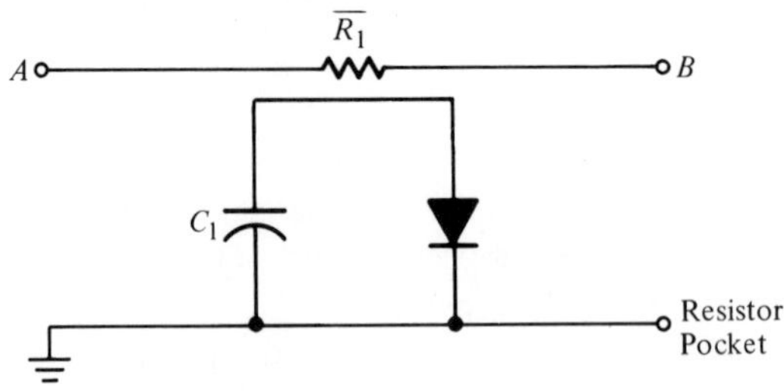

Figure 3.9 High frequency equivalent circuit of a diffused resistor.

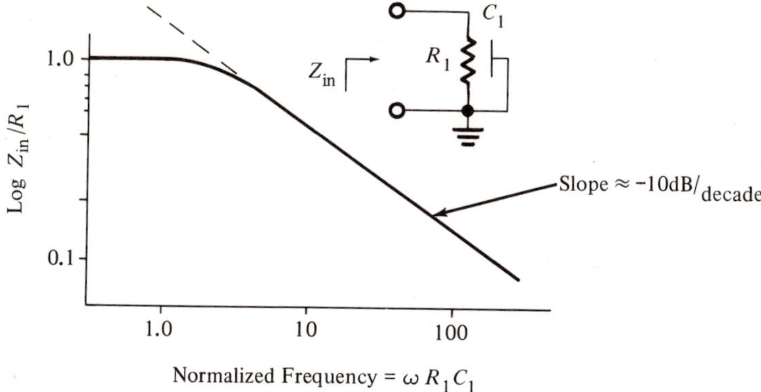

Figure 3.10 Typical frequency response of a diffused resistor.

associated with R_1. Capacitor C_1 appears distributed along the entire length of R_1. At high frequencies, the net effect of C_1 is to shunt the ac signal to ground, and cause excessive phase lag. Figure 3.19 shows the typical frequency response of a diffused resistor, in terms of the magnitude of its driving point impedance.[4] Assuming that capacitance C_1 is uniformly distributed along R_1, one can show that the magnitude of the driving point impedance Z_{in} is down by approximately 3 dB at a frequency f_1 given as

$$f_1 \approx \frac{1}{3\ R_1 C_1} \tag{3.12}$$

Replacing the total resistor and capacitor values by the sheet resistance R_s and the capacitance C_0 per unit area of the junction, Eq. (3.12) can be written as

$$f_1 \approx \frac{1}{3\ R_s C_0 L^2} \tag{3.13}$$

Note that f_1 decreases as the square of the resistor length. For a typical 10 kΩ diffused resistor structure ($R_s = 200\ \Omega$, $L = 50$ mils, $W = 0.5$ mil), f_1 is approximately 10 MHz.

Resistor Layout and Tolerances

To obtain a given resistor (L/W) ratio with efficient use of the chip area often requires that the resistor should follow a folded or zigzag structure. This can be achieved by making square or round bends in the lateral layout of the resistor, as shown in Fig. 3.11. The equivalent values of the L/W ratios for such bends (measured between lines A and B) are also

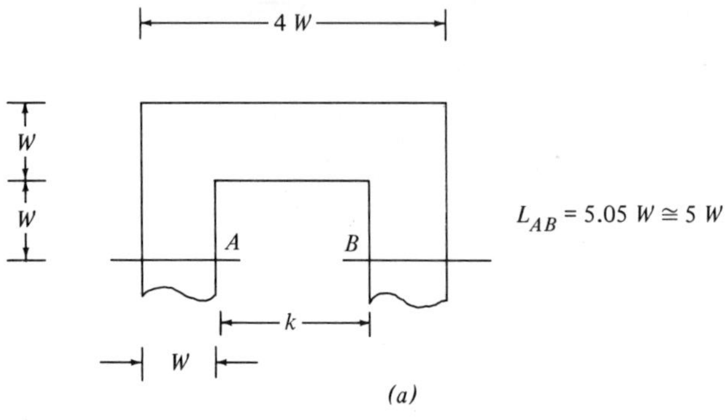

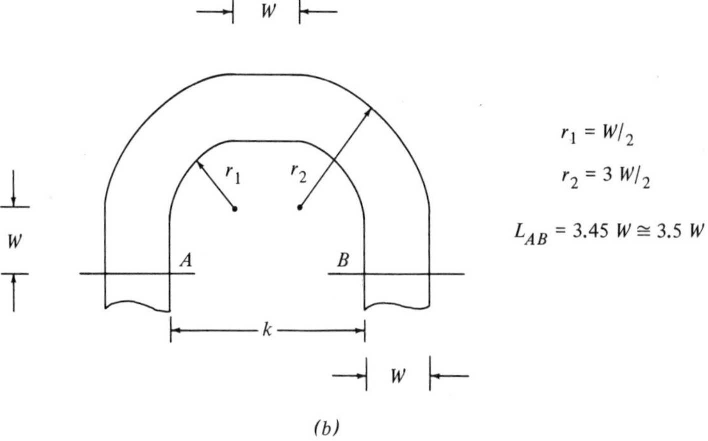

Figure 3.11 Calculation of effective resistor length around corners: (*a*) square corners; (*b*) rounded corners.

shown in the Figure. The inner separation in the resistor fold, shown as the distance (k) in the Figure is set by the side-diffusion effects (see Fig. 1.7) and is normally no less than 20 μ. The lower limit of the resistor width W is set by the masking tolerances, and is approximately 5 to 7 μ production circuits. The best matching of resistor values is obtained for the values of W in the range of 20 to 50 μ.

Table 3.3 shows typical values of the absolute value and ratio tolerances for diffused resistors fabricated during the base diffusion cycle. Typical values of resistance R that can be obtained from a sheet resistance

TABLE 3.3 TOLERANCES ASSOCIATED WITH A DIFFUSED RESISTOR

Resistor Width	Absolute Value Tolerance	Ratio Tolerance	
		1:1	5:1
W = 7 μ	±15%	±2%	±5%
W = 25 μ	±8%	±0.5%	±1.5%

R_s are in the range of

$$\frac{R_s}{4} < R < 10^4 R_s \tag{3.14}$$

for a base-diffused resistor. In the lower end of this range, the practical limit is set by the contact resistances associated with the resistor contact windows. The practical limit at the upper end is set by the available chip area.

3.4 BULK RESISTORS

The bulk resistance of the *n*-type epitaxial layer can be used in some applications to form a noncritical high value resistor. This can be done by using a structure as shown in Fig. 3.12. The resulting structure, known as a "bulk resistor" has a sheet resistivity given as

$$R_s = \frac{\rho_e}{d} \tag{3.15}$$

where ρ_e is the resistivity and d is the epitaxial layer. For a 5 Ω-cm epitaxial layer of 10 μ thickness, this results in a sheet resistivity of approximately 5 kΩ.

Since the bulk resistor is formed by the epi-isolation junction, its breakdown voltage is also substantially higher than that for the diffused resistor structure. The sidewalls of the bulk resistor are formed by the deep isolation diffusion which diffuses sideways as well as downward during the diffusion cycles. When calculating the effective cross-section of the bulk resistor, the side diffusion effects should be taken into consideration.

The temperature coefficient of the bulk resistor is significantly higher than the diffused resistor, due to the strong temperature dependence of mobility at low impurity concentrations. Typical values of bulk resistor temperature coefficients are in the range of +3500 to +5000 ppm/°C for 1 Ω-cm and 5 Ω-cm epitaxial layers.

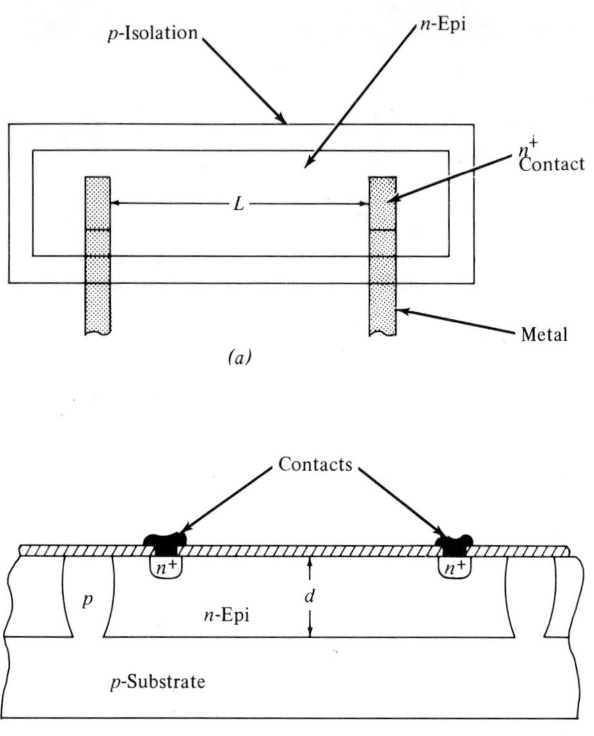

Figure 3.12 Bulk resistor: (*a*) lateral geometry; (*b*) cross-section.

The absolute value tolerances for a bulk resistor are quite loose, due to relatively poor control of the epi resistivity (± 20 percent) and epitaxial layer thickness (± 10 percent). Thus, practical absolute value tolerance for a bulk resistor is typically ± 30 percent.

3.5 PINCHED RESISTORS

The sheet resistivity of a semiconductor region can be increased by reducing its effective cross-section area. In a pinched resistor structure this technique is used to obtain a high value sheet resistance from the ordinary base-diffused resistor. Figure 3.13 shows such a resistor structure formed by placing an n^+-type emitter diffusion over the *p*-type diffused resistor. The emitter diffusion greatly reduces the effective cross-sectional area of the *p*-type resistor and consequently raises its sheet resistivity. The resulting value of R_s can be expressed as

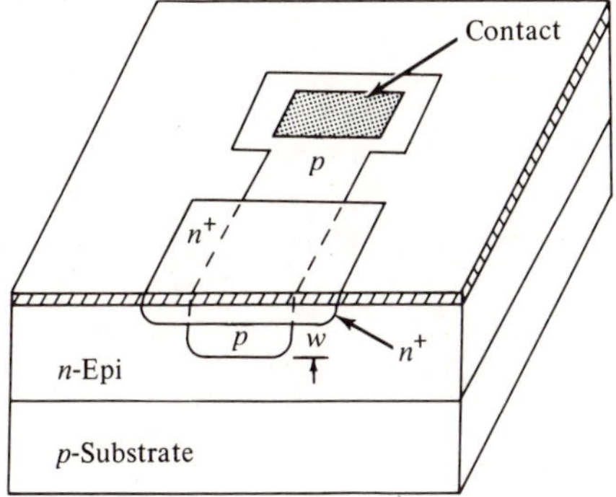

Figure 3.13 Structural diagram of a pinched resistor.

$$R_s = \frac{\rho}{x_b - x_e} = \frac{\rho}{W_b} \tag{3.16}$$

where x_b and x_e are the depths of the base and the emitter diffusions. The average resistivity ρ for the effective cross section of the resistor can either be calculated from the knowledge of the diffusion profiles, or can be estimated from the data of Fig. 3.7. It should be noted that the effective depth of such a pinched resistor structure is equal to W_b, the *npn* transistor base width. Typical values of R_s of a pinched resistor are in the range of 5 to 10 kΩ.

Figure 3.14 shows the circuit designation and the current-voltage characteristics for a pinched resistor. Normally, the n and n^+ regions surrounding the pinched resistor are shorted together and connected to the resistor terminal, which is maintained at a more positive potential than the rest of the resistor structure. Such a connection as shown in Fig. 3.14(*a*) ensures that all the junctions forming the resistor are reverse-biased. The current-voltage characteristics of the pinched resistor are linear only for small voltage drops across the resistor. In this range of operation, the device behaves as a linear resistor, with a sheet resistivity R_s given by Eq. 3.16. Application of a higher dc voltage results in an increase of reverse bias between the *p*-type resistor body and the surrounding *n*-type island. This reverse bias causes the junction depletion layer to extend into the resistor, and "pinch" the effective resistor cross-section, in a manner similar to the case of a junction-gate FET. Consequently, for increasing voltages, the pinched resistor current I_p saturates at a value I_0, as shown in Fig. 3.16(*b*). Since the top portion of the pinched resistor is comprised of the heavily doped emitter diffusion, the resistor exhibits a low breakdown voltage, equal to the transistor emitter-base breakdown, BV_{EBO} (typically 6 to 8 V).

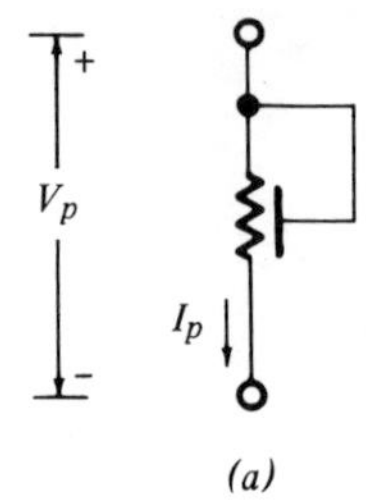

(a)

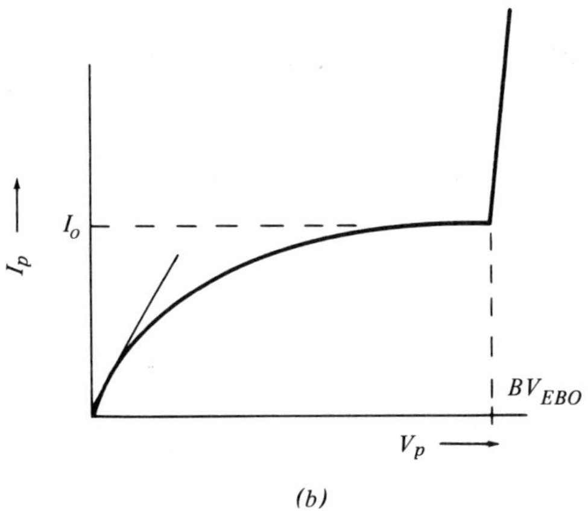

(b)

Figure 3.14 Pinched-resistor: (a) its electrical symbol; (b) current-voltage characteristics.

The sheet resistance of the pinched resistor structure exhibits a strong temperature coefficient, of the order of +3000 to +5000 ppm/°C. The saturation current I_0 is also strongly temperature dependent. Its temperature coefficient is almost identical to that of sheet resistance, but opposite in polarity. Absolute values of the pinch resistors are difficult to control in fabrication. Typical absolute value tolerances associated with R_s and I_0 are −30 to +50 percent range. However, matching and tracking of identical pinched resistor structures on the same chip can be held to within ±6 percent. Since the effective thickness of the pinched resistor is the same as the base width of the *npn* transistor, the variations in the absolute values of the pinched resistor tend to track the transistor β_0 variations.

The "buried FET" structure of Fig. 3.15 provides an alternate method of obtaining high value resistors without the breakdown limitations of the base-diffused pinch resistor. In this case, the channel of the buried FET,

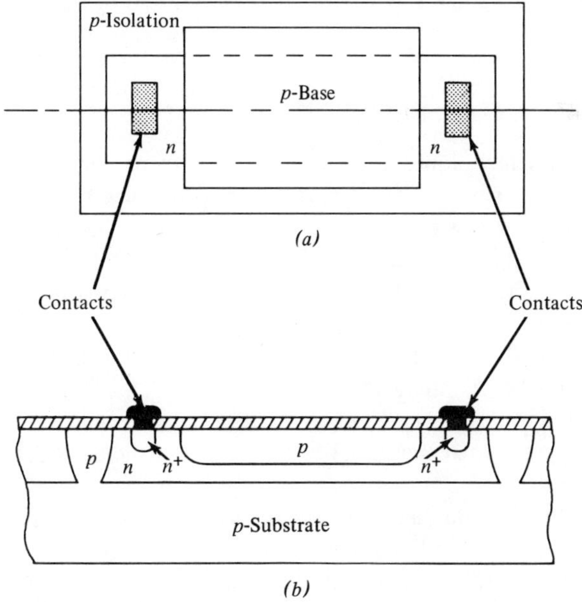

Figure 3.15 Buried FET structure: (*a*) device layout; (*b*) cross-section.

which constitutes the body of the resistor, is formed by the *n*-type epitaxial region surrounded on all sides by the *p*-type substrate, isolation, and the base diffusion. The sheet resistivity of the resulting device can be expressed as

$$R_s = \frac{\rho_e}{d - x_b} \tag{3.17}$$

where x_b is the depth of the base diffusion, and ρ_e and d are the resistivity and the thickness of the epitaxial layer. For resistivity ranges available in analog circuits, the values of R_s for a buried FET are in the range of 4 to 8 kΩ with a temperature coefficient of approximately +4000 ppm/°C.

Since all *pn* junctions forming the buried FET structure are lightly doped, the breakdown voltage associated with the device is quite high, typically of the order of transistor base-collector breakdown voltage. Thus, such a resistor structure is particularly suited for high voltage application. Unlike the base-diffused pinched resistor, the pinch-off voltage associated with the buried FET structure is a function of the lateral dimensions of the device as well as the epitaxial layer thickness. Therefore, it can be designed relatively independently of the *npn* bipolar transistor parameters. The absolute value and the matching tolerances associated with the buried FET structure are comparable to those of the *p*-type pinched resistor.

3.6 ION-IMPLANTED RESISTORS

Ion-implantation techniques can be utilized to form resistor structures on the semiconductor surface. The fundamental principles of ion implantation technology were briefly described in Chapter 1 (see Section 1.6) and are well covered in the literature.[5,6] With this technique the impurities are introduced into the silicon lattice by bombarding the wafer surface with high energy ions.

The implanted ions lie within a very shallow layer (typically of the order of 0.1 to 0.8 μ) along the silicon surface. Thus, for similar doping levels, the implanted layers yield a sheet resistivity which is roughly 20 times higher than a correspondingly doped diffused layer of 2 to 4 μ thickness.[6]

In fabricating ion-implanted resistors, normally boron-implanted *p*-type resistor structures are used. To define the resistor geometries, either a thick oxide layer or a metal layer (typically aluminum) can be used as a mask. After the implantation step, the wafer is annealed at a low temperature (typically 10 to 20 min at 500 to 600°C) to anneal out the structural damage sustained by the silicon lattice during the implantation step.

The sheet resistivity of the ion-implanted resistors is inversely proportional to the implantation dose. The final value of the sheet resistance is also effected by the post-implant annealing time and temperature, since this heat treatment determines the electrical activity of the implanted dopant atoms. By proper choice of implantation dose and annealing temperatures, sheet resistivity ranges of 500 Ω/square to 20 kΩ/square can be obtained. Both the implantation dose and the post-implant anneal cycles can be very accurately controlled. Thus, the absolute value of the sheet resistivity can be controlled to approximately ±6 percent and a uniformity or matching tolerance of ±2 percent can be obtained across the wafer.[7]

Figure 3.16 shows the planar layout and the cross-section diagram of a *p*-type ion implanted resistor. Due to the shallow depth of the ion-implanted resistor, it is difficult to obtain a good ohmic contact to the implanted region. Therefore, *p*-type diffused beds are used at the contact areas of the resistor, as shown in the Figure.

At high values of sheet resistivity, i.e., $R_s > 10$ kΩ, the implanted resistors exhibit JFET-like characteristics due to the pinching off of the resistor by the resistor-substrate depletion layer. At lower sheet resistance ranges this effect is negligible.

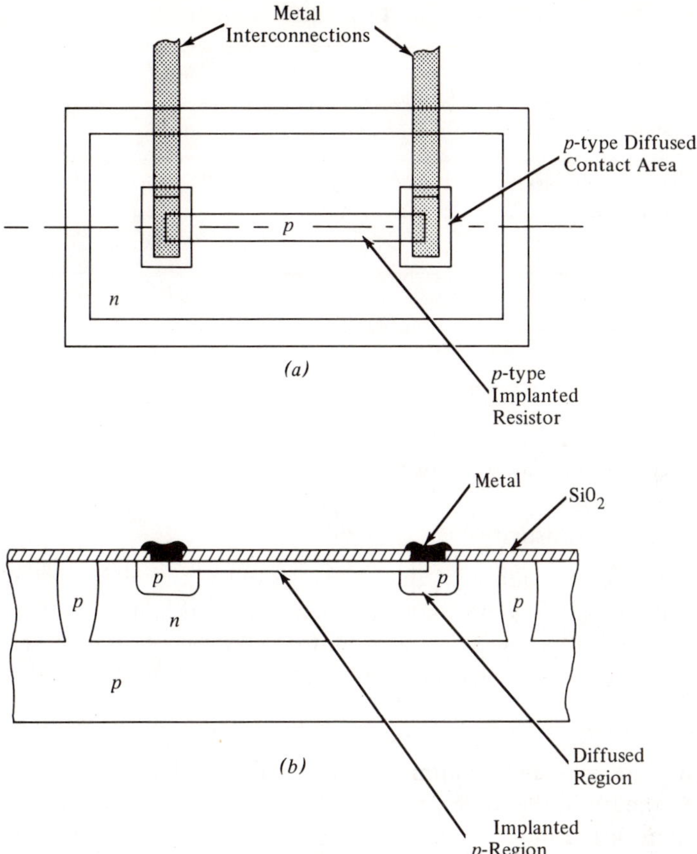

Figure 3.16 Ion-implanted resistor: (*a*) lateral geometry; (*b*) cross-section.

Ion-implanted resistors also exhibit a lower temperature coefficient of resistance than diffused resistors. The temperature coefficients of ion-implanted resistors in the 1 to 2 kΩ/square range are approximately a factor of four lower than a typical base-diffused resistor.[7]

The low temperature coefficient, high sheet resistivity and good matching properties make ion-implanted resistors readily suited for precision attenuators and ladder networks, as well as for micropower circuits.

3.7 THIN FILM RESISTORS

Using the film deposition techniques outlined in Chapter 1, resistive thin-film layers can be deposited and patterned on the silicon surface. Compared with the diffused resistors, thin films offer the following advantages:

TABLE 3.4 CHARACTERISTICS OF THIN-FILM RESISTORS

Characteristic	Ta	Ni-Cr	SnO_2
Sheet resistance (Ω)	200–5000	40–400	80–4000
Temperature coefficient (ppm/°C)	100	100	0 to −1500
Absolute value tolerance	±5%	±5%	±8%
Matching tolerance	±1%	±1%	±2%
Breakdown voltage	200 V	200 V	200 V

1. Low temperature coefficient
2. Tighter absolute value control
3. Lesser parasitics
4. Higher sheet resistivity

The main disadvantage of thin-film resistors is the additional process steps required in their fabrication. In some cases, in addition to the basic deposition and patterning steps, an additional SiO_2 deposition is necessary to stabilize the resistor structure by sealing it off from the ambient atmosphere.

The most commonly used thin-film resistors in integrated circuits are tantalum (Ta), nickel-chromium (Ni-Cr), and tin oxide (SnO_2) Table 3.4 gives a summary of the basic properties of each of these thin-film resistors. Of the three listed in the Table, Ta and Ni-Cr films are by far the most widely used.

In the circuit layout of thin-film resistors, zigzag geometrics with sharp corners are avoided since these etch poorly during the resistor masking step. Instead, a circular fold as shown in Fig. 3.11(*b*) is preferred. An alternate way of eliminating sharp corners in thin-film resistors is to lay out a number of parallel strips which can, then, be connected in series by aluminum interconnections at each end of the strips. The thin-film resistors should also be laid out over a smooth region of the surface oxide layer, with no steps or sudden changes of the oxide layer.

REFERENCES

1. A. B. Phillips, *Transistor Engineering,* McGraw-Hill, New York, 1962, Chap. 5.
2. H. Lawrence and R. M. Warner, "Diffused Junction Depletion Layer Calculation," *Bell System Tech. J.,* **39** (1960): 389–404.

3. J. C. Irvin, "Resistivity of Bulk Silicon and Diffused Layers in Silicon," *Bell System Tech. J.,* **41** (1962): 287–410.

4. A. B. Grebene, "A Practical Method for Reducing the Effects of Parasitic Capacitances in Integrated Circuits," *Proc. IEEE* (Letts.), **55** (2) (1967): 235–236.

5. J. T. Burrill, W. J. King, S. Harrison, and P. McNally, "Ion Implantation as a Production Technique," *IEEE Trans. Electron Devices,* **ED–14** (1), (Jan., 1967): 10–17.

6. J. D. Macdougall and K. E. Manchester, "Implanted Components in Microcircuits," *Proc. Natl. Electronics Conf.,* **25** (Dec., 1969), 140–145.

7. H. H. Stellrecht, D. S. Perloff and J. T. Kerr, "Precision Ladder Networks Using Ion Implanted Resistors," presented at WESCON 1971 (Aug., 1971).

4 General Design Considerations

For a linear circuit designer trained in the area of discrete circuits, the basic constraints and the limitations of integrated circuit technology often pose a difficult challenge. Some of the basic disadvantages of monolithic circuits which can render many of the discrete designs useless for integration are the following:

1. Poor absolute value tolerances
2. Poor temperature coefficients
3. Limitations on component values
4. Lack of integrated inductors
5. Limited choice of compatible active devices

On the other hand, integrated circuit fabrication methods offer a number of unique and powerful advantages to the circuit designer:

1. Availability of a large number of active devices
2. Good matching and tracking of component values
3. Close thermal coupling
4. Control of device layout and geometry

By making efficient use of these basic advantages associated with integrated circuits, it is often possible to come up with designs which can exceed the performance of similar discrete component circuits.

The purpose of this chapter is to outline some basic design guidelines for analog integrated circuits which make efficient use of the advantages of monolithic technology, and manage to circumvent most of its shortcomings. The basic circuit configurations to be discussed in this chapter deal with the dc design of a monolithic circuit, particularly with regard to voltage and current bias levels and their drift with temperature. These basic circuit configurations form a set of useful building blocks which in turn serve as the starting point for the design of larger and more complex functional circuits.

4.1 CONSTANT-CURRENT STAGES

In a constant-current stage, the reference current in one branch of the circuit is accurately reproduced in a second branch, relatively independent of the absolute values of device parameters. Such a circuit configuration is a particularly useful building block for analog circuit design, since it provides a means of establishing the dc bias levels within the circuit, within the accuracy of matching or tracking properties of the monolithic components.

A constant-current stage effectively simulates a current generator with one of its terminals connected to an ac ground point in the circuits. In general, this ground point is the positive or the negative supply. Depending on the polarity of the bias current flowing into a given circuit node from the constant-current stage, the latter can be classified as a "current source" or a "current sink," as shown in Fig. 4.1. From the circuit theory or network analysis point of view, the difference between the current source and the current sink is merely one of polarity. However, in monolithic circuits where the complementary devices are not readily available, this difference also implies that each kind would require a different polarity of device, i.e., *pnp* or *p*-channel FET for the current source and *npn* or *n*-channel FET for the current-sink circuits. Since the integrated circuit technology is built around the *npn*-bipolar process, the current-sink configuration is

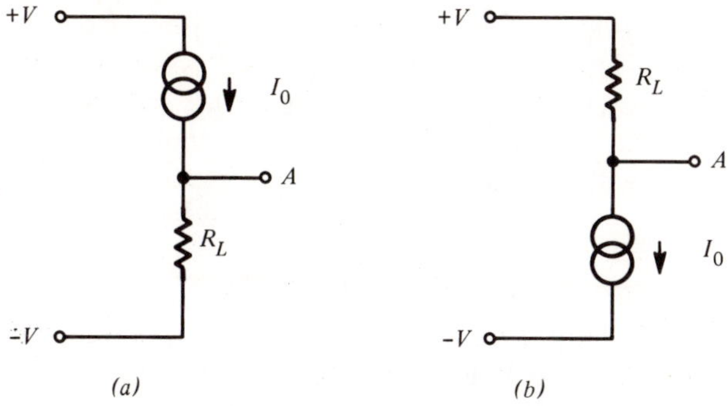

Figure 4.1 Circuit diagram of a current source (*a*); and a current sink (*b*).

by far the easier one to obtain. The analysis of this section deals with the current sink topology. However, the basic results and conclusions are directly applicable to the current source configuration by reversing the polarity of the transistors and the supply voltages.

Diode-Biased Current-Sink

As discussed in Chapter 2, the emitter current I_E of a transistor is related to the base-emitter voltage, V_{BE}, by an expression of the form

$$V_{BE} = V_T \ln (I_E/I_0) \tag{4.1}$$

where $V_T = kT/q$ = thermal voltage
I_0 = reverse saturation current

The reverse saturation current is related to the area of the emitter-base junction A as

$$I_0 = \gamma_E A \tag{4.2}$$

where the constant of proportionality γ_E depends on the intrinsic semiconductor parameters such as the minority carrier diffusion lengths and the concentration on either side of the junction. From Eqs. (4.1) and (4.2), if two transistors T_1 and T_2 are operated with the same base-emitter voltage, their emitter currents will be related as the ratio of their emitter areas, i.e.:

$$\frac{I_{E1}}{I_{E2}} = \frac{A_1}{A_2} \tag{4.3}$$

Since the collector current I_C differs from I_E by the base current I_{B1} for large values of β_0, i.e., $\beta_0 \gg 1$, the emitter current in Eqs. (4.1) and (4.3) can be readily replaced by I_C, without introducing any significant error. Figure 4.2 shows how the base-emitter voltage drop of one transistor can be used to set the current level in the next one.[1] Assuming for the moment that the base current drawn by T_2 is negligible, and that the transistors have emitter area ratios of A_1 and A_2 respectively, the collector currents of the two devices will be related as

$$\frac{I_2}{I_1} = \frac{A_2}{A_1} \tag{4.4}$$

Similarly, taking into account finite base-emitter voltage drops within the transistors current-sink, current I_2 can be determined as

$$I_2 = \left(\frac{V_{CC} - V_{BE}}{R_1}\right) \cdot \left(\frac{A_2}{A_1}\right) \approx \frac{V_{CC}A_2}{R_1 A_1} \tag{4.5}$$

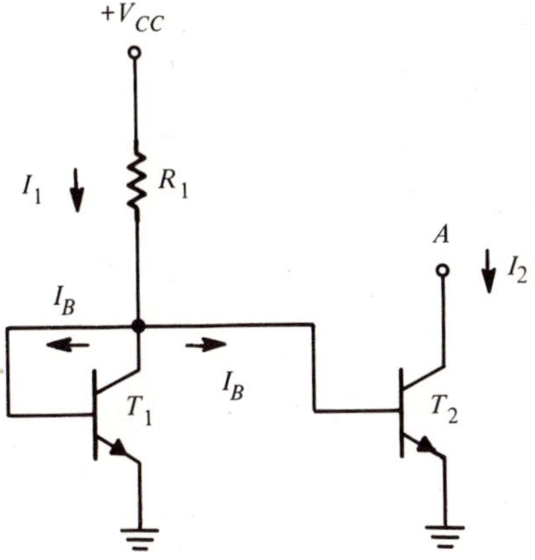

Figure 4.2 Diode-biased current sink.

The basic current-voltage equations, (4.1) thru (4.4), hold over six orders of magnitude in current, and over a broad temperature range (typically from -60 to $+130°C$). Thus the diode-biased current sink configuration of Fig. 4.2 provides a means of obtaining a current reference level, independent of implicit device parameters, which can be scaled by proper choice of the emitter areas of the two transistors.

Assuming that transistors T_1 and T_2 have identical geometry (i.e., $A_1 = A_2$), the effect of finite base current I_B into T_2 can be readily calculated to show that

$$I_2 = I_1 - 2\, I_B \tag{4.6}$$

Similarly, if the base-emitter voltages of the two transistors have a finite V_{BE} mismatch, the sink current I_2 can be related to the reference current I_1 as

$$I_2 = I_1 \left(1 - \frac{2}{\beta_0} + \frac{\Delta V_{BE}}{V_T}\right) \tag{4.7}$$

where $\Delta V_{BE} = V_{BE2} - V_{BE1}$. For a typical analog integrated circuit transistor, with $\Delta V_{BE} \approx \pm 1$ mV, $\beta_0 \approx 100$, the sink current I_2 is within ± 5 percent of I_1 over a wide range of temperature or current levels.

Figure 4.3 shows a modified version of the diode-biased current sink configuration which greatly reduces the β_0 dependence in the matching of currents I_1 and I_2.[2] In this configuration, the base current of T_2 which is extracted from the reference current is then resupplied to the base of the reference transistor T_1, thus keeping the current levels in T_1 and T_2 uneffected by the base current changes. The V_{BE} drop across the diode con-

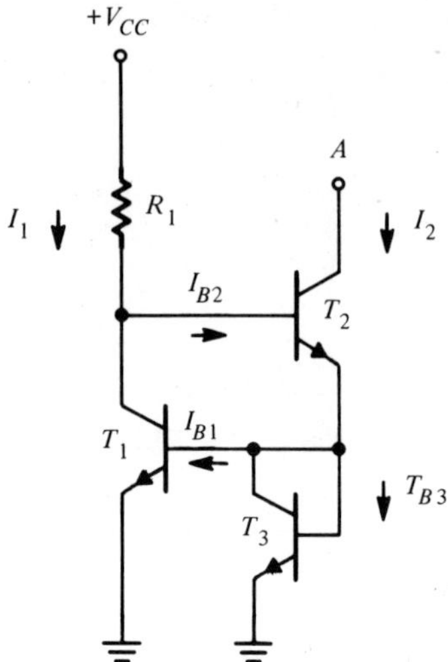

Figure 4.3 An improved version of the diode-biased current sink.

nected transistor T_3 sets the bias level for the reference transistor T_1, which in turn sets the current level in T_2. It can be shown by straightforward nodal analysis that the sink current I_2 is related to the reference current I_1 as

$$I_2 = I_1 + (I_{B1} + I_{B3} - 2\ I_{B2}) \tag{4.8}$$

Note that the additional terms within the parenthesis represent the third-order errors due to the base current mismatches, and reduce to zero if the transistor β_0's are matched. If the transistor β_0's are matched to better than ± 20 percent, the difference between I_2 and I_1 is less than 0.5 percent, for a typical β_0 value of 100.

Further insight can be gained into the operation of the improved constant-current stage of Fig. 4.3 by treating it as a special case of the shunt-series feedback, with a unity feedback connection in the shunt branch. This corresponds to the unity current feedback condition from the emitter of T_2 back to the base of T_1. As a consequence of this feedback arrangement, the output impedance of the resultant current sink stage is greatly increased. This is illustrated in the current-voltage characteristics of Fig. 4.4. Also, since there is practically no base-current error, the temperature tracking of I_1 and I_2 is nearly perfect.[3]

Resistor-Biased Current Sink

Figure 4.5 shows a modified version of the basic diode-biased current sink which uses resistor ratios rather than emitter areas to set current levels.

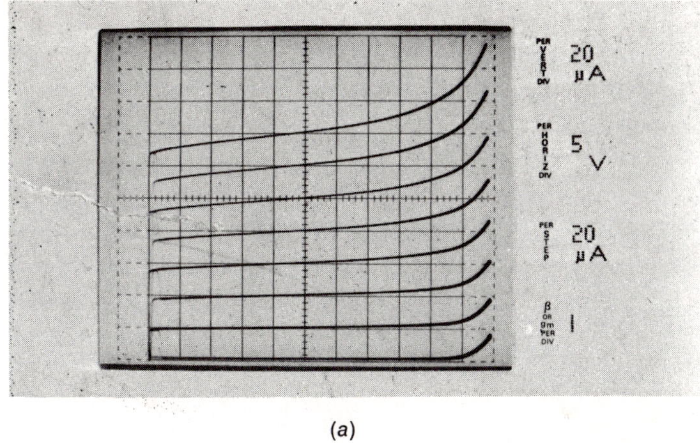

(*a*)

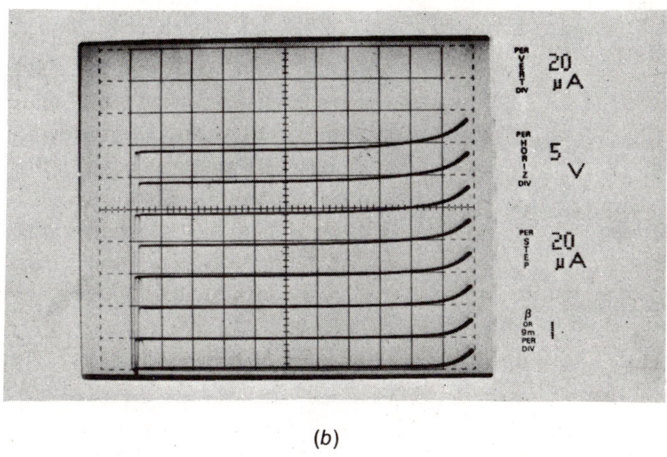

(*b*)

Figure 4.4 Current-voltage characteristics of constant-current stages shown in (*a*) Fig. 4.2; and (*b*) in Figure 4.3.

Neglecting the base current of T_2, current levels through each of the two transistors are related as

$$I_1R_1 + V_{BE_1} = I_2R_2 + V_{BE_2} = V_A \tag{4.9}$$

The base-emitter voltage drop difference between two identical transistors operating at respective collector currents I_1 and I_2 can be written as

$$\Delta V_{BE} = V_{BE_2} - V_{BE_1} = V_T \ln (I_2/I_1) \tag{4.10}$$

Thus from (4.9) and (4.10), the ratio of the currents can be expressed as

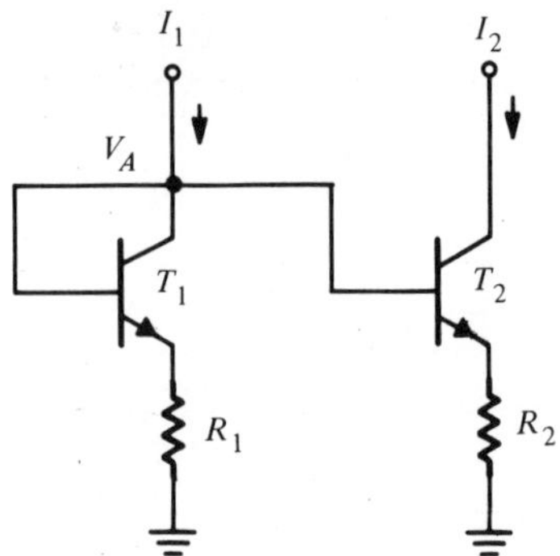

Figure 4.5 Resistor-biased current sink.

$$\frac{I_2}{I_1} = \frac{R_1}{R_2}\left[1 - \frac{V_T \ln (I_2/I_1)}{R_1 I_1}\right] \tag{4.11}$$

If the voltage drop across R_1 is made comparable to V_{BE}, then the second term within the brackets becomes negligibly small and the two currents are closely related by the resistor ratio, i.e.,

$$I_2/I_1 \approx R_1/R_2 \tag{4.12}$$

For $I_1R_1 \geq V_{BE}$, the two currents follow the resistor ratio of Eq. (4.12), with a maximum error of less than ± 10 percent over two orders of magnitude in current, i.e.,

$$(1/10 < (I_2/I_1) < 10)$$

independent of temperature.

The resistor-biased constant-current stage is preferred over the simple diode-biased current-sink of Fig. 4.2 for the cases where the ratio of (I_2/I_1) is significantly different than unity, since the resistor ratios can be varied over a broader range of values than the emitter areas.

Figure 4.6 shows a special case of the resistor-biased constant-current stage, with R_1 set equal to zero.[1] Such a configuration is particularly useful for obtaining very low values of sink current with relatively large values of the reference current. Assuming matched device geometries, T_1 operates at a higher current density than the base-emitter diode of T_2, and the voltage drop V_2 across R_2 is constrained to be

$$V_2 = I_2R_2 = V_{BE_1} - V_{BE_2} \tag{4.13}$$

Solving for R_2 from Eqs. (4.10) and (4.13), we can write

$$R_2 = V_T/I_2 \ln (I_1/I_2) \tag{4.14}$$

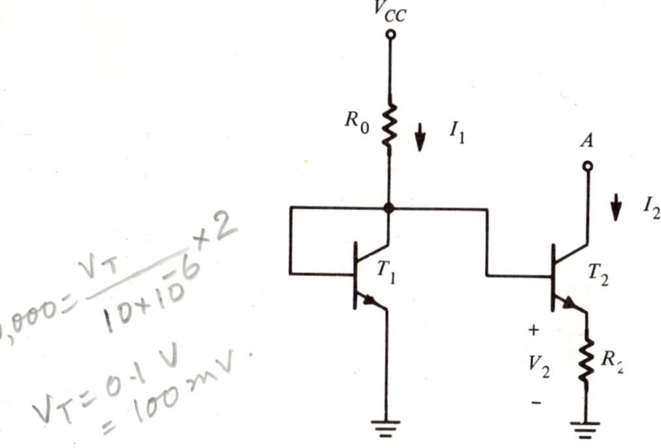

Figure 4.6 A constant-current stage for low current levels.

Since R_2 is proportional to the logarithm of the current ratio, a high degree of current mismatch can be obtained without requiring extremely high resistor values. For example, letting $R_2 = 20\ \text{k}\Omega$, one obtains a 100:1 ratio between the two current, with $I_2 = 1$ mA and $I_2 \approx 10\ \mu$A.

An added advantage of the constant-current circuit of Fig. 4.6 is that the sink current I_2 is relatively independent of the supply voltage, i.e.,

$$I_2 = V_T/R_2 \ln\left[\frac{V_{CC} - V_{BE}}{R_0 I_2}\right] \tag{4.15}$$

Thus, for $I_1 \gg I_2$, I_2 varies as the logarithm of the supply voltage. As will be discussed further in the next chapter, such a low-current sink circuit is particularly useful for biasing the input stage of an operational amplifier, where it can be used to set the quiescent current level of the stage independent of the supply voltage changes.

By differentiating (4.15) with respect to temperature, one can show the temperature coefficient of I_2 can be nominally reduced to zero if the temperature dependence of R_2 is such that

$$\frac{dR_2}{R_2} = \frac{dT}{T} \tag{4.16}$$

At room temperature, this corresponds to a resistor temperature coefficient of about +3300 ppm/°C. A monolithic resistor having such a temperature coefficient can be fabricated using either a high resistivity p-diffusion or an n-type bulk resistor structure, as described in Chapter 3.

4.2 VOLTAGE SOURCES

In a variety of circuit applications, it is necessary to establish a low impedance point within the circuit which can serve as an internal voltage supply. Ideally, such a voltage reference point in the circuit is required to have both a very low ac impedance and a very stable dc voltage level which is insensitive to power supply and temperature variations. In most applications however, only one of these two requirements, i.e., either the low impedance or the dc voltage stability is of prime importance. The circuits which primarily fulfill the low impedance requirements are known as voltage sources; whereas those specifically designed to provide a constant voltage independent of the supply or the temperature changes are called voltage references.

The voltage source stages are normally used to provide independent bias levels within the circuit. In such an application, the low ac impedance of the voltage source is necessary to buffer or decouple adjacent gain stages. An example of such an application is the use of voltage source stages to form a common bias point for the inputs of a differential gain stage. Figure 4.7 shows some of the practical voltage source configurations for integrated circuits. In the circuit of Fig. 4.7(*a*), the low output impedance of an emitter-follower stage is used to simulate a low impedance voltage source with an output voltage level V_L, given as

$$V_L \simeq V_{CC}\left(\frac{R_2}{R_1 + R_2}\right) \tag{4.17}$$

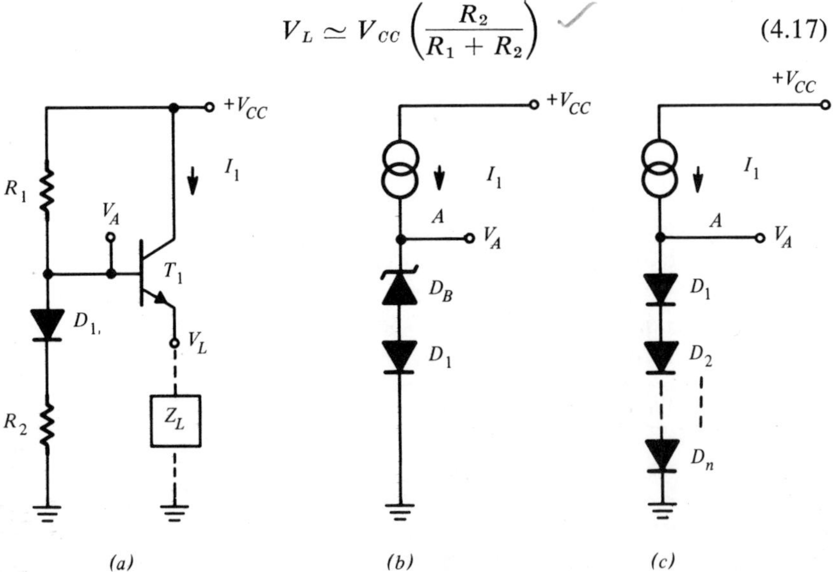

Figure 4.7 Practical voltage source configurations: (*a*) common-collector stage; (*b*) temperature-compensated avalanche diode; (*c*) diode string.

Note that in this configuration, the diode D_1 in the bias string is used to offset the dc value and the temperature dependence of the V_{BE} drop across T_1. The load, Z_L, represents the rest of the circuitry biased by the current through T_1. Using the hybrid-π model for the transistor (see Fig. 2.10), the impendance level R_0 looking into the emitter of T_1 can be expressed as

$$\mathrm{R}_0 \simeq \frac{V_T}{I_1} + \frac{R_1 R_2}{\beta_0(R_1 + R_2)} \tag{4.18}$$

Due to the resistive bias string in Fig. 4.7(*a*), the value of the bias voltage V_A and the output voltage V_L are both dependent on the supply voltage V_{CC}. This dependence can be avoided using the bias circuits of Fig. 4.7(*b*) and (*c*). In each of these circuits, the impedance looking into the bias terminal, A, is low enough for most applications to eliminate the need for an additional emitter-follower stage. The current source I is normally simulated by a resistor connected between V_{CC} and V_A, with the requirement that this resistor value be significantly higher than the impedance level measured looking into node A.

The circuit of Fig. 4.7(*b*) provides an output voltage level,

$$V_A = V_B + V_{BE} \tag{4.19}$$

where V_B is the breakdown voltage of the base-emitter avalanche diode D_B. The forward diode D_1 is used to provide partial compensation for the positive temperature coefficient of V_B (see Section 2.2). In a monolithic structure, the D_B and D_1 combination can be conveniently designed as a single transistor structure with two separate emitters, as shown in Fig. 2.21. The impedance level measured looking into the output terminal is

$$R_0 = R_B + \frac{V_T}{I_1} \tag{4.20}$$

where R_B is the dynamic impedance of the base-emitter breakdown diode. For typical integrated device structures, R_B is in the range of 40 to 100 Ω. Since the base-emitter avalanche breakdown voltage is set by the integrated circuit fabrication process V_A in the circuit of Fig. 4.7(*b*), it is restricted to be within the 6.5 to 9 V range.

Figure 4.7(*c*) shows how a number of diodes or diode-connected transistors can be cascaded to simulate a low impedance output voltage level V_A, given as

$$V_A = nV_{BE} \tag{4.21}$$

and an impedance level of

$$R_0 = \frac{nV_T}{I_1} \tag{4.22}$$

where n is the number of diodes in the string. Such a voltage source has a strong negative temperature coefficient, given as

$$\frac{\partial V_A}{\partial T} = n \frac{\partial V_{BE}}{\partial T} \simeq -2\, n\, mV/°C \tag{4.23}$$

Since each diode in the string requires a separate collector pocket, a string involving a large number of diodes may occupy a significant portion of the chip area.

In some analog circuit applications, it may be necessary to provide multiple voltage sources which are biased from the same reference voltage but are buffered from each other so that the ac signals in one source would be relatively isolated from those in the other. In conventional circuit design using discrete devices, this can be accomplished using two emitter-follower stages as shown in Fig. 4.8(*a*). However, using the design and layout advantages of monolithic circuits, the same circuit can be designed using a multiple emitter transistor as shown in Fig. 4.8(*b*).

A low impedance voltage source can also be obtained by using a transistor in a shunt feedback configuration as shown in Fig. 4.9. The principle of operation of such circuits can be best understood by considering the basic circuit of Fig. 4.9(*a*): The voltage drop across R_2 is constrained to be equal to the transistor V_{BE} drop. Assuming that the base current of T_1 is negligible, the current through R_2 is the same as that through R_1. There-

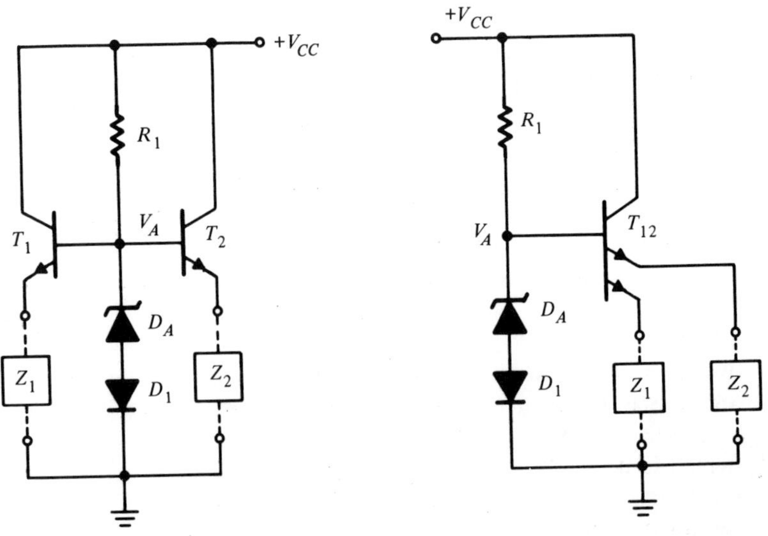

Figure 4.8 Multiple voltage sources biased from the same reference voltage: (*a*) discrete design (*b*) its monolithic counterpart using multiple-emitter transistors.

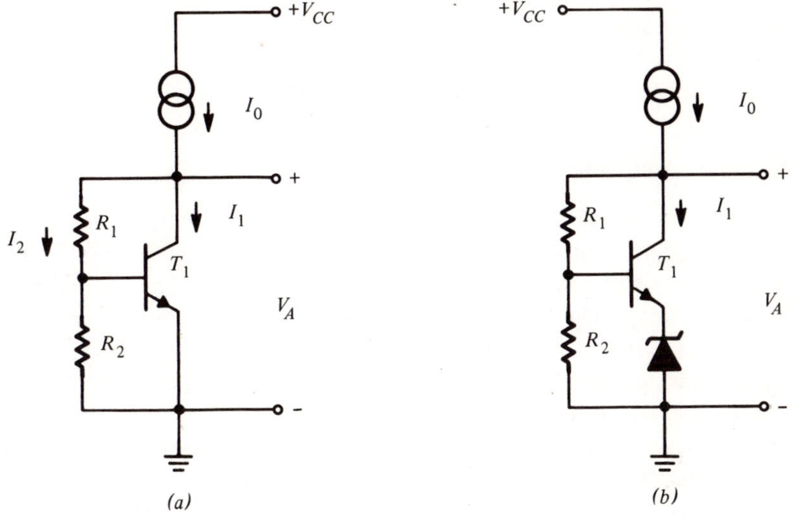

Figure 4.9 Additional voltage source configurations.

fore, the output dc level can be readily related to the transistor V_{BE} as

$$V_A = I_2(R_1 + R_2) = V_{BE}\left(1 + \frac{R_1}{R_2}\right) \tag{4.24}$$

Note that, due to the shunt feedback provided by R_1, the transistor current I_1 automatically adjusts itself to maintain I_2 and V_A relatively indepedent of the supply voltage. The circuit configuration of Fig. 4.9(*a*) provides a convenient substitute for the diode string voltage source of Fig. 4.7(*c*) when a relatively large number of diode drops are needed. Using the hybrid-π model for the transistor, the impedance level at node A can be readily calculated to be

$$R_0 = \frac{R_1}{\beta_0} + \frac{R_1 + R_2}{g_m R_2} \tag{4.25}$$

where g_m is the transconductance of T_1. For most applications, the value of R_0 is in the range of 50 to 200 Ω.

Fig. 4.9(*b*) shows a modified version of the basic shunt-feedback circuit, which is useful for obtaining high value voltage sources without requiring avalanche diodes with high breakdown voltages. Assuming that D_B is a base-emitter diode with a breakdown voltage V_B, the voltage level at the output terminal can be expressed as

$$V_A = (V_B + V_{BE})(1 + R_1/R_2) \tag{4.26}$$

The circuit configuration of Fig. 4.9(*b*) is particularly useful for high voltage integrated circuits where it can be used either as a high value voltage source (20 V $< V_A <$ 100 V) or can be substituted for a high voltage avalanche diode for overvoltage protection.

4.3 VOLTAGE REFERENCES

In the design of various analog circuits such as low drift amplifiers or voltage regulators, it is often necessary to establish an internal voltage reference within the circuit. Unlike the case of voltage sources, the main emphasis in a voltage reference stage is not on the low output impedance but on the thermal stability of the reference voltage. Temperature stability requirements for a reference voltage are typically $\leq$100 ppm/°C. Temperature coefficients of most monolithic circuit components or devices are much greater than this amount. However, by making use of the matching and tracking properties of integrated components, and the close thermal coupling on the chip, it is possible to compensate the thermal drifts to a few parts per million level. Figure 4.10 shows a practical circuit configuration for generating a reference voltage V_R with a very low temperature coefficient.[4] The base-emitter avalanche diode D_B is supplied by a constant current I_o and provides a bias voltage V_B with a positive temperature coefficient (typically $\approx$ +3 mV/°C). The temperature dependence of the V_{BE} drop across T_1 and D_1 results in a temperature coefficient of about +7 mV at the cathode of D_1. Similarly, the thermal variation of the voltage drop across D_2 creates a temperature coefficient of $\approx$ −2 mV/°C at the anode of D^2. Thus, by tapping the resistor string, R_1 and R_2 connecting these two points of opposite temperature drift, a voltage reference V_R can be made to have nominally zero temperature coefficient. The voltage level of V_R is given as

$$V_R = \frac{R_2 V_B + V_{BE}(R_1 - 2\,R_2)}{R_1 + R_2} \tag{4.27}$$

The temperature coefficient of V_R can be nominally set to zero by setting the resistor ratio as

$$\frac{R_1 - 2\,R_2}{R_2} = -\frac{(\partial V_B/\partial T)}{(\partial V_{BE}/\partial T)} \tag{4.28}$$

For typical values of $(\partial V_B/\partial T)$ and $(\partial V_{BE}/\partial T)$ associated with integrated circuit components, the nominal value of V_R for zero temperature coefficient is in the range of 1.7 to 2.5 V. If a different output dc level is required, either D_1 or D_2 (or both) can be replaced by a diode string.[5] Using the temperature compensation scheme outlined above, the tempera-

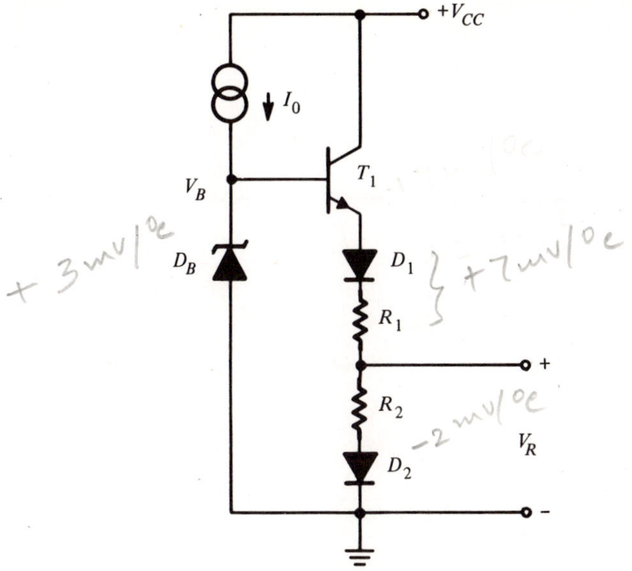

Figure 4.10 A circuit for generating temperature-independent voltage reference.

ture coefficient of V_R can be kept within less than ± 40 ppm/°C over a -55 to $+125$°C temperature range.

As indicated by the design example described above, the basic approach in a voltage reference design is to utilize the close thermal coupling between the monolithic elements, and to compensate for the known thermal drifts by introducing an opposing drift source of equal magnitude. An alternate design approach for a voltage reference stage which uses a similar temperature compensation scheme is shown in Fig. 4.11.[6] In this circuit, T_1 sets the current level in transistor T_2 which is biased in the current sink configuration of Fig. 4.6. The voltage across R_3 is equal to the base-emitter voltage drop difference ΔV_{BE} where

$$\Delta V_{BE} = V_{BE_1} - V_{BE_2} = V_T \ln (I_1/I_2) \tag{4.29}$$

Assuming $\beta_0 \gg 1$, the net voltage drop across R_2 is equal to V_2, given as

$$V_2 = (R_2/R_3)\,\Delta V_{BE} \tag{4.30}$$

The output voltage level V_R is then equal to the base emitter drop of T_3 plus V_2, i.e.,

$$V_R = V_{BE} + (R_2/R_3)\,\Delta V_{BE} \tag{4.31}$$

As described earlier in this section, V_{BE} decreases with temperature at a rate of ≈ -2 mV/°C. However, ΔV_{BE} has a positive temperature co-

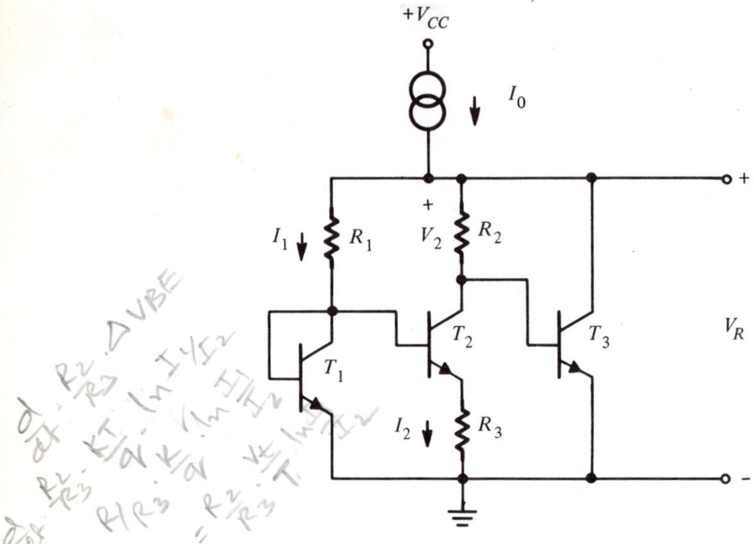

Figure 4.11 An alternate circuit for temperature-compensated voltage reference.[6]

efficient given as

$$\frac{\partial(\Delta V_{BE})}{\partial T} = \frac{V_T}{T} \ln\left(\frac{I_1}{I_2}\right) \tag{4.32}$$

Thus, by proper choice of current levels I_1 and I_2 and the resistor ratio (R_2/R_3), first-order temperature dependence of V_R can be reduced to zero. Letting $(R_2/R_3) = 10 \approx (I_1/I_2)$, one can obtain a nominal zero temperature coefficient for V_R at $V_R \approx 1.2$ V.

One basic drawback of the temperature compensated voltage reference circuits shown in Figs. (4.10) and (4.11) is that the dc level of V_R is directly related to the desired temperature coefficient, i.e., the values of V_R and $\partial V_R/\partial_T$ cannot be chosen independently. An integrated voltage-reference configuration which does not have this drawback is the balanced bridge-type reference circuit shown in Fig. 4.12.[7] The bridge is balanced by letting $R_1 = R_2$ and $I_1R_1 = nV_{BE}$. In this mode of operation, the current flow through the center leg of the bridge is negligible; and the dc level of the reference voltage V_R is set independent of R_A or R_B as

$$V_R = nV_{BE} \tag{4.33}$$

However, the temperature coefficient of V_R depends on where the output tap is taken along the center leg of the bridge, and can be expressed as

$$\frac{\partial V_R}{\partial T} = n\,\frac{\partial V_{BE}}{\partial T}\left(\frac{R_B}{R_A + R_B}\right)\left[1 - \frac{R_A R_1}{R_B(R_1 + 2\,R_0)}\right] \tag{4.34}$$

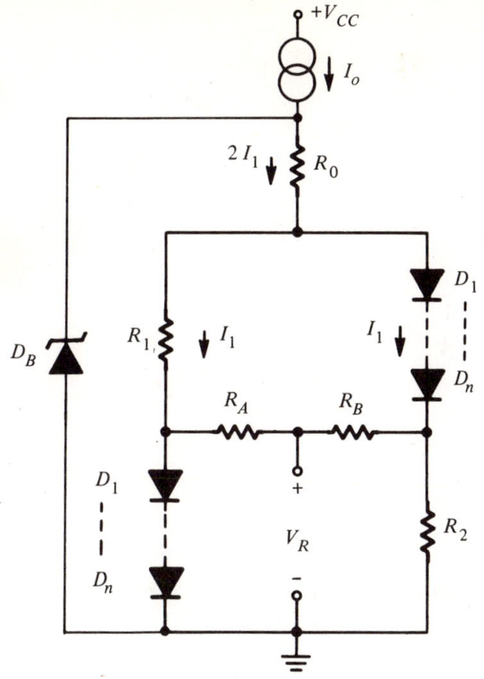

Figure 4.12 A balanced bridge-type voltage reference.

Thus, the temperature coefficient of V_R can be reduced to zero by choosing

$$\frac{R_B}{R_A} = \frac{R_1}{R_1 + 2\,R_0} \tag{4.35}$$

independent of the number of diodes in the bridge.

4.4 DC LEVEL-SHIFT STAGES

Since large value coupling capacitors are not available in monolithic circuits, all broadband gain stages need to be dc-coupled. This means that the output dc level of a gain stage should be compatible with the dc level at the input of the next stage. In an *npn* common-emitter gain stage, the output dc level is always higher than the dc level of the input. Therefore, if a number of such gain stages are cascaded, the output dc level rapidly builds up toward the positive supply voltage. This in turn limits amplitude and the linearity of the available output swing. Ideally, such a dc level buildup can be avoided by using complementary *pnp-npn* gain stages. However, the *pnp* transistors available in monolithic form have relatively poor frequency response and current gain characteristics.

If an analog integrated circuit comprised of a cascade of *npn* gain stages, this positive dc level buildup can be overcome by using a level-shift

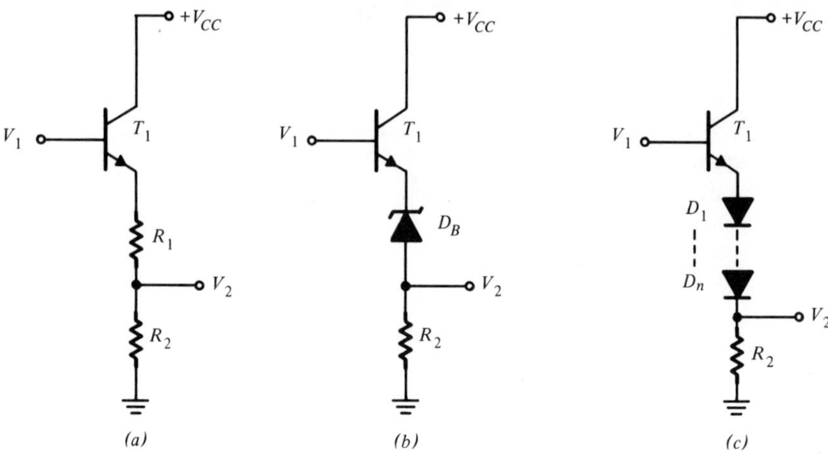

Figure 4.13 Some practical dc level-shift stages: (*a*) resistive; (*b*) avalanche-diode; (*c*) diode-string.

stage between each gain stage, to shift the output dc level toward the negative supply, with minimum attenuation of the ac signal. In general, such a stage also serves as a unilateral buffer between successive gain stages; therefore it is required to have a high input impedance and a relatively low output impedance to prevent interstage loading. Figure 4.13 shows some practical dc level-shift stages for monolithic circuit applications. In each case, a common-collector stage is used at the input to avoid loading the output of the gain stage connected to V_1.

The resistive level-shift stage of Fig. 4.13(*a*) provides a simple means of shifting the input level V_1 to a more negative dc level V_2, where

$$V_2 = (V_1 - V_{BE}) \frac{R_2}{R_1 + R_2} \tag{4.36}$$

The main drawback of the resistive level-shift stage is the attenuation of the ac signal along with the dc level shift. The ac voltage gain A_v for the resistive level-shift stage is less than unity, given as

$$A_v = \frac{R_2}{R_1 + R_2} < 1 \tag{4.37}$$

Thus, as the value of R_2 is decreased to improve the net dc level shift from V_1 to V_2, the ac gain of the stage deteriorates rapidly. The output impedance of such a stage is also relatively high, being equal to the shunt combination of R_1 and R_2.

The avalanche-diode level-shift stage of Fig. 4.13(*b*) provides an alternate means of shifting the dc level by an amount

$$V_1 - V_2 = V_{BE} + V_B \tag{4.38}$$

where V_B is the breakdown voltage of the avalanche diode D_B. Reverse breakdown characteristics of the base-emitter junction is used to form the avalanche diode ($V_B \approx$ 6–9 V). If the bulk resistance of D_B is negligible compared to R_2, the voltage gain for the stage is approximately unity. The two main disadvantages of the avalanche-diode level-shift stage are the limitations on the values of V_B available in integrated circuits and the excess noise generated by the breakdown diode D_B. Therefore, such a level-shift scheme is not suitable for low level ac signals.

The diode-string level-shift stage of Fig. 4.13(*c*) provides a net dc level shift $V_1 - V_2$ given as

$$V_1 - V_2 = (n + 1)V_{BE} \tag{4.39}$$

where n is the number of diodes in the string. Normally, the diodes D_1 through D_n would be formed by diode-connected transistors. Assuming that the dynamic impedance of the diodes is negligible compared to R_2, the voltage gain of the stage is approximately unity. The output impedance R_o for the circuit is quite low, given as

$$R_0 \approx (n + 1)\frac{V_T}{I_1} \tag{4.40}$$

where V_T is the thermal voltage. The diode string level-shift stage has two disadvantages: (1) since each diode requires a separate isolation pocket, such a stage may take up sizable chip area and have appreciable shunt capacitance to the substrate; (2) the output dc level shows a strong temperature dependence due to the change of the diode voltage V_{BE} with temperature.

Figure 4.14 shows an alternate circuit for dc level-shift applications which uses the voltage source configuration of Fig. 4.9(*a*). The net dc level change across the stage is

$$V_1 - V_2 = V_{BE}(2 + R_1/R_2) \tag{4.41}$$

Assuming that $\beta_0 \gg 1$ the net voltage gain across the stage can be expressed as

$$A_v \approx \frac{R_3R_2g_m}{R_3R_2g_m + 1} \approx 1.0 \tag{4.42}$$

where g_m is the transconductance of T_2. Similarly, the output impedance R_o for the stage can be calculated to be

$$R_0 = R_3 \,\Big/\!\!\Big/ \left(\frac{R_1}{R_2g_m}\right) \tag{4.43}$$

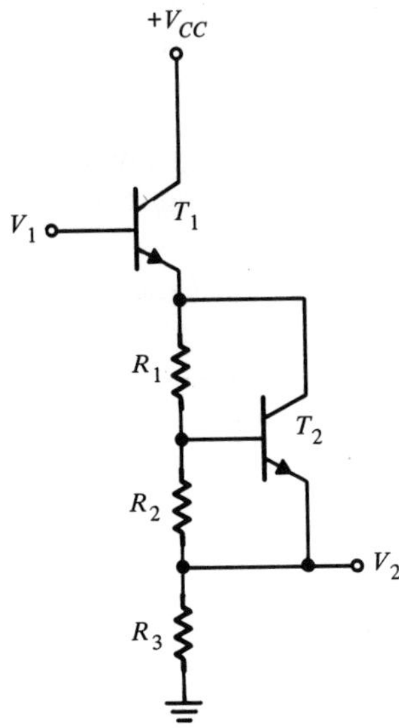

Figure 4.14 A dc level-shift configuration using the circuit of figure 4.9.

The main disadvantage of the level-shift circuit of Fig. 4.14 is the strong temperature dependence of V_2 due to the V_{BE} change with temperature.

A practical level-shift stage often used in bipolar analog circuits is the resistor and current sink stage combination shown in Fig. 4.15. Since the

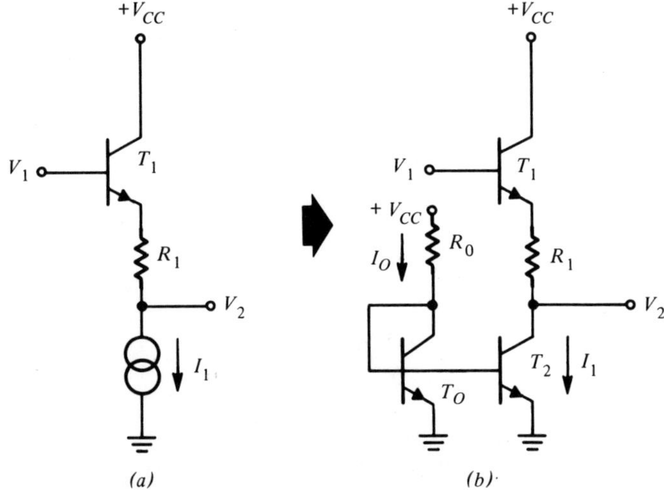

Figure 4.15 Resistor and current sink level shift stage: (*a*) simplified diagram; (*b*) actual circuit implementation.

dynamic impedance of the current sink is much higher than R_1, the ac voltage gain of the stage is very close to unity. However, the output dc level is shifted toward the negative supply (or ground) by an amount

$$V_1 - V_2 = V_{BE} + I_1 R_1 \tag{4.44}$$

The main disadvantage of this level-shift stage is its relatively high output impedance. This output impedance, which is approximately equal to R_1, makes it susceptible to capacitive loading at the output and can limit its frequency response. This drawback can be partly circumvented by adding an emitter-follower stage to the output of the level-shift circuit.

Figure 4.15(*b*) shows an actual circuit implementation of the resistor-current-sink level-shift scheme by using the diode-biased current sink circuit of Fig. 4.2. Note that if the transistors T_o and T_2 have the same emitter area, than the currents I_o and I_1 are equal, i.e.,

$$I_0 = I_1 = \frac{V_{CC} - V_{BE}}{R_0} \tag{4.45}$$

Therefore, the net dc level shift across the stage can be related to the supply voltage as

$$V_1 - V_2 = V_{BE} + (R_1/R_0)(V_{CC} - V_{BE}) \tag{4.46}$$

or, for $V_{CC} \gg V_{BE}$, as

$$V_1 - V_2 \approx V_{CC}(R_1/R_0) \tag{4.47}$$

Thus, using a circuit configuration similar to Fig. 4.15(*b*), the dc level shift provided by the stage can be made to track the supply voltage changes, and be insensitive to the absolute value tolerances associated with the bias current levels and the resistor values.

The voltage gain of the simple resistor and current-sink level-shift stage can be increased above unity by means of a positive feedback scheme[8] shown in Fig. 4.16. In this configuration, transistor T_3 provides a low-impedance output for the level-shift stage comprised of T_1 and T_2. The positive feedback is provided to the emitter of T_2 via R_3. Assuming $\beta_0 \gg 1$, the output dc level of the circuit can be expressed as

$$V_2 = (V_1 - 2\,V_{BE})\frac{R_3}{R_3 - R_1} - V_{CC}\frac{R_1(R_2 + R_3)}{(R_3 - R_1)(R_0 + R_2)} \tag{4.48}$$

The ac voltage gain A_v for the circuit can be written as

$$A_v = \frac{1}{1 - R_1/R_3} \tag{4.49}$$

As indicated by Eqs. (4.48) and (4.49), $R_3 > R_1$ is a design requirement

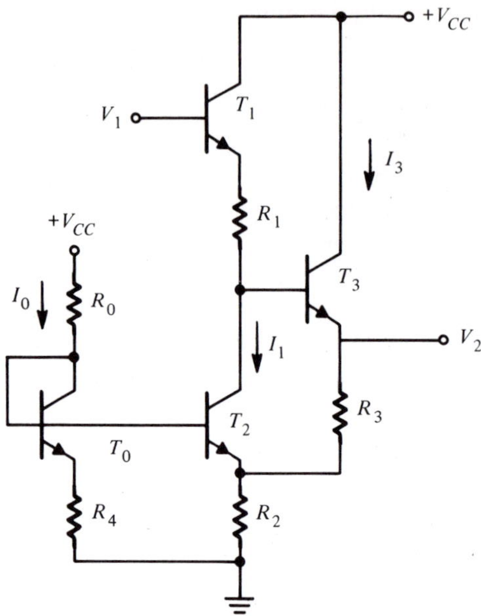

Figure 4.16 A dc level-shift stage with ac voltage gain.

for stable operation of the circuit. The frequency response of the circuit is relatively poor ($\approx$200 kHz with $A_v = 4$) due to the narrow-banding effect of the positive feedback.

The level shift scheme of Fig. 4.16 can result in a negative real part of the input impedance seen at the base of T_1. In order to avoid potential instability, it is necessary that the positive feedback stage be driven from a low impedance source, such that the effective source, impedance R_s at the base of T_1 must satisfy the inequality

$$R_s \ll \frac{\beta_0 R_3}{A_v} \tag{4.50}$$

Conventional design practice of cascading complementary gain stages to avoid level shifting can also be applied to monolithic circuits, using lateral *pnp* transistors. In many cases, however, the low value of the current gain and the poor frequency response characteristics associated with the lateral *pnp* make it a poor substitute for discrete (nonintegrated) *pnp* devices in a conventional design. By interconnecting a vertical *npn* and a lateral *pnp* transistor, it is possible to form a composite device which has gain characteristics superior to those of a lateral *pnp*. This composite device has the polarity of the *pnp* and can be used for a variety of level-shifting or current source applications. A detailed description of the applications of such composite devices is presented in the following section.

4.5 COMPOSITE *pnp-npn* CONNECTIONS

The relatively poor current gain characteristics of the lateral *pnp* transistor can be partially compensated by interconnecting a lateral *pnp* with a vertical *npn* transistor to form a composite device. As stated, the resulting structure has the current gain of the *npn* transistor but retains the polarity of the *pnp*. Since the lateral *pnp* has a much poorer frequency performance than the *npn*, the high frequency capability of the composite device is still limited by the *pnp* performance. Therefore, main applications for the composite, *pnp-npn* stages are in the design of dc level-shift stages or current-source stages. The current source stages formed by the lateral *pnp* transistors can be used either for biasing purposes or they can be employed as active loads to obtain large voltage amplification from a single gain stage. This latter application will be discussed further in a later section, in connection with operational amplifier design.

Figure 4.17 shows a commonly used dc level-shift circuit which utilizes a lateral *pnp*-vertical *npn* combination. Note that the interconnection of the devices shown in the Figure is the same as that mentioned earlier in Chapter 2 (see Fig. 2.16). The output dc voltage of the level-shift stage shown in Fig. 4.17 can be expressed as

$$V_2 = \frac{R_2}{R_1}(V_{CC} - V_{BE} - V_1) \tag{4.51}$$

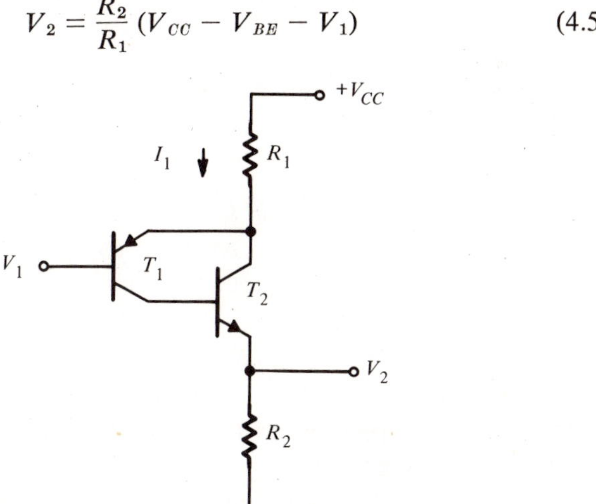

Figure 4.17 A pnp-npn level-shift stage.

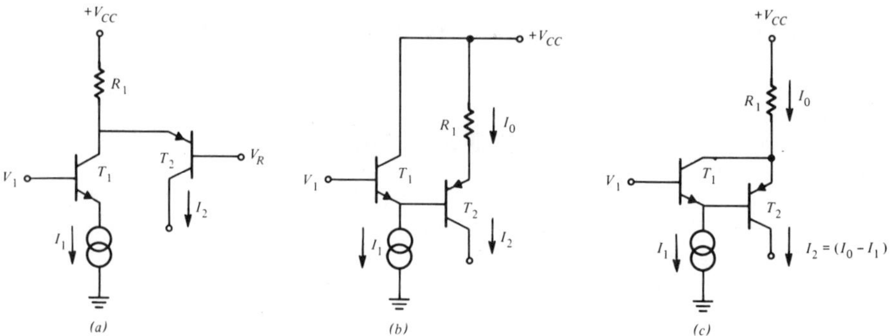

Figure 4.18 Current source circuits using composite *npn-pnp* transistors.

Similarly, the ac voltage gain of the circuit can be expressed as

$$A_v \approx -(R_2/R_1) \tag{4.52}$$

pnp-npn combinations are often utilized in the design of integrated current-source stages. Some of the basic composite device connections for this application are shown in Fig. 4.18. In the circuit connection of Fig. 4.18(a), the lateral *pnp* being used in common-base configuration with a conventional *npn* transistor. The base of the *pnp* is biased from a fixed-bias reference V_R. Resistor R_1 and the collector impedance of T_1 simulate a constant-current drive for the emitter terminal of T_2. The collector current I_2 of the *pnp* can be expressed as

$$I_2 = \alpha_p(I_0 - I_1) \tag{4.53}$$

where α_p is the common-base current gain for the lateral *pnp;* and the bias current I_o is given as

$$I_0 = (1/R_1)(V_{CC} - V_{BE} - V_R) \tag{4.54}$$

Note that the current generator I_1 can be formed by the collector of the *npn* gain stage, and thus be a controlled source. In this case, the entire *pnp-npn* composite stage functions as a current amplifier, with a current gain A_I given as

$$A_1 = \frac{\partial I_2}{\partial I_1} = -\alpha_p \tag{4.55}$$

with a 3-dB bandwidth equal to f_α of the lateral *pnp* (i.e. 3 to 5 MHz).

The circuit of Fig. 4.18(*b*) shows an alternate composite device to form a current-source stage. In this case, the *npn* is used to provide a low impedance voltage drive into the base of the lateral *pnp*. The current sink I_1 provides a bias path for the emitter and base currents of T_1 and T_2 respectively. The output current I_2 can be expressed as

$$I_2 = \alpha_p I_0 = \alpha_p \frac{V_{CC} - V_1}{R_1} \tag{4.56}$$

where the V_{BE} drops across the *pnp* and the *npn* emitters are assumed to be approximately equal. Note that, for an ac signal applied to the terminal V_1, the current source stage of Fig. 4.18(*b*) functions as a transconductance amplifier with a conversion gain G_m given as

$$G_m = \frac{\partial I_2}{\partial V_2} = -\frac{\alpha_p}{R_1} \tag{4.57}$$

which again has a 3-dB bandwidth equal to f_α of the *pnp*.

The α_p dependence of the output current can be compensated by incorporating a small amount of positive feedback into the composite device interconnection.[5] As shown in Fig. 4.18(*c*), this can be done by circulating some of the bias current I_0 through the *npn* input transistor. Assuming that the current gain of T_1 is high (i.e., input bias current is negligible), the sum of the three branch currents I_o, I_1, and I_2 must be zero. Therefore, the current continuity through the circuit requires that

$$I_2 = (I_0 - I_1) \tag{4.58}$$

which is the same as Eq. (4.53) except for the lack of the α_p factor. Note that I_o is still given by Eq. (4.54). The circuit of Fig. 4.18(*c*) can also be used as a transconductance amplifier with a conversion gain equal to $(-1/R)$. Since, to a first order, α_p dependence is eliminated, the 3-dB bandwidth of the conversion gain is significantly increased. Experimental results indicate that 3-dB bandwidths of G_m up to 3 f_α are achievable in this manner.

The physical structure of the lateral *pnp* also provides an additional design and layout advantage: since the *pnp* is formed by simultaneous emitter and collector diffusions into a *n*-type epitaxial pocket (see Fig. 2.14), the *pnp* transistors having the same base and emitter regions but separate collectors do not require separate isolation islands. Therefore, unlike the vertical *npn, a* lateral *pnp* transistor having multiple collectors can be readily formed by splitting the collector region of the *pnp*. This is schematically shown in Fig. 4.19 for the case of a two-collector *pnp*. The split-collector lateral-*pnp* is particularly advantageous for current source circuits of Fig. 4.18 since it allows the design of a number of independent current sources from a single bias reference, with a minimum increase of chip area. In a multiple collector *pnp* structure, the total collector current I_2 is split between the individual collectors, in direct proportion to the peripheral length of the respective collector regions facing the emitter in the device layout. Using the symmetrical structure similar to that

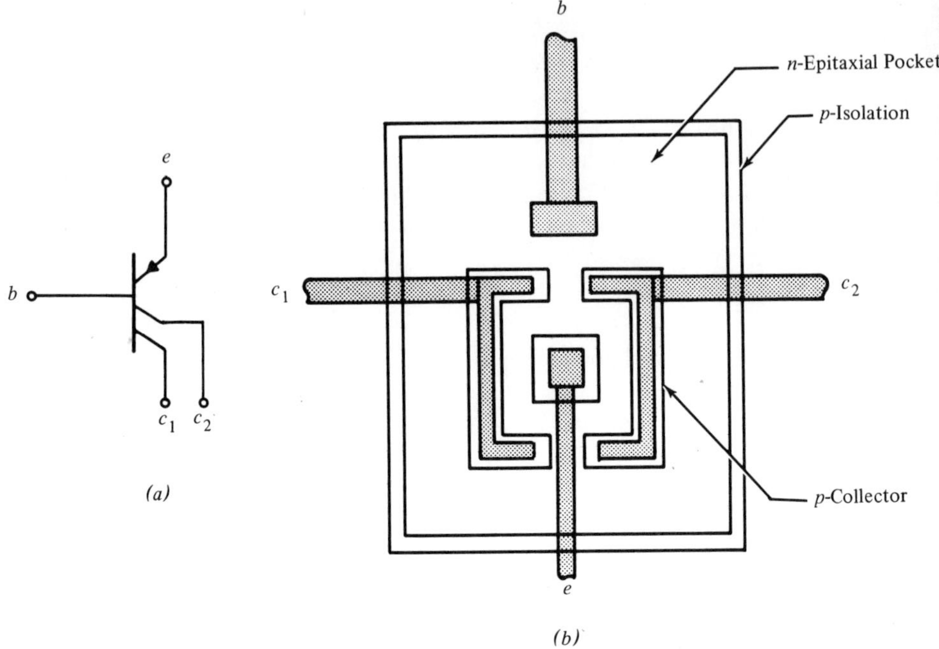

Figure 4.19 Multiple-collector lateral *pnp* transistor.

shown in Fig. 4.19, the collector currents of a two-collector lateral *pnp* transistor can be matched to better than ±5 percent over three orders of magnitude of current.

4.6 INPUT STAGE BIASING

In conventional circuit design using discrete devices and components the dc bias point of common-emitter gain stages can be determined by use of series feedback, in the form of an emitter degeneration resistor. Then for the ac operation of the circuit, this degeneration is bypassed by a bypass capacitor, and the full voltage gain of the stage is utilized. Figure 4.20 shows the circuit diagram of such a gain stage, where the capacitor C_E is chosen large enough to provide efficient ac bypassing of R_E in the frequency range of interest. In the design of integrated circuits where such large bypass capacitors are not available in monolithic form, alternate biasing schemes can be devised which rely on the matching and thermal tracking properties of integrated components to stabilize the operating point of a gain stage. Figure 4.21 shows two such biasing schemes which are derived from the diode-biased current sink configuration of Fig. 4.2.

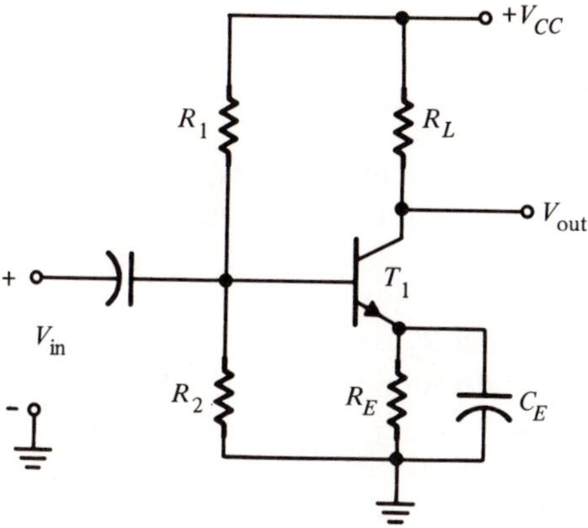

Figure 4.20 Conventional bias scheme for common-emitter gain stage.

In the circuit of Fig. 4.21(*a*) a low impedance transformer secondary is used to couple the input signal into the circuit without changing the dc bias conditions.[1] If transistors T_1 and T_2 are matched, then collector currents I_1 and I_2 are matched, and by choosing $R_2 \simeq R_1/2$, the output dc level is set at approximately $V_{CC}/2$, depending only on the matching of the circuit components. Note that the full voltage gain of the circuit can now be obtained without requiring any bypass elements. In some applications, the input signal may be derived from a low impedance floating (ungrounded) signal source such as a transducer or a sensor. In such a case, the need for a transformer coupling can be eliminated and the signal source can be connected directly across terminals B_1 and B_2 of the Figure.

With a minor modification, the diode biasing scheme can lend itself to capacitive coupling,[1] as shown in Fig. 4.21(*b*). As compared with the

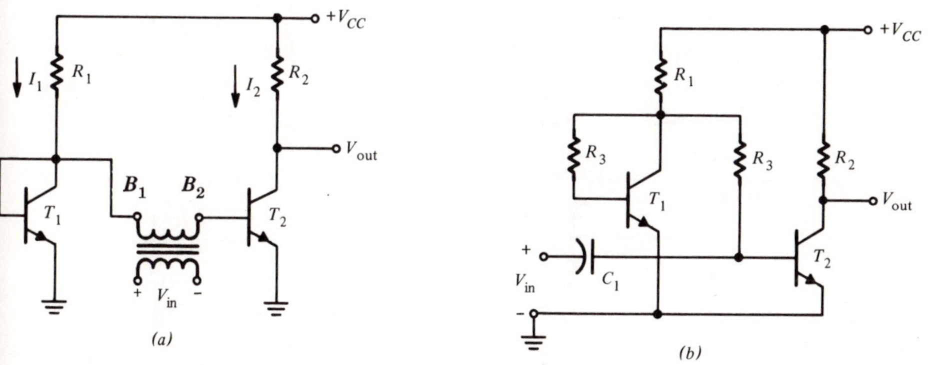

Figure 4.21 Diode-biased input stage configurations: (*a*) transformer-coupled; (*b*) R-C coupled.

basic diode bias circuit of Fig. 4.2, the addition of a pair of matched resistors R_3 does not change the dc balance of the circuit; yet it raises the impedance level at the base of T_2 to a sufficiently high level that the capacitive coupling of the input signal becomes feasible. As in the case of transformer coupling, the full voltage gain of the stage is again utilized without requiring any bypass elements. If desired, the voltage gain of the circuits in Fig. 4.21 can be controlled by using balanced emitter degeneration for transistors T_1 and T_2.

In the design of discrete amplifier circuits, the dc-biasing of each stage is normally done separately, and the overall circuit is completed by ac-coupling the individual gain stages. In monolithic circuits where large value coupling capacitors are not available, such a design approach is not feasible. Instead, an alternate design approach is to utilize overall feedback around a number of gain stages to establish the desired bias levels and bias stability. One particular circuit configuration often used for this purpose is the shunt-series feedback pair shown in Fig. 4.22(*a*). In this case, resistors R_2 and R_4 form the shunt and the series branches, respectively, of the feedback network around the two-stage current amplifier. Assuming that $\beta_0 \gg 1$ for each of the transistors, the overall current gain A_I for the circuit is set by the ratio of the resistors in the shunt and the series branches,[9] i.e.,

$$A_I \approx \frac{R_2}{R_4} \tag{4.59}$$

Since the accurate resistor ratios can be easily realized in integrated circuits, this gain is highly predictable and stable.

In the particular circuit under discussion the base current required to keep the first transistor in its active region is derived from the emitter cur-

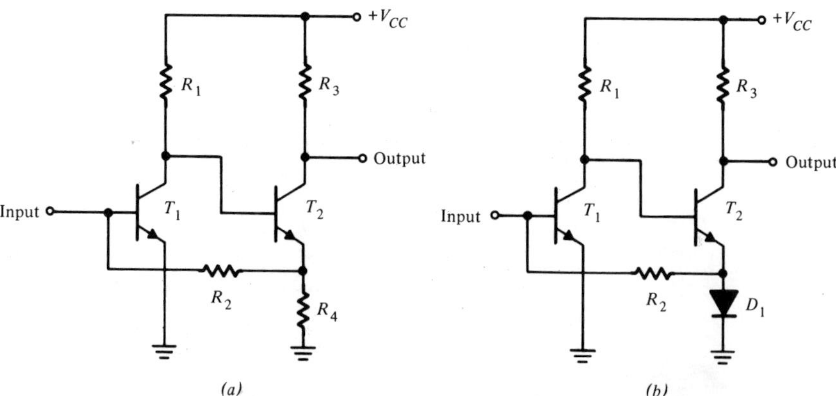

Figure 4.22 Shunt-series feedback pairs: (*a*) resistor-ratio feedback; (*b*) diode-resistor feedback.

rent of the second transistor. Thus the value of R_4 would be chosen so that the emitter current of T_2 caused a voltage drop of V_{BE} across R_4. Similarly, for a desired dc level at the output, the load resistor R_3 at the collector of T_2, is determined by the amount of collector current in T_2. Even though R_3 and R_4 might have large variations in their absolute resistances, a predictable and constant ratio assures that the amplifier is correctly biased.

The effective dc drift referred to the input, due to V_{BE} variations with temperature, is unaffected by the feedback arrangement, and is of the order of -2 mV/°C. Since extreme resistor ratios are impractical and inaccurate in integrated circuits, the maximum value of A_I obtainable from such a pair is limited to ≤ 1000.

For small signal operation, the gain of a shunt-series feedback pair can be further increased by utilizing the dynamic impedance r_d, of a forward-biased diode, D_1, to replace R_4, as shown in Fig. 4.22(*b*). An interesting and highly useful property of this scheme is its tolerance tracking ability. If, for example the resistors are doubled in value, both collector currents would be cut in half. The diode, therefore, runs at half the current and its dynamic resistance is increased by a factor of two. Thus, the ratio of R_2 to r_d remains unchanged, providing the same amount of gain for the amplifier.

The shunt-series feedback pairs shown in Fig. 4.22 exhibit a low input impedance due to the nature of the feedback loop.[9] If a relatively high input impedance level is required (i.e., R_{in} in the range of several kilo-ohms) a series-series feedback circuit can be used in place of the shunt-series feedback pair, as shown in Fig. 4.23. Here, the base current required

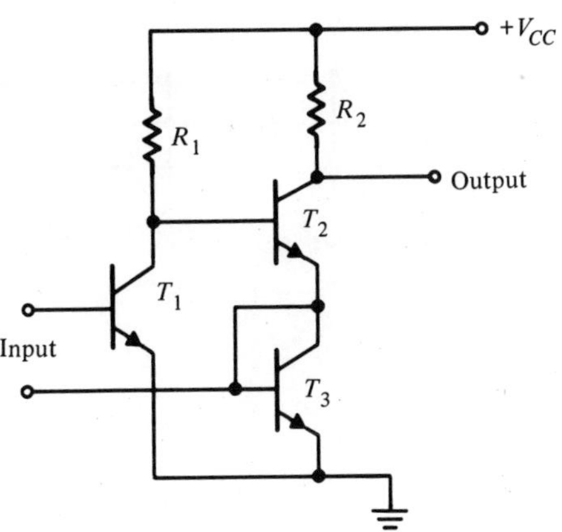

Figure 4.23 Series-series feedback pair.

for the first transistor flows through the source. The source, therefore, must have a reasonably low dc resistance and must float above ground, resembling the transformer coupled gain stage of Fig. 4.21(*a*).

The series-series feedback circuit of Fig. 4.23 offers the same matching and tracking advantages previously discussed in connection with the shunt-series feedback pair, and provides a voltage gain given by

$$A_v \approx R_2 r_d \tag{4.60}$$

where r_d is the dynamic impedance of T_3. Since r_d is inversely proportional to the current level through it, the circuits of Figs. 4.22(*b*) and 4.23 are useful only for small signal operation. At high signal levels the nonlinear behavior of r_d can result in a significant harmonic distortion in the output signal.

In a variety of circuit applications, capacitively coupled high input impedance amplifier stages are required. In monolithic circuit design, where large bias resistor values are impractical, the high input impedance requirement can be met by use of a "bootstrap" bias configuration, as shown in Fig. 4.24.[10] In this circuit, T_1 and T_2 form a Darlington-connected common-emitter gain stage providing a voltage gain A_V given as

$$A_v \approx -R_1/R_4 \tag{4.61}$$

The branch of the circuit formed by diode-connected transistors T_3 and T_4 provides a bias reference level at the collector of T_4 which is at the

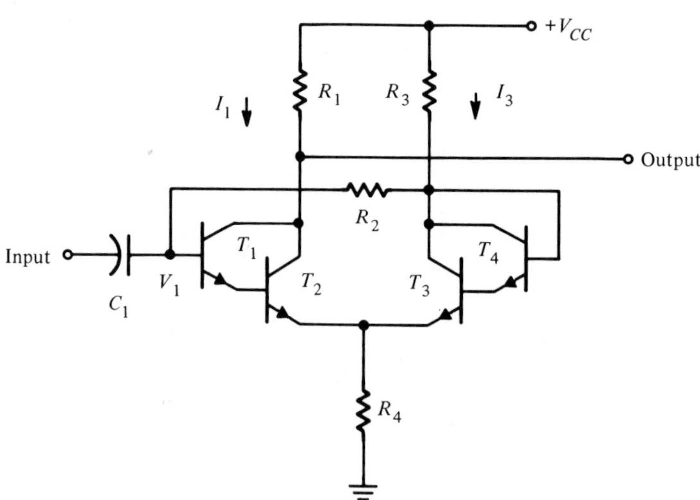

Figure 4.24 A self-biased high input impedance gain stage.

same dc potential as the input. Thus, upon connecting a resistor R_2 between this point and the input, the input stage can be made self-biasing. Assuming that the voltage drop across R_2 is negligible, and that $V_{CC} \gg V_{BE}$, currents I_1 and I_3 are equal, and the dc bias level V_1 at the input is constrained to be

$$V_1 \approx V_{CC} \frac{2\, R_4}{R_3 + 2\, R_4} + 2\, V_{BE} \tag{4.62}$$

Similarly, the output dc level V_2 is set by the choice of R_1 as

$$V_2 = V_{CC}\left(1 - \frac{R_1}{R_3 + 2\, R_4}\right) \tag{4.63}$$

The maximum output swing can be obtained from the circuit by setting $R_1 = (\frac{1}{2})R_3$.

The circuit offers a very high input impedance due to the "bootstrap" action of the self-bias resistor R_2. This can be explained as follows: Neglecting the dynamic diode or emitter impedances, the voltage gain between the input and the collector points of T_3 and T_4 is that of an ideal emitter-follower, i.e., exactly unity. Thus, the voltage levels at both ends of R_2 change in the same manner, and R_2 behaves as a virtual open circuit. More specifically, considering the dynamic impedances of the transistors, the effective value of R_2' for small signal operation can be written as

$$R_2' \simeq \frac{R_2}{1 - \dfrac{R_3 R_4}{(r_d + R_3)(r_d + R_4)}} \tag{4.64}$$

where r_d is the dynamic diode or emitter resistance of the transistors in the circuit. The total input impedance of the circuit can be then expressed as

$$R_{\text{in}} = R_2' \,\Big/\!\!\Big/ \left[\frac{(\beta_0)^2\, R_3 R_4}{R_3 + R_4}\right] \tag{4.65}$$

where the term in brackets is the input impedance seen looking into the base terminal of T_1.

Using the "bootstrap" bias circuit of Fig. 4.24, self-biased input impedance levels as high as 1 MΩ can be obtained with practical diffused resistor values.

4.7 STABILIZATION OF SUBSTRATE TEMPERATURE

The close thermal coupling between the integrated components on a monolithic chip allows the circuit designer to control the thermal environment of the circuit by maintaining the substrate at a steady temperature. Because of the small thermal capacity of the chip, such a stabilization scheme can be achieved without requiring excessive amounts of power dissipation. Regulation of the substrate temperature can be done by incorporating a temperature sensor-controller circuit on the same monolithic substrate as the circuit to be stabilized.[11,12] The temperature regulator circuit is basically a controlled heater unit whose function is to maintain the substrate at a constant elevated temperature, with minimum dependence on the ambient variations.

Figure 4.25 shows a functional block diagram of a temperature sensor and controller circuit for stabilizing the substrate temperature. All the necessary circuit elements to form such a temperature regulator system are readily available in the form of integrated components. The predictable temperature dependence of the transistor V_{BE} can be utilized as the temperature-sensing element, and a power transistor can be used as the heating element. In order to minimize the thermal gradients through the chip, the circuit to be stabilized is laid out symmetrically with respect to the heating and the sensing elements, and the most critical components, such as the input stage of a high gain amplifier, are located nearest to the sensor. The threshold level of the heater unit is set such that the heater is operative over the entire temperature range of interest, and keeps the chip temperature at a relatively constant level, above the highest ambient temperature to be encountered. The power dissipation level of the heater unit is set by the lowest ambient temperature limit and by the thermal resistivity of the circuit package. Normally, to provide the desired chip temperature regula-

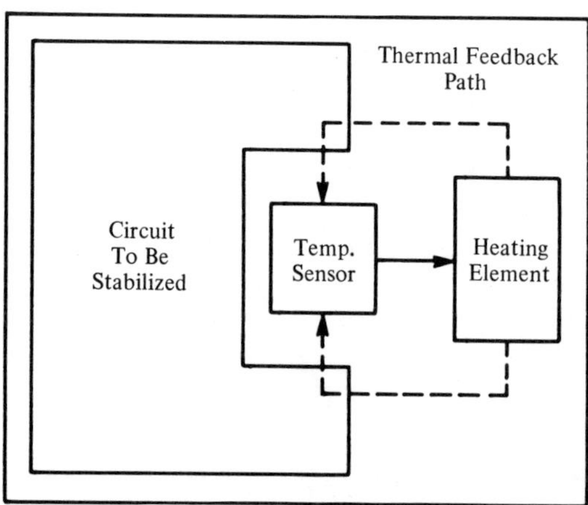

Figure 4.25 Block diagram for substrate temperature stabilization.

tion with a minimum amount of additional power dissipation, thermal insulation is provided between the chip and the package. This insulator is normally a ceramic layer between the latter two. Using the thermal stabilization technique discussed above, it is possible to control the substrate temperature constant to better than $\pm 5°C$ over a 180°C change in the ambient temperature (i.e., from -55 to $+125°C$), with an added power dissipation of <500 mW.[13]

At this point, it is instructive to examine some of the typical offset and thermal drift levels in a monolithic chip to have an idea of the close thermal coupling between the components. Without controlling the substrate temperature, the thermal drift of the V_{BE} mismatch between two adjacent devices on the chip can be made as low as 10 $\mu V/°C$. Since the V_{BE} varies with temperature at a rate of -2 mV/°C, this implies that the temperature differential between two adjacent transistors under identical bias conditions is $\leq 0.002°C$ per degree change of temperature. With substrate heating, this V_{BE} offset drift can be reduced to less than 0.5 $\mu V/°C$ of ambient change, indicating that the temperature differential between adjacent transistors can be kept as low as $10^{-4}°C$ per degree change of ambient temperature.

As mentioned in the beginning of this chapter, the excellent matching of integrated components, as well as the close thermal coupling between them, provides the integrated circuit designer with a diverse set of design techniques. With imaginative application of these unique circuit techniques —some of which are outlined in this chapter—it is possible to design monolithic analog circuits which offer superior performance characteristics over their discrete counterparts.

REFERENCES

1. R. J. Widlar, "Some Circuit Design Techniques for Linear Integrated Circuits," *IEEE Trans. Circuit Theory,* **CT–12** (Dec., 1965): 586–590.
2. G. R. Wilson, "A Monolithic Junction FET-NPN Operational Amplifier," Int. Solid State Ckts. Conf., *Digest Tech. Papers,* **11** (1968): 20, 21.
3. H. R. Camenzind and A. B. Grebene, "An Outline of Design Techniques for Linear Integrated Circuits," *IEEE J. Solid State Ckts.,* **SC–4** (1969): 110–122.
4. R. J. Widlar, "A Versatile Monolithic Voltage Regulator," *Natl. Semiconductor Application Note AN–1,* 1967.
5. T. M. Frederiksen, "A Monolithic High-Power Series Voltage Regulator," *IEEE J. Solid State Ckts,* **SC–3** (1968): 380–387.

6. R. J. Widlar, "New Developments in IC Voltage Regulators," Int. Solid State Ckts. Conf., *Digest Tech. Papers,* **13** (1970): 158–159.
7. A. B. Grebene, "A Monolithic Frequency-Selective AM/FM Demodulator," *Proc. Natl. Electronics Conf.,* **26** (1970): 519–524.
8. R. J. Widlar, "A Unique Circuit Design for a High Performance Operational Amplifier Especially Suited to Monolithic Construction," *Proc. Natl. Electronics Conf.,* **21** (1965): 85–89.
9. M. S. Ghausi, *Principles and Design of Linear Active Circuits,* McGraw-Hill, New York, 1965.
10. M. J. Hellstrom, *et al.,* "An Integrated Circuit Preamplifier with Nonlinear Bootstrapped Input Impedance," *Proc. Natl. Electronics Conf.,* **23** (1967): 321–324.
11. S. P. Emmons and H. W. Spence, "Very Low Drift Complementary Semiconductor Network DC Amplifier," *IEEE J. Solid State Ckts.,* **SC–1** (1966): 13–18.
12. T. F. Prosser, "An Integrated Temperature Sensor-Controller," *IEEE J. Solid State Ckts.,* **SC–1** (1966): 8–13.
13. J. D. Lieux, "A New Approach to Low Drift Amplifiers," *Fairchild Semiconductor Application Note 20–BR–0021–48,* 1968.

5 Differential and Operational Amplifiers

Differential amplifiers represent a broad class of circuits whose basic function is to amplify the difference between two input signals. For this reason, they are also referred to as "difference amplifiers." The bias levels and the gain characteristics of a differential stage by and large depend on the symmetry between the two branches of the circuit. This balanced nature of the differential amplifier makes it ideal as a gain block for integrated circuits, since close matching is inherent to the monolithic components. In fact, since the matching and the temperature tracking properties of monolithic components are far better than their discrete counterparts, the performance characteristics of an integrated differential gain stage is, in general, superior to that of a nonintegrated one. The differential gain stage is also the basic building block in the design of operational amplifiers. Therefore, monolithic operational amplifiers also enjoy similar advantages over their discrete counterparts.

Today, the most widely accepted class of analog integrated circuits is the monolithic operational amplifier. In the design of such an amplifier, it is possible to go to a much higher order of circuit complexity than would have been used in discrete types. Since all the monolithic components are fabricated simultaneously, this higher level of circuit design complexity greatly improves performance without significantly increasing fabrication cost. Therefore, the monolithic operational amplifier offers significant cost and performance advantage over its discrete counterpart in all but a few specialized applications. In fact, the cost of the monolithic operational type has reached such a low level that these devices are now used as simple circuit components in applications where an operational amplifier would not even have been considered a few years ago.

5.1 DIFFERENTIAL GAIN STAGES

The differential gain stage is a balanced amplifier circuit, designed to amplify only the *difference* between the two input signals. Figure 5.1 shows the circuit diagram of the basic differential amplifier stage. Assuming that the resistors and the transistors in both legs of the circuit are precisely matched, currents I_1 and I_2 in each branch of the circuit would be identical, for zero input voltage. Similarly, the output dc voltage levels V_{o1} and V_{o2} would be equal:

$$V_{O1} = V_{O2} \approx V_{CC} \frac{I_0 R_c}{2} \tag{5.1}$$

The differential amplifier belongs to a special class of circuits known as "symmetrical networks." Therefore, for small signal ac analysis, the performance of a differential amplifier can be analyzed by using a well-

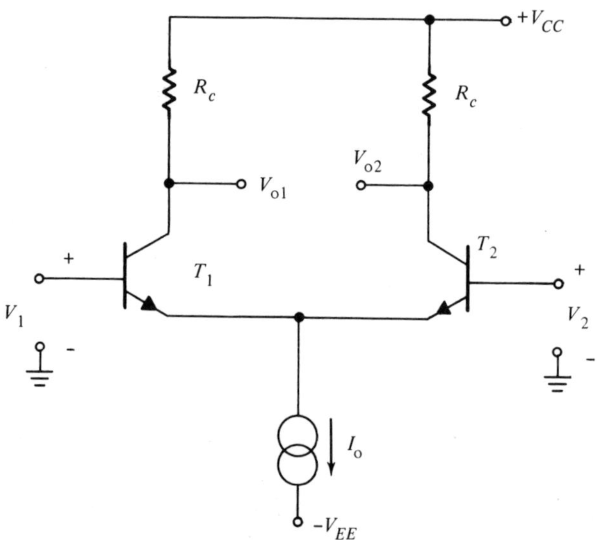

Figure 5.1 Circuit diagram of a differential gain stage.

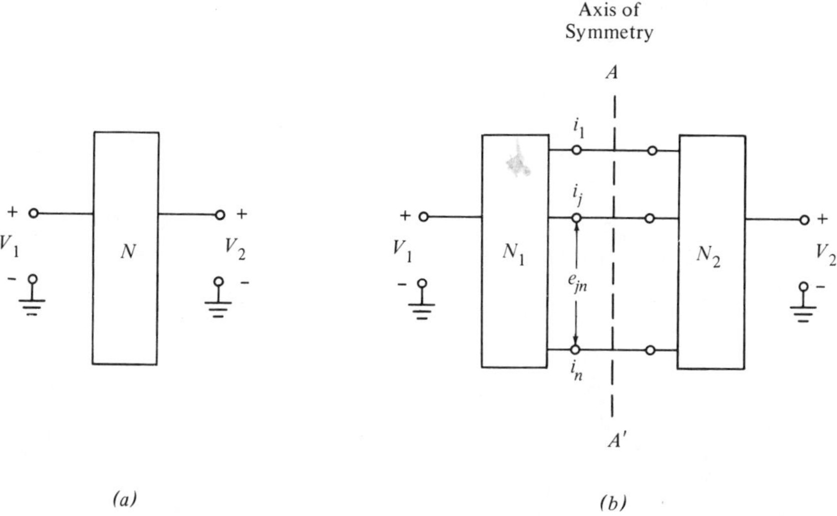

Figure 5.2 Separation of a symmetrical network N into an equivalent set of networks N_1 and N_2 about the axis of symmetry.

known circuit theorem, called the "bisection theorem."[1,2] This approach allows one to determine the overall ac performance by examining the behavior of only one half of the circuit for a given set of symmetrical (common-mode) or antisymmetrical (differential-mode) input signals.[3,4]

The symmetrical networks can be separated into two parts which are mirror-images of each other, about an axis of symmetry. This is known symbolically in Fig. 5.2 where a symmetrical network N, with two independent ac inputs v_1 and v_2 is redrawn as a combination of networks N_1 and N_2 about the AA' axis of symmetry.

The bisection theorem provides a powerful analysis technique for networks possessing such a symmetry. It states that, if a network such as that of Fig. 5.2(*b*) is excited with a set of symmetrical or common-mode (CM) inputs v_c, such that

$$v_1 = v_2 = v_c{}^* \tag{5.2}$$

the currents and the voltages through out the network are not disturbed if the branches joining the two parts of the network are open-circuited. In other words, all the branch currents i_1 through i_n of Fig. 5.2(*b*) due to a set of CM inputs are identically equal to zero. Similarly, the bisection theorem also states that when the network is energized by a set of anti-

* Lower case letters are used to imply ac small-signal quantities.

symmetric or differential-mode (DM) inputs such that

$$v_1 = -v_2 = v_d \tag{5.3}$$

the currents and the voltages in the interconnecting branches are not disturbed if these branches are cut and shorted together in each of the two parts, N_1 and N_2. In other words, all the relative node voltages e_{jk}, between the jth node and the kth node of Fig. 5.2(*b*) due to a set of DM inputs, are identically equal to zero. Thus, for differential input signals, a "virtual ground" exists along the axis of symmetry, since none of the node voltages, e_{jk}, along this axis is affected by DM inputs.

The proof of the bisection theorem can be readily derived by applying a linear superposition to the network response due to each of the independent inputs v_1 and v_2. A detailed discussion and the proof of this theorem is well covered elsewhere in the literature.[1,2]

Figure 5.3 is a redrawing of the basic differential amplifier stage shown in Fig. 5.1. In this case, the circuit schematic has been rearranged to emphasize the symmetrical nature of the circuit. For CM or DM inputs, the circuit of Fig. 5.3 can be analyzed by separating it into two sets of

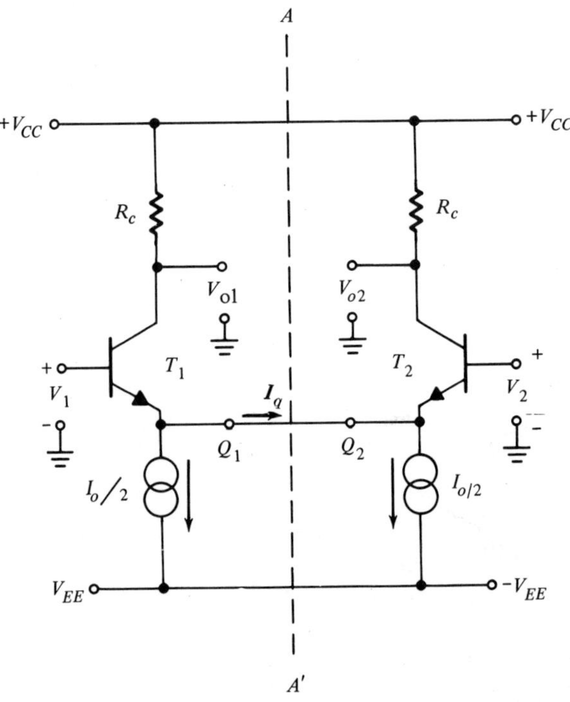

Figure 5.3 Axis of symmetry for a differential gain stage.

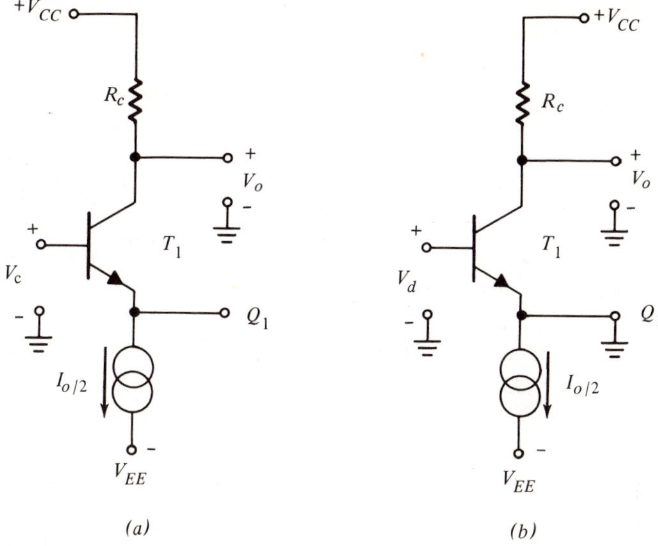

Figure 5.4 Equivalent half-circuits for (*a*) common-mode; (*b*) differential-mode input-signals.

equivalent half-circuits about the *A-A′* axis of symmetry, in accordance with the bisection theorem. These common-mode and differential-mode half-circuits are shown in Figs. 5.4(*a*) and (*b*), respectively.

If a set of identical common-mode inputs were applied to the inputs of Fig. 5.3 (i.e., $v_1 = v_2 = v_c$), the current and voltage levels in each symmetrical half of the circuit would vary in an identical manner. The relative voltage levels on either side of the interconnecting link $(Q_1 - Q_2)$ would remain unchanged and current I_q in this branch would be identically equal to zero. Therefore, for a CM input condition, the interconnecting link $(Q_1 - Q_2)$ can be left open-circuited, as predicted by the bisection theorem. This results in the CM equivalent circuit of Fig. 5.4(*a*). Under CM input conditions, the voltage gain of a differential amplifier is very low. From the CM half-circuit of Fig. 5.4(*a*), this voltage gain can be written as:

$$(A_v)_{\text{CM}} = \frac{v_{o1}}{v_c} = -\frac{R_c}{1/g_m + R_0} \tag{5.4}$$

where g_m is the transconductance of T_1 given as

$$g_m = I_o/2\ V_T \tag{5.5}$$

and R_0 is the impedance of the current source $I_o/2$. Since $R_0 \approx \infty$, common-mode voltage gain is very low.

If a set of differential-mode inputs were applied to the differential stage of Fig. 5.3 (i.e. $v_1 = -v_2 = v_d$), then the current and voltage levels in each symmetrical half of the circuit would vary *antisymmetrically*. Under these conditions, the voltage v_q at the interconnecting node $(Q - Q')$ would remain unchanged; and this point will behave as a virtual ground node, as predicted by the bisection theorem. Therefore, for DM input signals, the small signal performance can be analyzed using the DM half-circuit of Fig. 5.4(b) The voltage gain of the circuit under this condition is equal to that of a conventional common-emitter gain stage:

$$(A_v)_{\mathrm{DM}} = -R_c g_m \tag{5.6}$$

As indicated by Eq. (5.4) and (5.6), gain characteristics of a differential amplifier are quite different for differential and common-mode signals. The ability of the circuit to amplify differential-mode signals while rejecting common-mode ones is known as the common-mode rejection (CMR) property of the amplifier. CMR is defined as the ratio of DM gain to CM gain, i.e.,

$$\mathrm{CMR} = \frac{(A_v)\ \mathrm{DM}}{(A_v)\ \mathrm{CM}} = 1 + g_m R_o \tag{5.7}$$

and is normally expressed in decibels.

The analysis presented so far has been based on a set of symmetrical (CM) or antisymmetrical (DM) input voltages. This analysis can be extended to any *arbitrary* set of input voltages, v_1 and v_2, by separating them into their net common-mode and differential-mode components, as:

$$v_c = \frac{(v_1 + v_2)}{2} \tag{5.8}$$

$$v_d = \frac{(v_1 - v_2)}{2} \tag{5.9}$$

In most cases, a differential gain stage is driven with a single ended input, i.e., $v_{\mathrm{in}} = v_1$ and $v_2 = 0$. As indicated by Eq. (5.9), this corresponds to a net differential input v_d of $v_{\mathrm{in}/2}$; and the overall single-ended voltage gain of the stage is therefore equal to one half of that for a simple common-emitter stage, expressed as

$$A_v = v_{o1}/v_1 = -\frac{g_m R_c}{2} \tag{5.10}$$

The different behavior of a differential gain stage for common-mode and the differential-mode signals makes it ideal for integrated circuits. This is true because the basic absolute value tolerances and thermal drifts of component values appear as a common-mode variation, since they simultaneously appear at each branch of the circuit. Therefore, they do not affect the basic operation of the gain stage. On the other hand, component mismatches appear as an effective differential signal since they cause an asymmetry between the two branches of the circuit. Since the close matching of the components is an inherent property of monolithic circuits, the differential stage topology is nearly ideal for integration.

Drift and Offset Considerations

In spite of the inherent matching advantages of monolithic devices, small but finite mismatches still exist even between two adjacent, identical transistors on the same chip. In this section, the effects of these small mismatches will be examined, with regard to the current balance in a differential pair. In a differential amplifier stage, there are two dominant types of mismatches which arise from small nonuniformities in the masking and the diffusion steps:

1. *Input voltage offset,* ΔV_{io}*:* This is defined as the difference in the base-emitter voltages of the input transistors to obtain equal emitter currents.
2. *Input current offset,* I_{io}*:* This is defined as the difference in the base currents required for equal emitter currents.

In addition to the two basic device mismatches mentioned above, additional offset sources may exist in the circuit to cause equivalent input offsets comparable to inherent ΔV_{io} or ΔI_{io} mismatches. Some of these are possible mismatches in collector potentials or emitter resistors (when present). The mismatches in the collector potential may reflect to the input as a ΔI_{io} or ΔV_{io} offset, due to the "base width modulation" effects.

The effect of the input voltage or current offsets on the matching of the collector currents I_1 and I_2 can be readily calculated from Fig. 5.3 and 5.4(*b*). Any small change in the base-emitter voltage of a transistor can be related to the corresponding change in the collector current, by the transconductance g_m. Therefore, for the basic circuit of Fig. 5.1, the collector current mismatch for a given input offset voltage can be expressed as

$$(I_1 - I_2) = \Delta I_C \approx g_m(\Delta V_{io}) = (I_1/V_T)\,\Delta V_{io} \tag{5.11}$$

where $V_T = kT/q$ is the thermal voltage. Then, the relative, or per-unit, collector current mismatch can be written as

$$\frac{\Delta I_C}{I_1} = \frac{I_1 - I_2}{I_1} = \frac{V_{io}}{V_T} \tag{5.12}$$

for $V_{io} \ll V_T$. This means that a typical V_{BE} mismatch of 1 mV in the input transistors T_1 and T_2 can cause a collector current mismatch of approximately 4 percent.

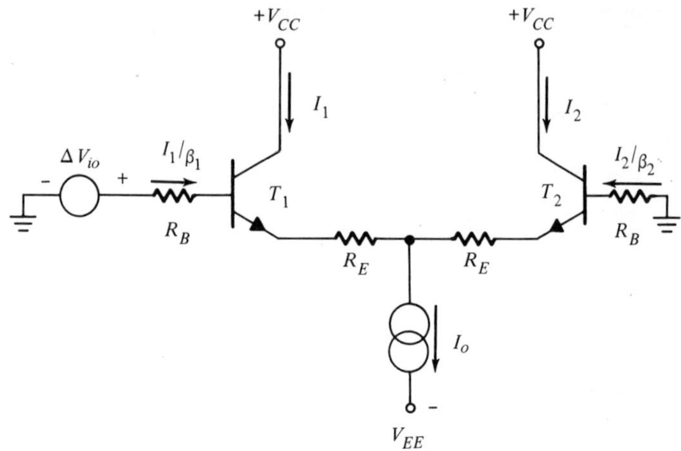

Figure 5.5 Offset sources in a current-biased differential amplifier.

Figure 5.5 shows the generalized circuit diagram of a current-biased differential amplifier stage. For completeness, emitter degeneration and the base bias resistors, R_E and R_B, are also included in the diagram. If, for the time being, one assumes that the external resistors R_B and R_E are well matched, one can derive a generalized expression for the collector current mismatch for a given input offset voltage and the beta mismatch as

$$\frac{\Delta I_C}{I_1} \approx \frac{\Delta V_{io} + \dfrac{R_B I_1}{\beta_1}(1 - \beta_1/\beta_2)}{V_T + R_E I_1 + \dfrac{R_B I_1}{\beta_1}} \tag{5.13}$$

If significant resistor mismatches were present in the circuit, then the collector current mismatch expression of Eq. 5.13 could be made to include these by redefining an effective input offset voltage as

$$(\Delta V_{io})_{\text{eff.}} = \Delta V_{io} + (\Delta R_E)I_1 + (\Delta R_B)I_1/\beta_1 \tag{5.14}$$

where ΔR_E and ΔR_B are the emitter and the base resistor mismatches. Even though it is fairly complicated, Eq. (5.13) gives a good insight to the circuit parameters which contribute to the current mismatch in a differential stage. The effect of the base resistors on the offset are in general negligible except where transistor β is very low, or when gross β mismatches are present between T_1 and T_2. The use of an emitter resistor R_E decreases the collector current mismatch by reducing the effective transconductance of the stage. However, if the emitter resistors are not well matched, then the increased effective offset due to the resistor mismatch (see Eq. 5.14) may deteriorate the collector current matching. For example, if one tries to reduce the collector current mismatch to less than 1 percent by using emitter degeneration, then the required matching of R_E should be significantly better than 1 percent.

In a well designed integrated circuit, the differential offset voltage due to the V_{BE} mismatch is in the range of 0.5 to 2 mV. Even though each transistor V_{BE} varies with temperature at an approximate rate of -2 mV/°C, the offset voltage variation with temperature is much less than this, due to the close thermal tracking of the transistor V_{BE} drops. Typical temperature drift of ΔV_{io} in a well designed circuit is in the range of ± 3 to ± 10 μV/°C.

5.2 DESIGN REQUIREMENTS FOR AN OPERATIONAL AMPLIFIER

An "ideal" operational amplifier can be defined as a voltage-controlled voltage amplifier circuit which offers infinite voltage gain with an infinite input impedance and zero output impedance. The advantage of such an idealized block of gain is that one can perform a large number of mathematical "operations," or generate a number of circuit functions, by applying passive feedback around the amplier. If the input and the output impedance levels are respectively high and low with respect to the feedback impedances, and if the voltage gain is sufficiently high, then the resulting amplifier performance becomes solely determined by the external feedback components. Figure 5.6 shows the feedback circuit connection for an inverting amplifier, with a voltage gain of $-A$. Neglecting the output impedence of the amplifier, the overall voltage gain A_v can be expressed as

$$A_v = V_o/V_{\text{in}} = -\frac{R_f}{R_s}\left[\frac{1}{1 + \frac{1}{A}(1 + R_f/R_s + R_f/R_{\text{in}})}\right] \tag{5.15}$$

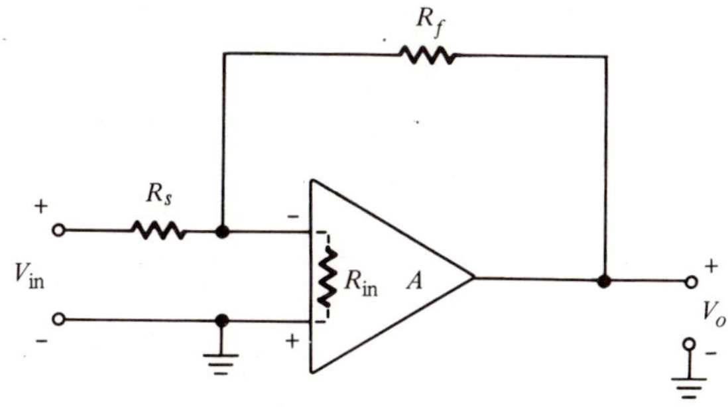

Figure 5.6 Basic feedback amplifier.

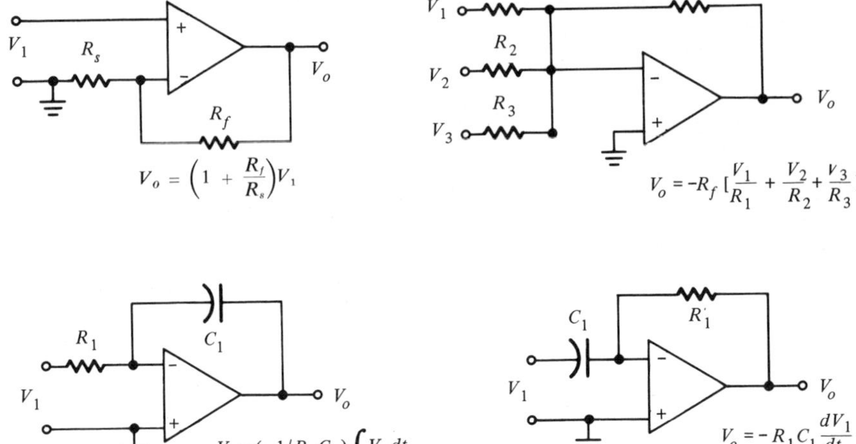

Figure 5.7 Some operational amplifier applications: (*a*) voltage-follower; (*b*) summing amplifier; (*c*) integrator; (*d*) differentiator.

where R_{in} is the input impedance of the amplifier. For large values of R_{in}, as $A \rightarrow \infty$ A_v becomes determined solely by the external components as

$$A_v = -\frac{R_f}{R_s} \tag{5.16}$$

Figure 5.7 shows some of the typical applications for an operational amplifier. The circuit of 5.7(*a*) is a voltage-follower which provides a very high degree of isolation between the output and the input and gives a non-inverting gain set by the feedback network. The circuit of 5.7(*b*) is a summing amplifier where the output is equal to the weighed sum of the individual inputs. Similarly, by putting reactive elements in the feedback circut, the operational amplifier can be made to perform the functions of integration or differentiation shown in Fig. 5.7(*c*) and (*d*). In addition to these basic applications, the operational amplifier can be used in connection with an analog multiplier to perform the mathematical operations of division and the square root extraction (see Section 7.4).

Since the operational amplifier has become a universal building block for circuit and system design, a number of widely accepted design terms have evolved which describe the comparative merits of various operational amplifier circuits. Some of these terms are defined below:

Input Offset Voltage: The input voltage which must be applied across the input terminals to obtain zero output voltage.

Input Offset Current: The difference of the currents into the two input terminals with the output at zero volts.
Common-Mode Range: Maximum range of input voltage that can be simultaneously applied to both inputs without causing cutoff or saturation of amplifier gain stages.
Common-Mode Rejection Ratio: Ratio of the differential open-loop gain to the common-mode open-loop gain.
Supply-Voltage Rejection Ratio: Input offset voltage change per volt of supply voltage change.
Slew Rate: Maximum rate of change of output voltage, for a step input. It is normally measured with unity gain, at the zero-crossing point of the output waveform.
Full-Power Bandwidth: Maximum frequency over which the full output voltage swing can be obtained.
Unity-Gain Bandwidth: Small signal 3-dB bandwidth, with unity gain close-loop operation.
Overload Recovery Time: Time required for the output stage to return to active region, when driven into hard saturation.

In the design of a monolithic general purpose operational amplifier it is impossible to optimize all of the performance characteristics associated with the device. For example, a high input impedance requirement may not be compatible with the low input offset voltage requirement, or the high slew rate or unity-gain bandwidth requirement may make the frequency compensation of the circuit difficult. Therefore, a large number of design compromises are necessary in the design of a monolithic operational amplifier. In the following sections, some of these basic design considerations will be examined separately.

5.3 INPUT STAGE DESIGN

The input stage is by far the most critical section of a monolithic operational amplifier since its design determines some of the key performance parameters of the amplifier. Over the years, a number of input stage topologies have been developed which efficiently utilize the inherent advantages of monolithic components in optimizing the circuit performance.[5-10] In this section, some of these input stage topologies will be reviewed, starting with the simplest and working toward the more complex circuit configurations.

To be suitable for operational amplifier design, the input stage has to fulfill the basic requirements listed below. Note that when applicable a typical upper or lower bound is also known in parenthesis, next to the parameter.

1. High input impedance ($>100\ \text{k}\Omega$)
2. Low input bias current (<500 nA)
3. Small input voltage and current offset
4. High common-mode rejection ratio (>60 dB)
5. High common-mode range ($\geq V_{CC}/2$)
6. High differential input range ($\geq V_{CC}/2$)
7. High voltage gain (>40 dB)

The first four requirements stem from basic ones associated with operational amplifier performance requirements discussed in the previous section. The fifth and the sixth requirements ensure the large input signal handling capability of the circuit. The last requirement, high voltage gain, is necessary for the input stage for two reasons: First, it decreases the gain requirements associated with the successive stages; second, it reduces the contribution of the second stage unbalance to the effective input offset. The small current and voltage offset requirement also dictates that the input devices should be well matched in size and geometry and be located in close proximity on the chip to ensure close thermal coupling.

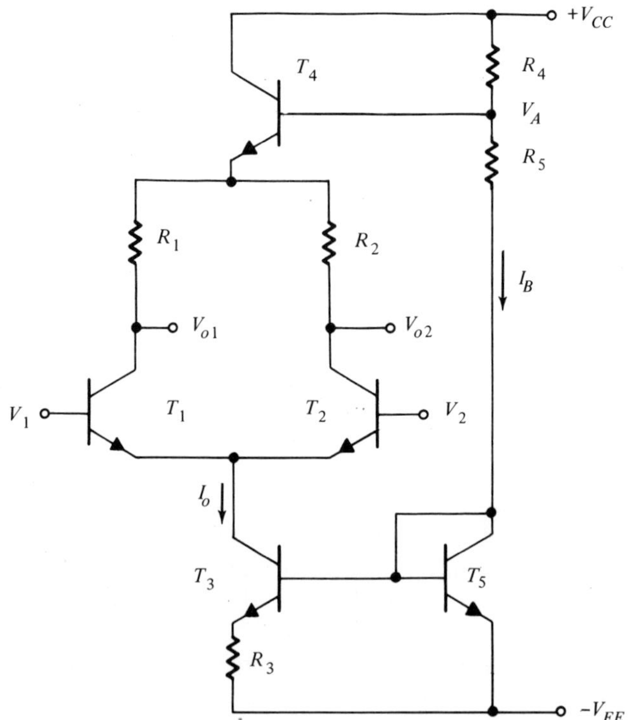

Figure 5.8 Typical input stage for a monolithic operational amplifier.

Fig. 5.8 shows a simplified circuit diagram for a typical monolithic operational input stage.[5] This circuit can meet the high input impedance and low input bias current requirement by operating input transistor T_1 and T_2 at very low current levels (typically 10 to 20 μA). T_3, T_5, and R_3 form a stable low value current-sink (see Fig. 4.6) to bias the stage. R_1 and R_2 form the matched collector load resistors, which set the gain of the stage as

$$A_v = -g_m R_1 = -\frac{I_0 R_1}{2\,V_T} \tag{5.17}$$

Since the collector currents of T_1 and T_2 are very small, the dc voltage drop across load resistors R_1 and R_2 is small. In order to avoid using excessively large resistor values to keep the output dc levels sufficiently below V_{CC}, an emitter follower stage, such as T_4 in the Figure, can be used to provide an internal voltage bias source. The main drawback of the input stage configuration is its low common-mode voltage range in the positive direction. The maximum positive common mode voltage V_C for such a circuit is

$$V_C = V_A - V_{BE} - \frac{I_0 R_1}{2} \approx V_{CC}/2 \tag{5.18}$$

The input stage configuration described above is typical of the so called "first-generation" monolithic operational amplifiers made popular by the well known 709 circuit. In this class of circuits, the input stage voltage gain is low ($\approx$30 to 40 dB). Therefore, to obtain the desired operational amplifier gain (typically $>$80 dB), at least two more gain stages are used. This large number of separate gain stages introduces too much excess phase delay into the circuit and makes the frequency compensation difficult for the operational amplifier.

An alternate design approach, used in the so-called "second-generation" operational amplifiers[6-8] is to use active loads to increase the voltage gain of the first stage to $\geq$60 dB. Figure 5.9 shows the circuit diagram for a practical input stage configuration which uses no resistors. The diode-connected transistor D_1 sets the bias level of the lateral *pnp* transistor T_3 and causes the differential current swing at the collectors of T_1 and T_2 to be converted to a single-ended voltage output. The gain of the stage is determined by the output impedance level of T_2 and the loading of the second-stage input, and is typically of the order of 50 to 60 dB. The main advantage of the input circuit of Fig. 5.9 over that of Fig. 5.8 is the improved input positive common-mode range, which is approximate equal to $V_{CC} - V_{BE}$. However, as compared to the other active-load input stages to be discussed shortly, the circuit of Fig. 5.9 has one fundamental drawback: with a balanced input voltage, the collector voltages of T_1 and T_2

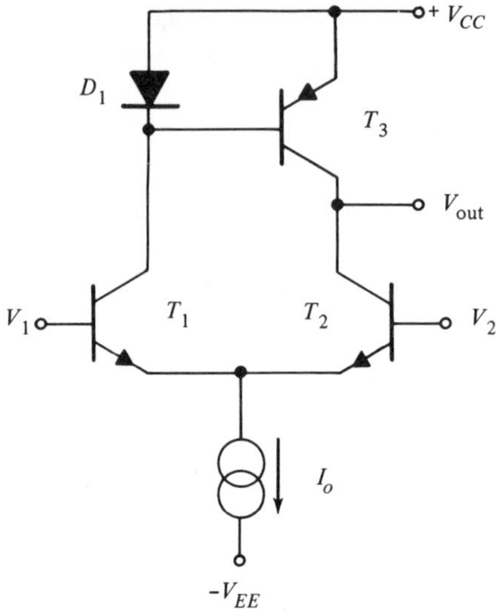

Figure 5.9 An input stage configuration using active loads.[8]

are quite different from each other, which can lead to an increased input offset voltage because of basewidth modulation effects. The low β of the *pnp* loads also causes a collector current mismatch which tends to increase the offset voltage. Typical offset voltages associated with the circuit of Fig. 5.9 are ≥ 3 mV.

Fig. 5.10 shows an alternate high gain input stage configuration used in most of the state-of-the-art operational amplifier circuits.[5-7] Even though the circuit of Fig. 5.10 is somewhat more complicated than that of Fig. 5.9, it has no contribution to offset voltage ΔV_{io} from the base width modulation effects. This is because both input transistors of Fig. 5.10 are at all times operated at the same collector voltage and current. Furthermore, the presence of the common-base *pnp* stage in series with input transistors T_1 and T_2 results in a higher input impedance level. Normally, lateral *pnp* devices are used for D_1, T_3, and T_4 and T_5.

The basic operation of the circuit can be explained as follows: Emitter-coupled transistor pairs T_1/T_3 and T_2/T_4 are equivalent to a common-emitter *pnp* pair with a low-frequency current gain equivalent to that of *npn* transistors T_1 and T_2. *pnp* transistors T_3 and T_4 are biased through a common constant-current sink I_B. Assuming that the base-emitter voltages of T_1 and T_2, T_7 and T_8 are well matched, the collector currents through each device are approximately equal. T_6 sets the bias level for T_7 and T_8, and also serves as a unity-gain stage, converting the collector-voltage swing of T_7 into a base-voltage drive for T_8. In this manner, the

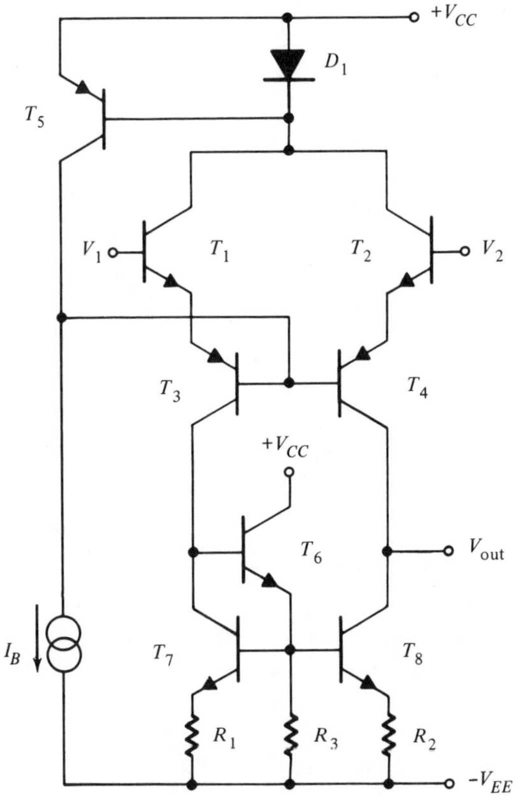

Figure 5.10 A high gain input stage configuration using active loads.

differential output of the stage is converted to a single-ended output at the collector of T_8. Using the circuit configuration of Fig. 5.10, voltage gains in excess of 60 dB can be obtained. The actual voltage gain is set by the second-stage loading and by the collector impedances of T_4 and T_8. The dc level of the output is close to the negative supply. This eliminates the need for dc level-shifting when coupling into the second stage.

Since the absolute values of the lateral p*np* transistor characteristics are not well controlled, a feedback arrangement is necessary to stabilize the operating points of T_3 and T_4. In the circuit, this feedback is provided by T_5 and the diode-connected lateral *pnp* transistor D_1. D_1 monitors the current level through T_1 and T_2, and correspondingly sets the current level through T_5 which in turn adjusts the base current of T_3 and T_4 by adding or subtracting current from the constant sink current I_B. This mode of feedback is known as a "common-mode" feedback since the feedback path is activated by the common-mode current changes in T_1 and

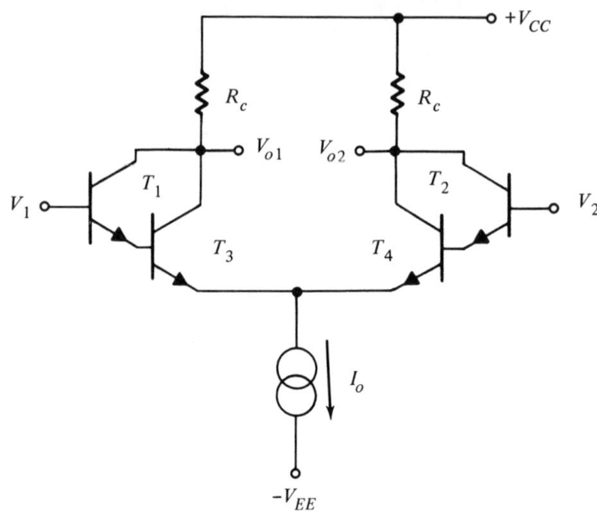

Figure 5.11 A Darlington-connected input stage.

T_2. In a differential stage, such a feedback simultaneously improves the bias stability and the common-mode rejection, without effecting the differential gain characteristics of the circuit.

In some applications, the input bias current and the input impedance levels may be respectively too high or too low to be acceptable. In such a case, a Darlington-connected set of input transistors can be used, as shown in Fig. 5.11. In such a configuration, each of the two input transistor pairs T_1-T_3 and T_2-T_4 are equivalent to a single transistors with an effective current gain, β_0 given as

$$\beta_0 \approx \beta_1 \beta_3 \tag{5.19}$$

Using a Darlington configuration, input impedance levels of 10 to 20 MΩ can be obtained with input bias currents of 5 to 10 nA. Unfortunately, the base currents and the V_{BE} drops of the Darlington-connected composite transistor pairs do not match or track as well as a single device. Therefore, input voltages and drifts are somewhat increased (typically ± 3 mV and ± 15 μV/°C, respectively). Furthermore, the beta variation of the composite transistor over the temperature range can cause the input bias currents and the impedance level to vary significantly with temperature.

A more practical method of achieving high input impedance and low bias currents is the use of "punch-through" transistors at the input stage. As described in Chapter 2, punch-through devices offer current gain values of several thousand, at base currents of a few nanoamps. They can be fabricated simultaneously with *npn* bipolars or the lateral *pnp* transistors, by use of one additional diffusion step (see Section 2.1). However, the collector-emitter breakdown voltages of punch-through transistors is very low (typically 3 to 4 V), and the total voltage swing across them must be limited. This can be done by using them in a composite connection along with a lateral *pnp* transistor, as shown in Fig. 2.13. Figure 5.12 shows a practical input stage configuration using punch-through transistors. Punch-through transistors, T_1 and T_2 are drawn with a thick base to distinguish them from the rest of the devices. The common-mode feedback network formed by the bias string T_5 and T_6 "bootstraps" the net voltage swing across T_1 or T_2 to less than one V_{BE} under all input conditions.[10] Thus, each of the cascade connected transistor pairs T_1-T_3 and T_2-T_4 correspond to a single *npn* device with the breakdown characteristics of the conventional *npn* (T_3) and the current gain of the punch-through *npn* (T_1).

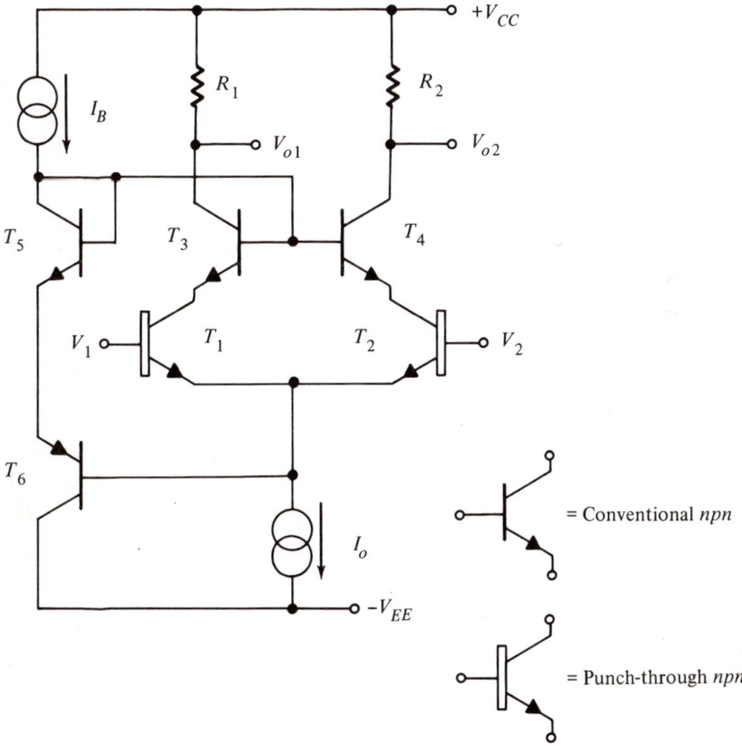

Figure 5.12 An input stage configuration using punch-through transistors.[10]

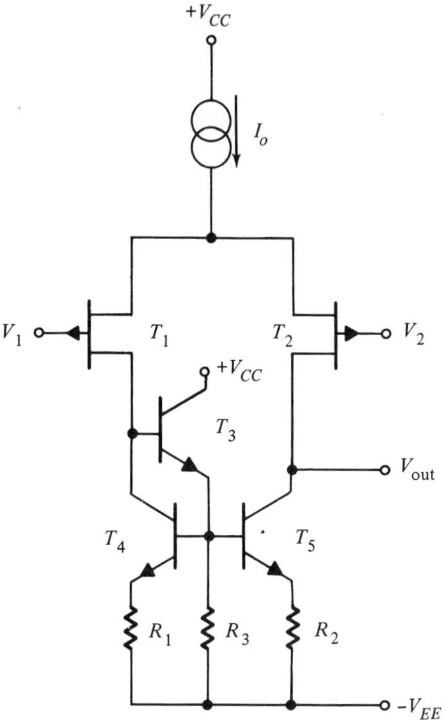

Figure 5.13 An FET-bipolar input stage.

Using punch-through transistors, the input bias current levels can be kept <10 nA, with typical offset voltages of ± 0.5 to ± 1 mV.

An alternate approach to the design of very high input impedance and low bias current input stages is to use JFET devices at the input. Figure 5.13 shows a practical JFET-bipolar input stage topology[11] where *p*-channel FETs are used as the input devices T_1 and T_2. As described in Chapter 2, *p*-channel JFETs can be fabricated simultaneously with *npn* bipolars or the lateral *pnp* transistors by use of one extra diffusion step. This results in an FET-bipolar structure shown in Fig. 2.26. Note that the bipolar portion of the circuit of Fig. 5.13 is similar to that described in connection with Fig. 5.10, where transistors T_4 and T_5 are used as active drain loads for T_1 and T_2.

Such a JFET-bipolar input stage can offer a very high input impedance ($>10^9\Omega$) with input bias currents in the picoampere range. However, the basic disadvantage of an FET input stage is the poor matching characteristics of JFETs as compared to bipolars. This leads to relatively large input offset voltages (typically 10 to 20 mV) and offset drifts (>40 μV/°C).

The different circuit configurations described in this section represent only a few possible and practical input stage topologies that are used in present-day monolithic operational amplifiers. A large number of alternate circuit schemes, as well as many variations of those described here, are available in the literature. The ultimate choice of the input stage configuration depends on the particular application for which the operational amplifier is intended.

5.4 OUTPUT STAGE DESIGN

The output stage of an operational amplifier should be able to deliver a substantial amount of power into a low impedance load. Therefore, it should have the following desirable properties:

1. Large output current swing capability
2. Large output voltage swing
3. Low output impedance
4. Low standby power

In addition to these four basic properties, the output stage should have built-in short-circuit protection such that the circuit will not burn out if the output terminal is shorted to $+V_{CC}$ or $-V_{EE}$. In earlier operational amplifier designs, a simple *npn* emitter-follower was used as the output stage.[12] Even though an emitter-follower stage can meet the low impedance and the large output swing requirements, it has very poor current swing capability and requires excessive standby power dissipation.

The four basic requirements for the operational amplifier output stage can be met by using a class-B or class-AB output stage configuration. One such possible circuit topology is the all-*npn* class-B amplifier stage shown in Fig. 5.14(*a*). This circuit configuration, often referred to as the "totem-pole" configuration was initially developed for digital integrated circuits, but is readily acceptable to analog integrated circuits. The operation of the circuit can be briefly described as follows: For positive swing of the input voltage, T_2 is in its active region, and the load current is sinked toward the negative supply through D_1 and T_2. The diode drop across D_1 assures that T_1 is off during this half-cycle. For negative going input signals, T_2 is cutoff and the load current is supplied from T_1. For standby condition, with no ac signal at the input, T_1 stays off, and T_2 is biased near cutoff with a small standby current I_1, set by the value of R_1, where

$$I_1 \approx \frac{V_{CC} + V_{BE}}{R_1} \tag{5.20}$$

In order to switch conduction from one transistor to the other, the

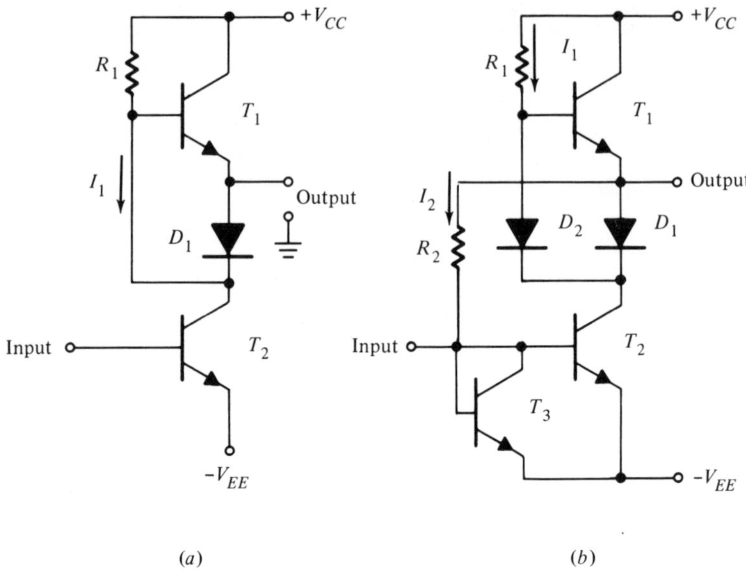

Figure 5.14 All *npn* class-B output stage: (*a*) basic circuit: (*b*) its actual implementation.

output voltage in the circuit of Fig. 5.14(*a*) has to change by two V_{BE} drops. This "dead-band" at the output is common to all class-B output stages, and results in a so-called "crossover distortion" at the output waveform. In the basic all-*npn* circuit of Fig. 5.14, this dead-band can be reduced by one V_{BE} by connecting an additional diode between the base of T_1 and the collector of T_2, as shown in Fig. 5.14(*b*). The circuit of Fig. 5.14(*b*) also demonstrates a practical method of biasing a totem-pole class-B stage for maximum output swing.[13] Diode-connected transistor T_3 sets the base bias for T_2 (see Fig. 4.2). Ignoring the error due to finite base currents, one can show that currents I_1 and I_2 are related as

$$I_1 = nI_2 \tag{5.21}$$

where n is the ratio of the effective emitter area of T_2 to T_3. Assuming symmetrical supply voltages, i.e., $V_{CC} = V_{EE}$, the output dc level is set at ground if

$$I_2 = (V_{EE} - V_{BE})/R_2 \tag{5.22}$$

In the quiescent state, bias current I_2 is supplied by T_1. Similarly, the collector current of T_2 is supplied from R_1 as

$$I_1 = (V_{CC} - V_{BE})/R_1 \tag{5.23}$$

The quiescent output voltage level can be located midway between the two supply voltages, over a wide range of symmetrical supply voltages and resistor values. The scale factor n is normally chosen to be >1 since T_2 is designed to handle large output currents. Within the constraints of the matching and tracking tolerances of integrated components, the value of n is normally chosen to be in the range of $1 < n < 10$.

Figure 5.15 shows a more conventional class-B output stage often used in operational amplifiers. The grounded-collector *pnp* transistor is normally designed as a substrate-*pnp* (see Fig. 2.17), which has a relatively low β but good high-current capability. Diodes D_1 and D_2 are diode-connected transistors whose V_{BE} drops match those of T_1 and T_2 respectively. Therefore, in the quiescent state, a bias current I_2 flows through both T_1 and T_2 such that:

$$I_2 \approx mI_1 \tag{5.24}$$

where m is the emitter area ratio of T_1 to D_1 or T_2 to D_2. Therefore, in a true sense, the circuit of Fig. 5.15 operates as a class-AB amplifier since both T_1 and T_2 can be "on" in standby condition. For positive going input signals, the load current is sinked to $-V_{EE}$ through T_2. Similarly, for negative going inputs, T_2 is off, and the load current is sourced from $+V_{CC}$ through T_1.

One possible failure mode for monolithic operational amplifiers is the destructive burn-out of the output stage due to an accidental short-circuit

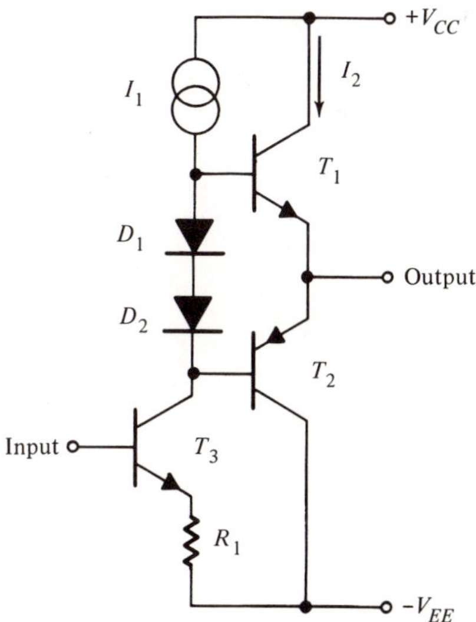

Figure 5.15 Class-B output stage using substrate *pnp*.

of the output terminal to one of the supplies. This is particularly true for the complementary class-B stage of Fig. 5.15 where T_1 or T_2 may burn out if the current through them is not limited to a safe value, with the output connected to $-V_{EE}$ or the $+V_{CC}$. This problem can be partially solved by putting series resistors R_E between the emitters of T_1 and T_2 and the output terminal. Then the maximum current through T_1 or T_2 would be limited to

$$I_{max} \approx V_{CC}/R_E \tag{5.25}$$

neglecting transistor saturation resistances. However, for most applications this may require an excessively large value of R_E, which will limit the available voltage swing at the output. In such cases, an alternate means of short-circuit protection can be provided using a circuit configuration similar to that of Fig. 5.16. In the circuit, T_4 and T_5 are normally off, except under output short-circuit conditions. As the output current increases, the voltage drop across R_{E1} or R_{E2} increases, causing T_4 or T_5 to turn on, and divert the base current available to T_1 or T_2. This causes the output current to limit at

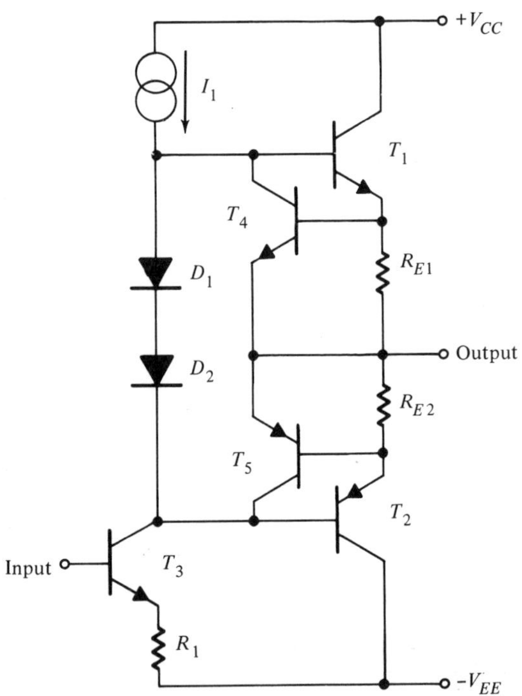

Figure 5.16 Class-B output stage with short-circuit protection.

$$I_{max} \approx V_{BE}/R_E \tag{5.26}$$

Using the active current limiting scheme of Fig. 5.16, the output current can at all times be maintained within safe limits, without impairing voltage swing capability. Typical values of I_{max} are in the range of 10 to 50mA for most commercially available monolithic operational amplifiers.

5.5 FREQUENCY COMPENSATION

In most circuit applications, the operational amplifier is required to be "unconditionally stable." In other words, an operational amplifier should not break into oscillations for any value of *resistive* negative feedback applied around it. In most operational amplifier circuits, excess phase shifts associated with the gain stages within the circuit cause the circuit to break into oscillations for increasing values of negative feedback. Therefore, in the design and the application of an operational amplifier, it is necessary to compensate the frequency performance of the circuit to avoid oscillations or peaking in the close-loop response.

For unconditional stability, it is necessary for an operational amplifier to approximate a one-pole open-loop transfer function A(s), given as

$$A(s) = \frac{A_0}{1 + (s/\omega_0)} \tag{5.27}$$

where $s = \sigma + j\omega$ is the complex frequency variable, and ω_0 is the 3-dB bandwidth. The transfer function of Eq. (5.27) corresponds to a -6 dB/octave roll-off of the open-loop gain for frequencies in excess of ω_0. In terms of the basic feedback theory,[14] it can be shown that, for unconditional stability, the excess phase shift across the amplifier (neglecting the 180° phase reversal associated with the inverting input) must be less than 180° for all frequencies where $|A(j\omega)| \geq 1$. The margin of stability in an operational amplifier can be related to its open-loop gain and phase response by the so-called gain and phase margins, defined as follows:

Gain Margin (M_G): The amount by which the voltage gain is below the unity (0 dB) level, at the frequency where the *excess* phase shift across the amplifier is exactly 180°. It is measured in decibels, and must be positive for unconditional stability.
Phase Margin (M_P): 180° minus the excess phase shift at the frequency where $|A(j\omega)| = 1$. It is measured in degrees and must be positive for unconditional stability.

Figure 5.17 shows the asymptotic open-loop magnitude and phase response for a typical monolithic operational amplifier, both with and with-

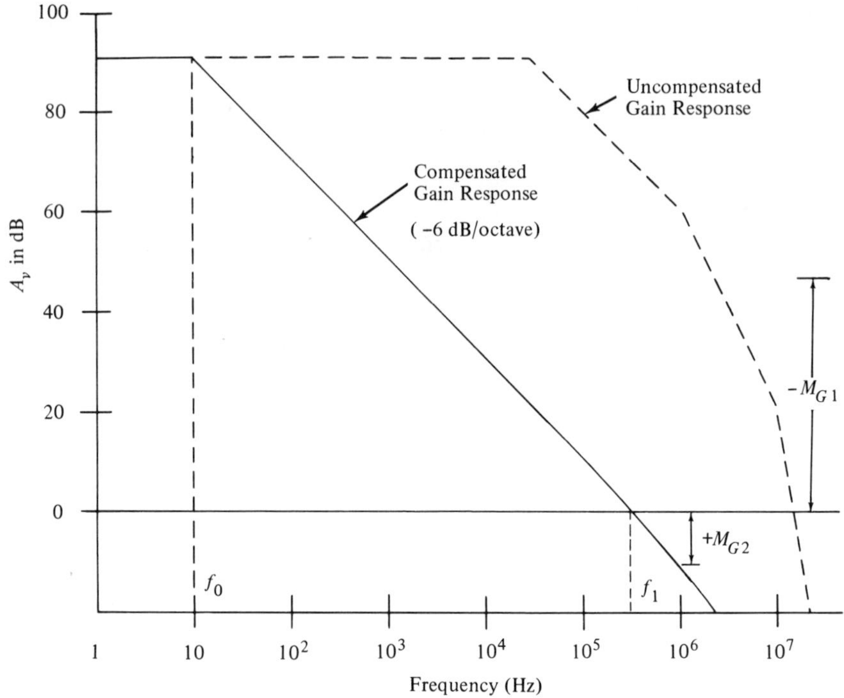

Figure 5.17 (*a*) Open-loop magnitude response with and without compensation.

out frequency compensation. The uncompensated magnitude response (shown with dotted lines) exhibits at least three break points, and -18 dB/octave high frequency roll-off. This in turn results in excessive phase lag at the unity-gain crossover frequency, and gives a negative gain margin (M_{G1}) and a negative phase margin (M_{P1}). The solid lines show the open-loop frequency response for the same amplifier after it has been frequency-compensated for unconditional stability. As shown in Fig. 5.17(*a*), the compensated amplifier exhibits a -6 db/octave roll-off (i.e., single-pole response) down to at least the unity-gain frequency, $\omega_1 = 2\,\pi f_1$. The corresponding gain and phase margins, M_{G2} and M_{P2}, are both positive quantities.

Frequency compensation is normally achieved by reducing the 3-dB bandwidth of the amplifier by introducing a dominant pole at low frequencies. This is often done by adding a *phase-lag* network into the amplifier circuit to narrow-band the dominant gain stage. In an operational amplifier, each gain stage introduces at least one dominant pole or break-frequency. Therefore, frequency compensation for unconditional stability

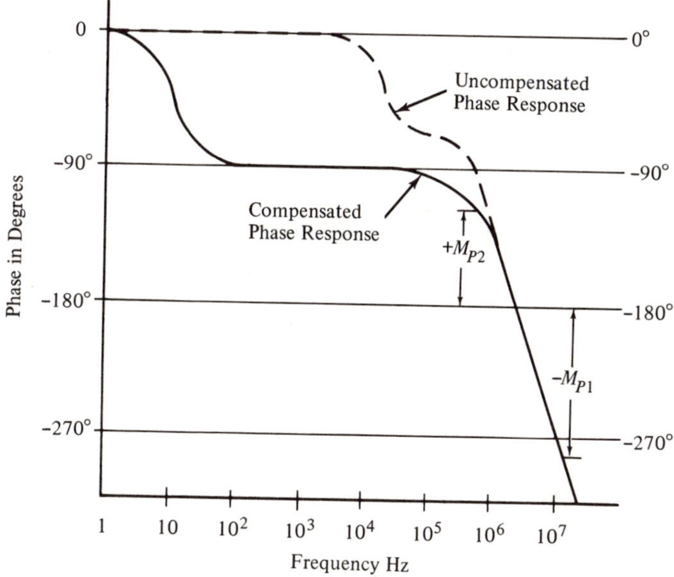

Figure 5.17 (*b*) Open-loop phase response with and without compensation.

becomes more difficult as the number of gain stages is increased. In earlier operational amplifier designs using three or more gain stages, simple lag-compensation is generally not sufficient due to the excess phase shifts introduced by the additional gain stages. Instead, a more complicated compensation scheme using a combination of lag and lag-lead networks may be necessary.[4,11]

In the so-called "second-generation" monolithic operational amplifiers, only two gain stages are used to achieve the desired overall gain. This results in only two dominant poles in the open-loop frequency response, and the unconditional stability can be achieved by a single compensation capacitor. In many cases, such a capacitor is small enough to be fabricated monolithically on the same chip as the rest of the circuit[7] (see Fig. 5.23).

Figure 5.18 shows a simplified ac equivalent circuit of a two-stage operational amplifier. Note that T_1 and T_2 represent the differential half-circuit for the input stage topology discussed earlier in connection with Fig. 5.10, and T_3, T_4 represent the second-stage. The collectors of T_2 and T_4 provide a set of convenient points across which a frequency compensating capacitor C_p can be connected. Assuming large second stage gain, the T_3 and T_4 combination along with the active load formed by the current-source I_2 serve as a nearly ideal integrator, as shown in the equivalent circuit of Fig. 5.18(*b*). Thus, the compensated open-loop response can be

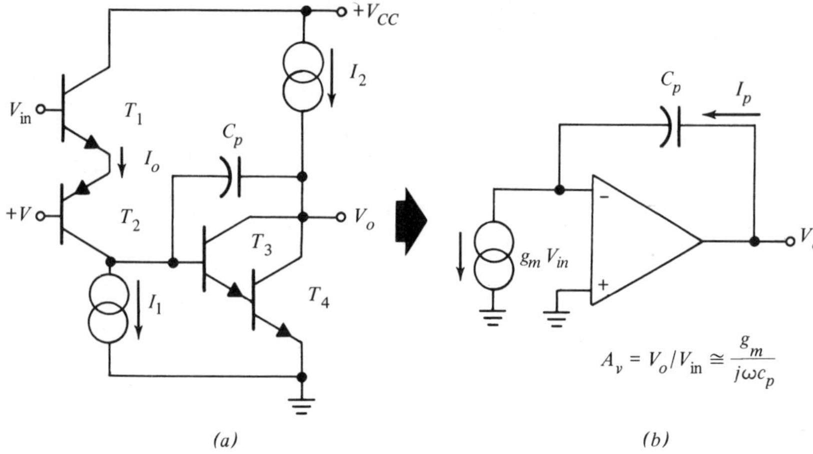

Figure 5.18 Frequency compensation in two-stage operational amplifiers.

approximated as

$$A(j\omega) = V_o/V_{\text{in}} \approx \frac{g_m}{j\omega C_p} \tag{5.28}$$

where g_m $(= I_o/V_T)$ is the input stage transconductance. Note that the unity gain crossover frequency ω_1 can be related to the compensation capacitor C_p by setting $|A(j\omega)| = 1$, as

$$\omega_1 = \frac{I_o}{V_T C_p} \tag{5.29}$$

Normally, the value of ω_1 is determined by the desired phase and gain margin considerations, and the value of C_p necessary to achieve the desired compensation becomes

$$C_p = g_m/\omega_1 \tag{5.30}$$

In most monolithic amplifiers, the value of ω_1 is determined by the alpha cutoff frequency of the lateral *pnp* devices in the gain stages. In most cases, the value of C_p is in the 15 to 30 pF range, making it readily compatible with the planar fabrication processes.

Slew Rate Considerations

The slew rate of an operational amplifier is defined as the time rate of change of the output voltage for a voltage step applied at the input. It is normally measured at the zero crossing point of the output voltage swing, with the amplifier compensated for unity gain. From the simplified circuit diagram for a two-stage operational amplifier shown in Fig. 5.18(*a*), it can be seen that the slew rate is minimum for a negative input step which can cause T_1 and T_2 to cut off, leaving the constant-current stage I_1 to sink the discharging current from the compensation capacitor. In this worst case, one obtains

$$\text{slew rate} = dV_o/dt = I_1/C_p \tag{5.31}$$

Assuming the current gain, α_p, of the *pnp* is nearly unity, currents I_o and I_1 are nearly equal; and since C_p is proportional to I_o (see Eq. 5.23), the slew rate expression becomes independent of current level:

$$\frac{dV_o}{dt} \approx V_T\omega_1 = (kT/q)\omega_1 \tag{5.32}$$

Typically, ω_1 is of the order of ω_α of the lateral *pnp* devices, i.e., in the 3 to 5 MHz range. Thus, a typical slew rate obtainable with the compensated gain stage of Fig. 5.18 can be estimated as

$$dV_o/dt \approx (25\ mV)(2\pi\cdot 5\cdot 10^6\ \text{rad/sec}) \approx 1V/\mu\text{sec} \tag{5.33}$$

This first-order approximation presented above demonstrates why the slew rate of most present-day general purpose operational amplifiers is in the range of 0.5 to 3 V/μsec. In the circuit of Fig. 5.18, the independence of the slew rate from the bias current level can be physically explained as follows: increasing the bias current level causes the loop gain to increase. This results in a proportionally larger value of C_p for compensation; thus the slew rate remains relatively unaffected.

The slew rate can be increased at the expense of increased power dissipation if the bias current levels, I_1 or I_o, can be changed without increasing the input stage transconductance. This can be done by inserting an emitter degeneration resistor R_E in series with the emitter of T_1 in the simplified circuit of Fig. 5.18(*a*) (or in series with emitters of T_1 and T_2 in the actual circuit of Fig. 5.10).[9] Then, the effective input stage transconductance, g_m', becomes

$$g_m' = \frac{1}{(1/g_m) + R_E} \simeq \frac{1}{R_E} \tag{5.34}$$

and the necessary value of C_p, from Eq. (5.24), becomes

$$C_p = \frac{g_m'}{\omega_1} = \frac{1}{R_E\omega_1} \tag{5.35}$$

Then, from Eq. (5.25), the slew rate can be written as

$$dV_o/dt = I_1/C_p = I_1R_E\omega_1 \tag{5.36}$$

which is proportional to the bias current level. Thus, the slew rate can now be increased at the expense of added power dissipation. For example, with $R_E = 1\ k\Omega$, $I_1 = 1$ mA, and $\omega_1 = 2\pi$ (5 MHz), one obtaines $(dV_o/dt) \simeq$ 40 V/μsec. However, the use of emitter degeneration at the input stage may not be desirable in all applications because it results in a reduced loop gain, increased noise, and offset voltage (see Eq. 5.14).

An alternate method to avoid slew rate limiting at the input is to use the cross-coupled input stage configuration shown in Fig. 5.19. For large input signals, the effective input transconductance, g_m', becomes determined by the cross-coupling resistors R_1 and R_2 as

$$g_m' = \frac{1}{R_1 + R_2} \tag{5.37}$$

which is independent of current. Thus, the slew rate can again be increased by increasing the operating current I_2 for the gain stage. Unity-gain slew rates in excess of 30 V/μsec have been reported using this input stage topology.[15]

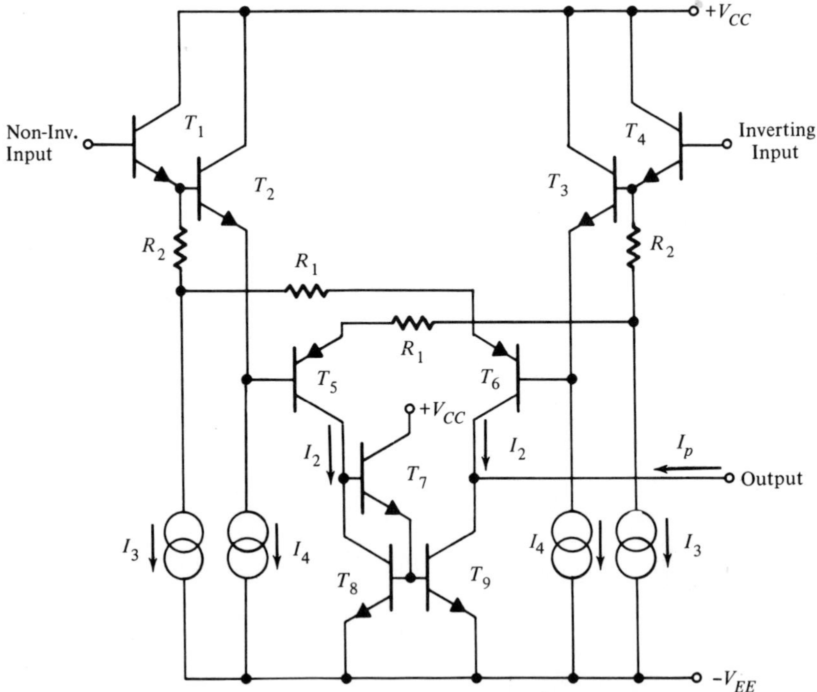

Figure 5.19 A cross-coupled input stage which avoids slew limiting.[15]

Full-Power Bandwidth (ω_p)

The full-power bandwidth is defined as the maximum frequency at which the peak output swing can be obtained. For a sinusoidal input, the output waveform is of the form,

$$v_o(t) = E_o \sin \omega t \tag{5.38}$$

The peak output voltage swing is obtainable as long as (dv_o/dt) does not exceed the slew rate. From (Eq. 5.38) one gets

$$(dv_o/dt)_{\max} = E_o\omega \tag{5.39}$$

Thus, the full-power bandwidth ω_p can be expressed from Eq. (5.39) by setting $\omega = \omega_p$ as

$$\omega_p = 2\pi f_p = \frac{\text{slew rate}}{\text{peak output swing}} \tag{5.40}$$

For most general purpose monolithic operational amplifiers using lateral *pnps* and no input emitter degneration, f_p is in the range of 5 to 50 kHz for a peak output swing of 10 V.

5.6 PRACTICAL OPERATIONAL AMPLIFIER CIRCUITS

Over 130 different monolithic operational amplifier circuits are commercially available at the present time. Some of these are special purpose types, designed to meet specific performance requirements such as low power dissipation, low noise, high slew rate, or high tolerance to radiation damage. However, by far the largest group of present-day monolithic operational amplifiers are designed as general purpose devices, suitable for a wide variety of applications. Although a large number of different designs exists for general purpose operational amplifiers, most of these can be related to one or the other of several basic designs which have become the industry standard. In this section, circuit design and layout and the electrical characteristics of some of these operational amplifiers will be examined.

The first monolithic operational amplifier to enjoy widespread acceptance is the "709" circuit, initially introduced by Fairchild Semiconductor, in 1965.[5] Figure 5.20 shows the basic circuit diagram for this operational amplifier. The input circuit uses the basic differential configuration shown in Fig. 5.8; with the input transistors operated at low collector currents ($\approx 20\ \mu$A) to maintain high input impedance and low bias currents. The first-stage voltage gain is achieved through the matched load resistors R_1 and R_2. T_8 simulates a low value positive voltage supply for the input

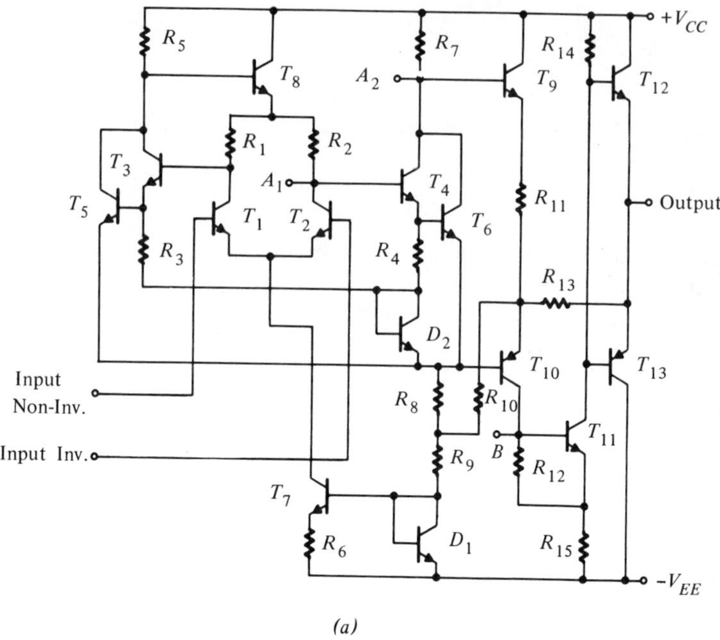

(a)

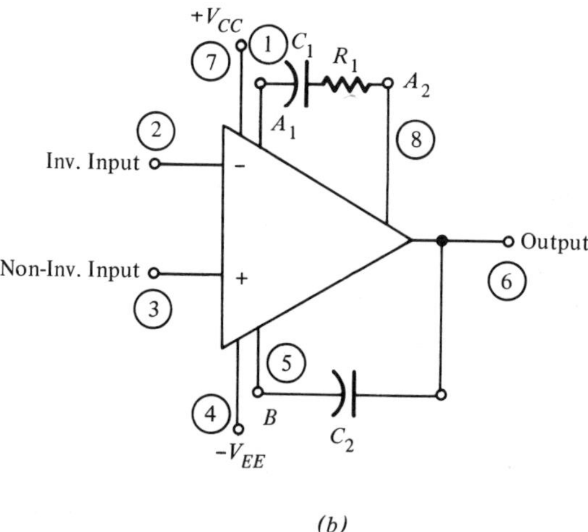

(b)

Figure 5.20 Circuit diagram and external connections for the 709 operational amplifier.

stage. T_5 functions as a unity-gain inverter and develops the full output swing of the input stage across the base-emitter junction of T_6. T_3 and T_4 are connected in Darlington configuration with T_5 and T_6 to minimize the second stage loading. R_3, R_4, and D_2 provide a low current "bleeder" path for the emitter currents of T_3 and T_4. The second stage gain is developed across the load resistor R_7. T_9 serves as a unity gain buffer for the second stage output, and in conjunction with R_{11} and the lateral *pnp* T_{10} shifts the dc level toward the negative supply. Since the *pnp* is used solely for level-shift purposes, the circuit can operate satisfactorily with lateral *pnp* current gains as low as 0.2.

The third gain stage of the 709 amplifier is formed by T_{11} and R_{14}, which drive the class-B complementary output stage (see Fig. 5.15) formed by T_{12} and T_{13}. T_{13} is a substrate-*pnp* transistor. R_{13} applies internal feedback across the third gain stage, and sets the stage gain as the ratio of the resistors R_{13} to R_{11}. The circuit does not have explicit provisions for output current limiting and short-circuit protection; however, the small geometry of the output transistors provides sufficient current limiting for most applications through the reduction of transistor emitter efficiency at high current levels. The 709 operational amplifier can be externally compensated by connecting a series (R-C) network between terminals A_1 and A_2 and a single capacitor C_2 between the output and terminal B of the circuit diagram shown in Fig. 5.20. The need for two separate frequency compensation networks arises from the three stage configuration of the 709 amplifier, as opposed to two stage designs encountered in later-generation operational amplifiers.

Since 709 was the first monolithic operational amplifier to gain wide acceptance, its external pin configuration has also set an industry standard, and all later monolithic amplifiers have their terminal pins numbered to be pin-compatible with the 709. The pin configuration for the 709 operational amplifier is shown in Fig. 5.20(*b*), including the external compensation networks R_1, C_1, and C_2.

The photomicrograph of the integrated circuit chip for the 709 operational amplifier is shown in Fig. 5.21. The dimensions of the circuit chip are 40 mils by 44 mils. The circuit is designed to be fabricated using base-diffused resistors and standard planar-epitaxial processing technology. The key components, as well as the circuit terminals are also identified in the Figure. The key features to note in the layout of the integrated circuit are the following:

1. Input transistors T_1 and T_2 are identical in geometry and are located in close proximity to ensure low offset voltage and currents.

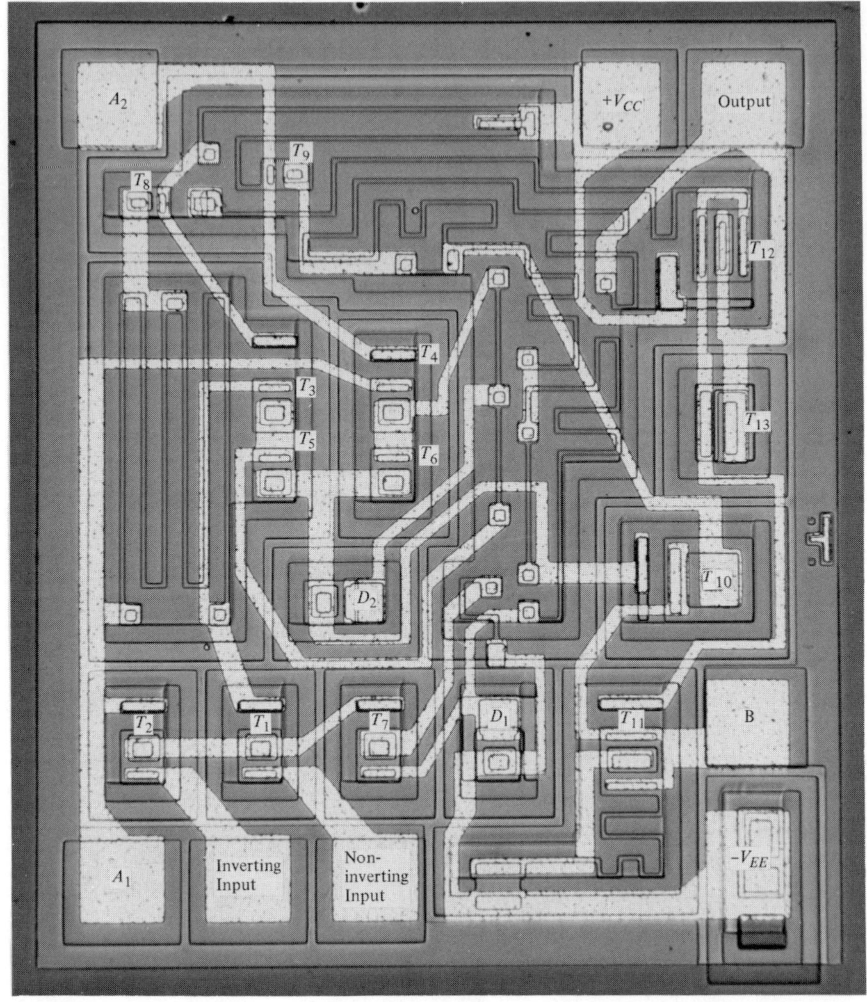

Figure 5.21 Circuit layout for the 709 operational amplifier. (*Photo: Fairchild.*)

2. Input and output terminals are placed as far away as possible on the chip to minimize parasitic coupling.
3. The matched resistors, such as R_1 and R_2 of the input stage, are placed in close proximity and are laid out with identical geometrics.
4. Common-collector transistors, such as (T_3, T_5), (T_4, T_6), (T_7, T_8), are located in the same isolation pocket to conserve chip area.

The basic shortcomings of the 709 operational amplifier are low differential and common-mode input swings. Input common mode swing is limited to approximately (⅔) V_{CC} (see Section 5.3), and maximum differential swing is limited to ≤ 7 V by the emitter-base breakdown of the input transistors. The frequency compensation is relatively complicated since it requires three compensating elements.

A more recent operational amplifier circuit which overcomes most of the shortcomings of the 709 circuit is the 741 operational amplifier.[7] The circuit schematic for the 741 operational amplifier is shown in Fig. 5.22. The circuit uses only two gain stages to obtain the desired open loop gain. Active, or current source loads are used in each gain stage, rather than resistive loads, to obtain higher voltage gains per stage. The input stage has the same configuration as that shown in Fig. 5.8 and provides a voltage gain of >60 dB. The second gain stage is formed by Darlington-connected transistors T_{16} and T_{17}, along with the multiple collector lateral *pnp* transistor, T_{13}, which serves as an active load for the stage. This stage provides a voltage gain of approximately 45 dB. The drive to the output stage is provided by T_{24}, with T_{19} and T_{21} setting up the bias levels for the class-B output stage. Effectively, these correspond to diodes D_1 and D_2 of Fig. 5.15.

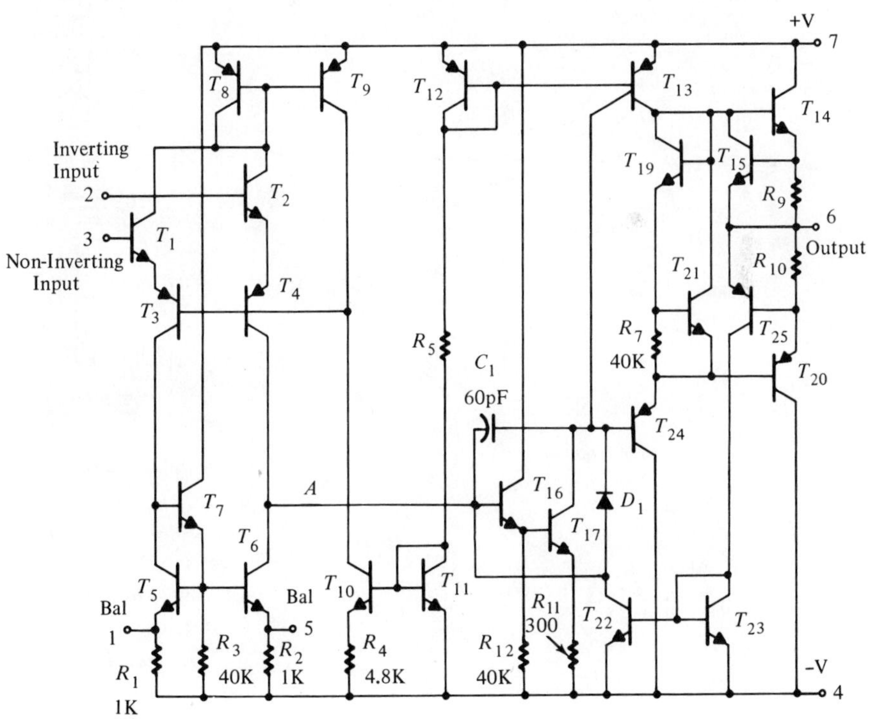

Figure 5.22 Circuit diagram of the 741 operational amplifier.

The class-B output stage is formed by complementary common-collector stages, T_{14} and T_{20}, where the latter is a substrate *pnp*. Transistors T_{15} and T_{25} along with resistors R_9 and R_{10} provide output short-circuit protection, as described in Section 5.4 (see Fig. 5.16).

The 741 operational amplifier is internally compensated for unconditional stability by a single roll-off capacitor C_1 connected across the second gain stage. This compensation technique is the same as that shown in Fig. 5-18 of the previous section (typically $C_1 \simeq 30$ pF). Typical electrical characteristics of the 741 operational amplifier are also listed in Table 5.1.

Figure 5.23 shows the photomicrograph of the 741 operational ampli-

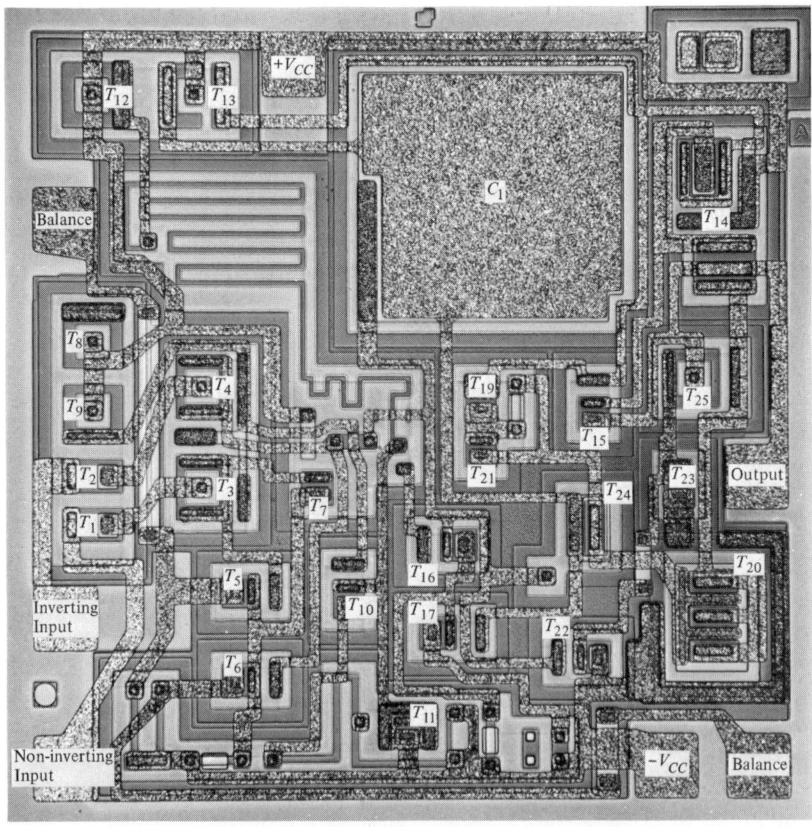

Figure 5.23 Photomicrograph of the 741 operational amplifier. Die size: 56 mils square. (*Photo: Fairchild.*)

fier chip, measuring 56 mils square. The compensating capacitor C_1 is realized as a MOS capacitor (see Fig. 3.4) with the n-type collector island and the aluminum interconnection layer serving as the plates of the capacitor. The capacitor is shown as the large metallized area on the upper right-hand side of the figure. The external terminals and the key devices within the circuit are also identified in the photomicrograph. Note that the basic layout rules outlined in connection with the 709 circuit are still adhered to. Except for the output stage, all the *pnp* devices in the circuit are of the lateral type. In the circuit layout, these can be identified by the wrap-around collector structure which encircles the emitter (see T_3 and T_4 of Fig. 5.23.)

The input impedance, and input bias current requirements of a monolithic operational amplifier can be significantly improved by using punch-through transistors at the input stage, as described in Section 5.3 (see Fig. 5.12). Since the punch-through transistors can be fabricated with β_0 values of approximately 5000 at 2 to 3 μA collector current levels, using the input stage topology of Fig. 4.10, it is possible to keep the input bias currents below 1 nA, and the bias current offsets below 0.1 nA. A good example of a practical circuit which makes use of punch-through devices at the input stage is the 108 operational amplifier whose circuit diagram is shown in Fig. 5.24.[10] For convenience, the punch-through devices are identified in the figure with rectangular base regions. The basic input stage configuration of the 108-type amplifier is similar to that shown in Fig. 5.12, where the cascade connected conventional bipolars T_5 and T_6 stand off the common-mode voltage across the punch-through input transistors T_1 and T_2. Thus, both input transistors are thus operated at nearly zero collector base voltage under all common-mode input conditions. Cross-coupled transistors T_3 and T_4 are normally off, and serve as input overvoltage protection. Diode connected lateral *pnp* transistors T_7 and T_8, along with matched load resistors R_4 and R_5, provide collector loads for the gain-stage. T_9 and T_{10} serve as the second gain stage and the transistor pair. T_{22} and T_{21} convert the differential output of the second stage to a single-ended output, similar to the differential to single-ended output conversion shown in Fig. 5.9. To make this conversion more efficient, T_{21} and T_{22} are designed as punch-through devices. The single-ended output is then applied to the base of T_{14}, which provides the base drive for output transistors T_{18} and T_{19}. T_{19} is a substrate *pnp,* similar to that described in the 709- or the 741-type output stage. T_{15} and T_{16} correspond to the biasing diodes D_1 and D_2 of Fig. 5.15. T_{29} simulates a low value current source ($\approx 6\ \mu$A) which sets the input stage current level; T_{27} and T_{28} provide the "bootstrap" action to operate the punch-through transistors at nearly zero col-

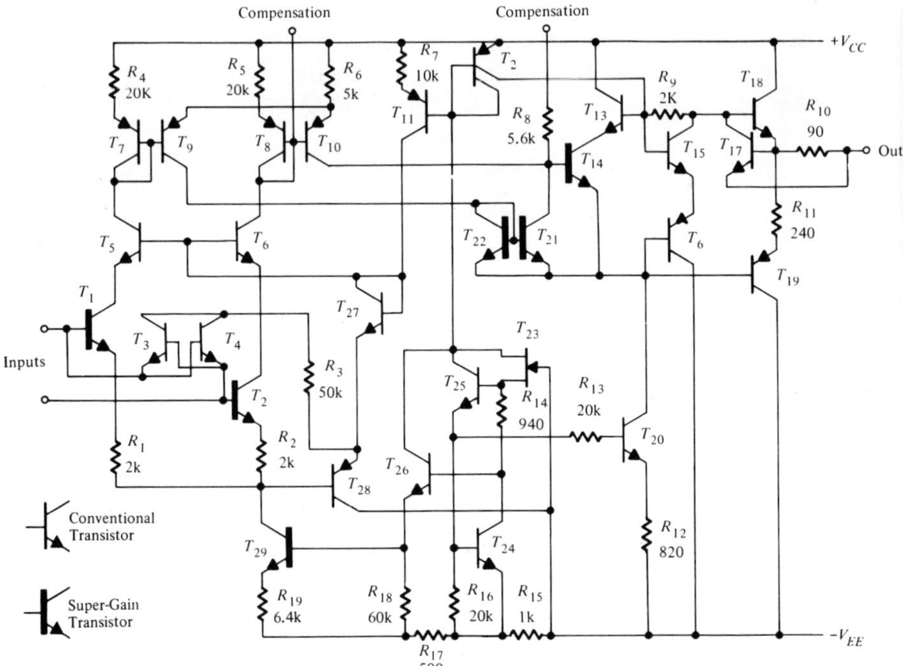

Figure 5.24 Circuit diagram for an operational amplifier using punch-through input transistors:[10] National Semiconductor LM 108.

lector voltage. *n*-channel FET T_{23} serves as a constant-current bias source to set the bias levels for T_2, T_{20}, and T_{29}. This *n*-channel FET is obtained using the *p*-type base diffusion as the gate, and the *n*-type epitaxial layer as the channel, similar to the structure of the device described in Fig. 2.25. The 108 circuit is compensated by connecting an external capacitor across the compensation terminals in Fig. 5.24. Internal resistor R_8, in series with compensation for a single external capacitor between A_1 and A_2. Typical electrical characteristics of the 108 operational amplifier are listed in Table 5.1 along with the other circuits described in this section.

Figure 5.25 shows the die photograph for the 108 operational amplifier. Various circuit terminals and the active devices are also identified in the layout.

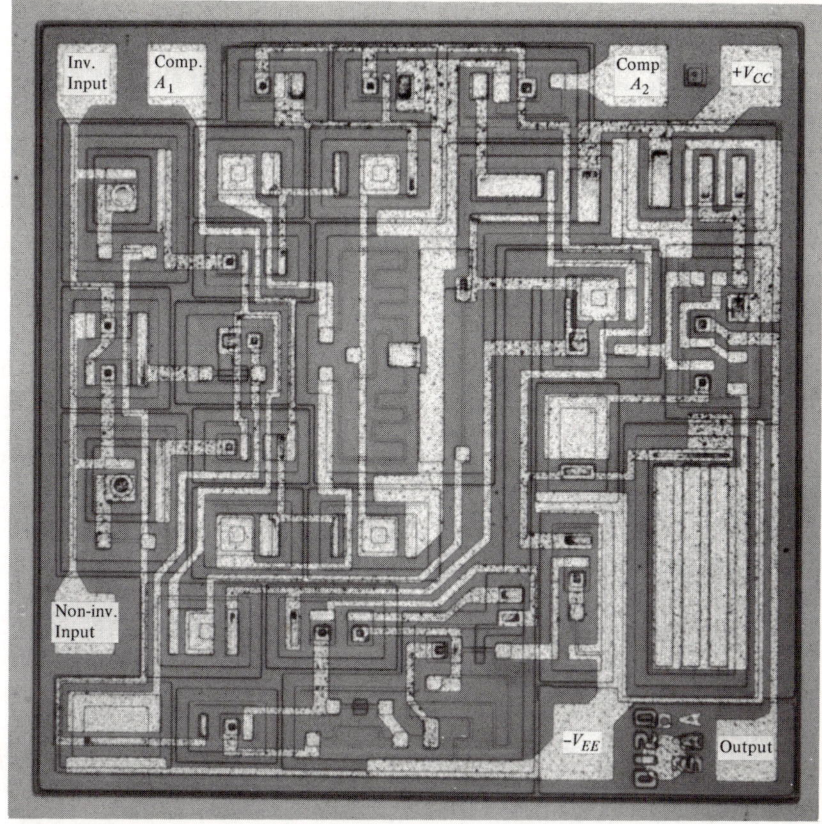

Figure 5.25 Circuit layout of the 108 operational amplifier. (*Photo: National Semiconductor.*)

5.7 OPERATIONAL AMPLIFIER BASED CIRCUITS

Comparators

The function of a voltage comparator is to compare the instantaneous value of a signal voltage at one input with a reference voltage on the other input, and produce a digital one or zero level at the output when one input is higher than the other. It is normally used to perform any one of the following classes of circuit functions:

1. Variable threshold detector
2. Pulse-height discriminator
3. Zero-crossing detector

4. High noise-immunity digital line receiver
5. Voltage level comparator (for analog to digital conversion)

A voltage comparator circuit is designed as a high gain differential input, single-ended output amplifier, with the output swing and the dc levels adjusted to be compatible with the conventional logic circuits. In terms of its basic circuit function, a comparator is very similar to a differential input operational amplifier; in fact, in many applications operational amplifiers can be used as comparators. The input impedance, voltage gain, and output voltage swing requirements are somewhat lower for a comparator; thus the circuit configurations for a comparator are quite a bit less complicated than for a general purpose operational amplifier.

Since a comparator is switched between two output states, it is required to have a rapid recovery from saturation, and a fast rise time.[16] The voltage gain is necessary only to reduce the differential input level change necessary to make the output swing from one extreme level to another. To interface with digital circuits, the required peak-to-peak output swing levels are in the range of 3 to 5 V. Therefore, a voltage gain of ≥ 1000 is usually sufficient to bring the input signal amplitude needed for full output swing to a level comparable with the input stage offset (see Section 5.3). Therefore, very high values of voltage gain are not required. Since the overall voltage gain is relatively low, and the comparator operates under open-loop conditions, it does not require frequency compensation.

Figure 5.26 shows the circuit schematic for a practical comparator circuit (Fairchild μA 710). The operation of the comparator circuit of Fig. 5.26 can be briefly described as follows: T_1 and T_2 form a differential input stage, with the balanced collector loads R_1 and R_2; T_5 serves as a virtual positive supply for the input stage. This basic circuit configuration is similar to that described earlier in Fig. 5.8. T_3 and T_4 form the second differential gain stage, with $(R_3 + R_4)$ and R_5 serving as the collector loads. The emitter-base breakdown diode Z_1 serves as the emitter bias for the second stage. Emitter follower T_7 provides a buffered low impedance output from the second stage. The second emitter-base breakdown diode Z_2 shifts the output dc potential to a level compatible with the logic circuitry. The diode D_1 connected across the second stage limits the positive output swing, both to increase speed and to provide level compatibility with integrated logic circuits. Neglecting the interstage loading effects, the overall voltage gain of the comparator can be expressed as:

$$A_v = (A_{v1})(A_{v2}) = (g_{m1}R_1)(g_{m4}R_5) \tag{5.41}$$

where A_{v1} and A_{v2} are the first and the second stage voltage gains; g_{m1} and g_{m4} are the transconductances of T_1 and T_4. For operation of the circuit with (+12) and (−6) volt supply voltages, $A_v \approx 1700$. The overload re-

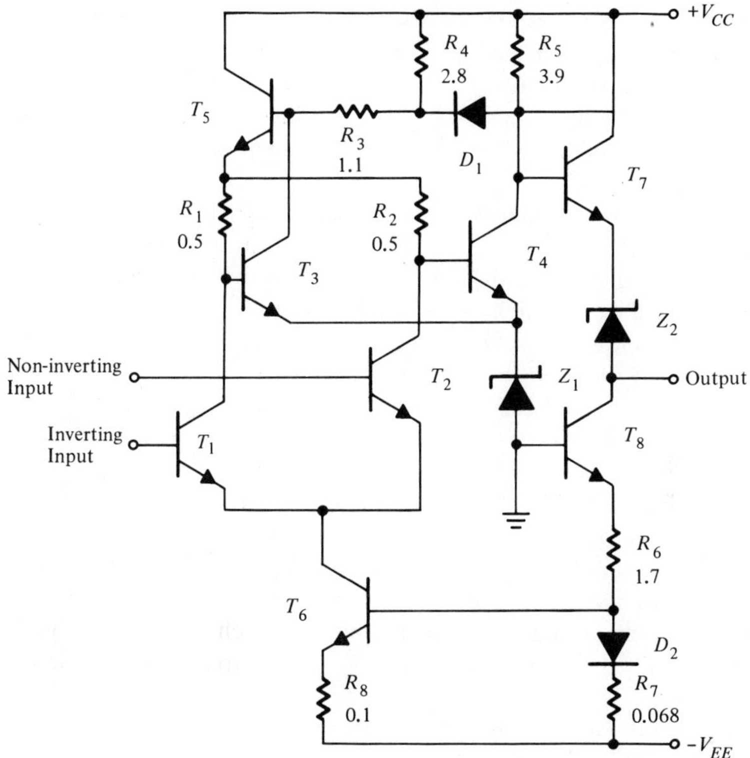

Figure 5.26 A high speed comparator circuit (μA 710) (resistor values in kilo-ohms).

covery time is approximately 30 nanoseconds for 10 mV input overdrive. The output dc levels for the circuit are +3.5 and −0.5 V, which make it compatible with most integrated logic circuits.

Sense Amplifiers

The basic function of a sense amplifier is to detect the presence and the polarity of low level signals, and transform these to high level logic signals. The sense amplifiers serve as the "read" elements in core or plated-wire memory systems. In a plated-wire or a ferrite core memory system, a closed flux path exists around each memory element. In the case of core memories, the ferrite body of the core forms the closed flux path; in the case of a plated-wire memory element, the plating serves as the closed path for the magnetic flux. The magnetic flux encircles the wire in one direction for a stored binary one, and in the opposite direction for a stored binary zero. To read the stored information, a word current pulse is passed through the word line which reverses or tilts the magnetization vector in the ferrite core or the plated film. This momentary change of magnetization causes a short voltage pulse to appear in a second wire, called the "sense wire." The function of the sense amplifier is to convert these small voltage pulses to high level digital signals compatible with integrated logic circuits.

In the case of core memory systems, input signals for the sense amplifier are relatively large (typically in the 20 to 50 mV range). In the case of plated-wire memories, the sense signals are in the 5 to 10 mV level, requiring a higher sensitivity and lower threshold for the sense amplifier. Along with accurate detection of input signals, the sense amplifier is also required to have a rapid differential and common-mode overload recovery time to facilitate rapid recycle times for the memory system.

Since a memory system is made up of a number of parallel information lines, monolithic sense amplifiers are often required to have more than one channel. In a multichannel sense amplifier system a decoder logic is also included on the monolithic chip which allows only one of the amplifier channels to be turned on at any given time. Figure 5.27 shows a functional block diagram for a 4-channel sense amplifier for a plated-wire memory system. The four independent input channels, designated A, B, C, and D, are controlled by a set of three control or decode logic inputs shown as S_1, S_2, and S_3. The digital one or zero signals applied to S_1 and S_2 allows only one of the four channels to be active at any one time. S_3 serves as an enable/disable terminal where a digital zero applied to this terminal disables all four channels. This is normally used during the "write" cycle to eliminate the overload transient. When the amplifier is enabled, the output of the selected channel is amplified by a second gain stage, A_o, and drives a NAND gate G_o to provide a binary one or zero at the output. The second input of the gate can be used as a "strobe" terminal to examine the output logic level at selected intervals of time.

Figure 5.28 shows a practical circuit implementation of the block diagram of the previous figure (Signetics SE 528). In this case transistor-transistor logic (TTL) gates are used for channel selection. The four input channels are comprised of the differential gain stages made up of transistor pairs (T_1, T_2), (T_3, T_4), (T_5, T_6), and (T_7, T_8). All these channels share a set of common collector loads, R_1 and R_2. T_{13} and T_{14} serve as unity-gain level shift stages and translate the output of the input channels to the input of the second gain stage. This stage A_o of Fig. 5.27, is formed by the differential gain stage made up of T_{15} and T_{16}. The output of this stage, through T_{19}, drives the output TTL gate.

From the circuit diagram of Fig. 5.28, operation of the channel-select logic can be explained as follows: All four amplifiers are biased from a common current source I_T, supplied through T_{12}. The control input S_3 can

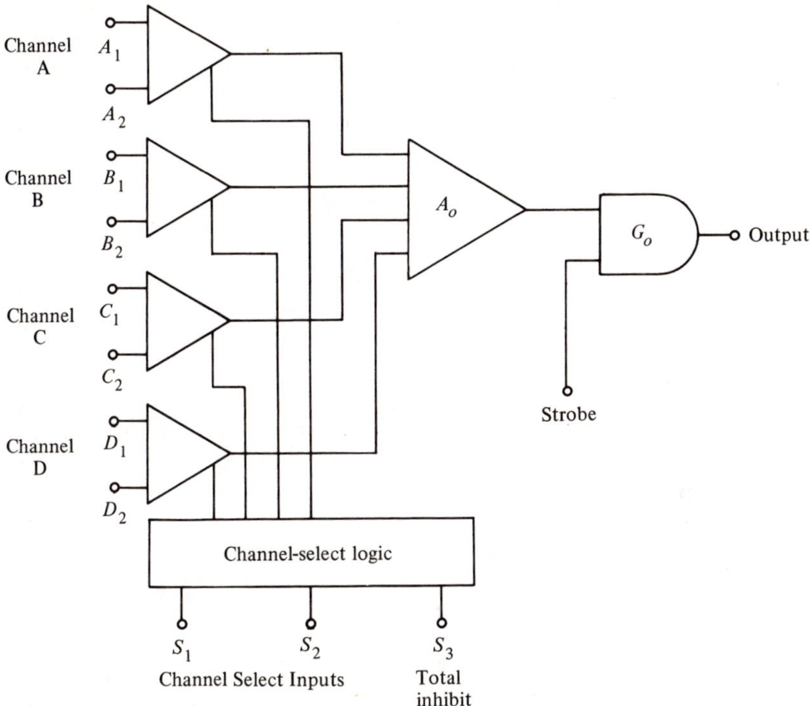

Figure 5.27 Functional diagram of a 4-channel plated-wire sense amplifier.

disable all four channels by steering this current to T_{11} and cutting T_{12} off. Similarly, with T_{12} conducting, S_2 steers the bias current to T_9 or T_{10}, and enables input channels A and B or C and D, respectively. S_1, in turn, steers the bias current to (T_{20}, T_{23}) or to (T_{21}, T_{22}), enabling the respective input channels A and D or B and C. Thus, the four channels of the amplifier can be controlled by the binary one or zero inputs to the channel select terminals as follows:

S_1	S_2	S_3	Active Channel
0	0	1	A
1	0	1	B
0	1	1	C
1	1	1	D
X	X	0	—

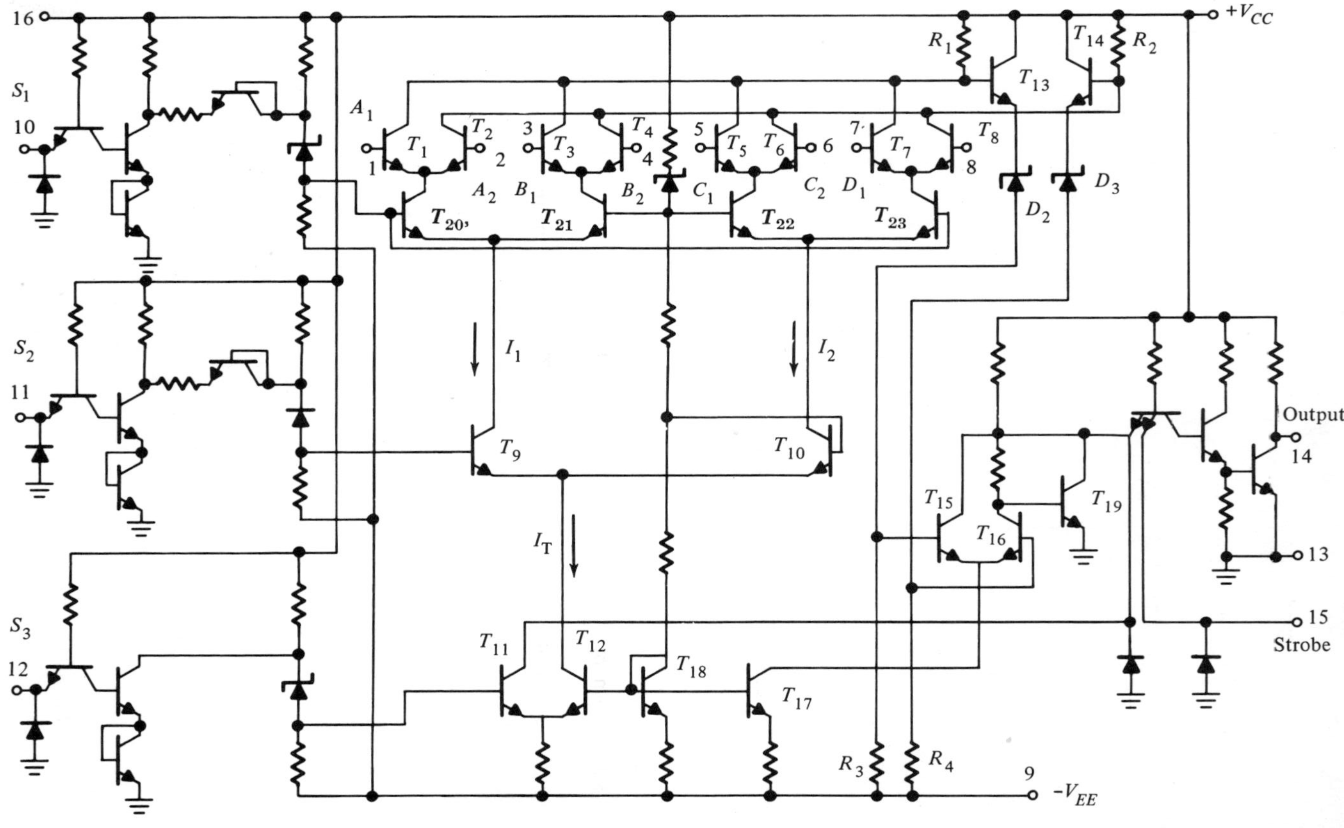

Figure 5.28 Circuit diagram of a 4-channel plated-wire sense amplifier (Signetics SE528B).

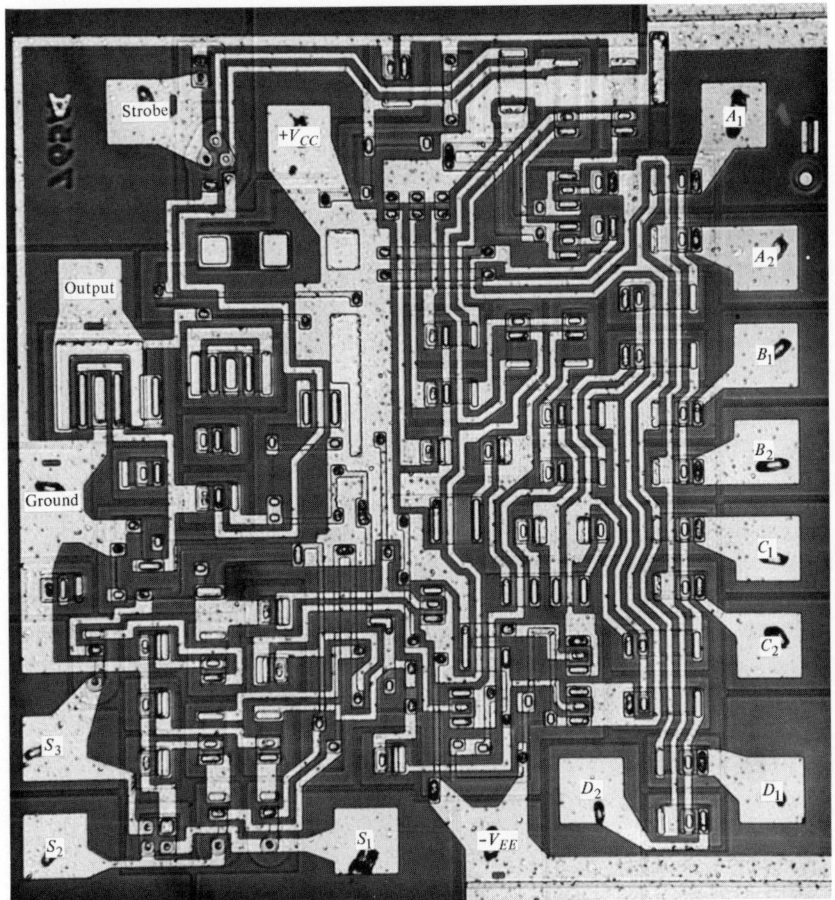

Figure 5.29 Topological layout of the circuit of Figure 5.28. (Chip size: 57 mils square). (*Photo: Signetics.*)

The 4-channel plated-wire sense amplifier circuit of Fig. 5.28 is designed to operate from ±5 V supplies, with an input threshold of 2 mV for a full-scale output swing. Typical propagation delay or channel select times for the circuit are of the order of 20 nanoseconds which is sufficient for high-speed memories having cycle times as low as 200 nanoseconds.

The photograph of the integrated circuit chip is shown in Fig. 5.29. Note that the analog and the digital sections of the circuit are laid out as far apart as possible from each other since they operate at drastically different signal levels. All the input channels are laid out symmetrically, and within close proximity of each other to take advantage of the close matching and tracking of integrated components.

REFERENCES

1. W. R. Le Page and S. Seely, *General Network Analysis,* McGraw-Hill, New York, 1952.
2. E. J. Angelo, Jr., *Eelctronic Circuits,* 2d ed., McGraw-Hill, New York, 1964, Chap. 4.
3. R. D. Middlebrook, *Differential Amplifiers,* Wiley, New York, 1963.
4. L. J. Giacoletto, *Differential Amplifiers,* Wiley, New York, 1970.
5. R. J. Widlar, "A Unique Circuit Design for a High Performance Operational Amplifier Especially Suited to Monolithic Construction," *Proc. Natl. Electronics Conf.,* **21** (1965): 85–89.
6. R. J. Widlar, "Monolithic Op. Amp. with Simplified Frequency Compensation," *EEE Magazine,* **15**(7) (1967): 58–63.
7. D. Fullagar, "A New High Performance Monolithic Operational Amplifier," *Fairchild Semiconductor App. Brief,* May, 1968.
8. M. J. Hellstrom and W. R. Harden, "An Integrated Low Voltage Operational Amplifier," Int. Solid State Ckts. Conf., *Digest Tech. Papers,* **12** (1969): 16–17.
9. J. E. Solomon, W. R. Davis and P. L. Lee, "A Self-Compensated Monolithic Operational Amplifier with Low Input Current and High Slew Rate," IEEE Int. Solid State Ckts. Conf., *Digest Tech. Papers,* **12** (1969): 14–15.
10. R. J. Widlar, "Design Techniques for Monolithic Operational Amplifiers," *IEEE J. Solid State Ckts.,* **SC–4**(9) (1969): 184–191.
11. G. R. Wilson, "A Monolithic Junction FET-NPN Operational Amplifier," Int. Solid State Ckts. Conf., *Digest Tech. Papers,* **11** (1968): 20–21.
12. R. J. Widlar, "A Monolithic High-Gain DC Amplifier," *Proc. Natl. Electronics Conf.,* (1964): 169–174.
13. H. R. Camenzind and A. B. Grebene, "An Outline of Design Techniques for Linear Integrated Circuits," *IEEE J. Solid State Ckts.,* **SC–4**(3) (1969): 110–122.
14. H. W. Bode, *Network Analysis and Feedback Amplifier Design,* Van Nostrand Reinhold, New York, 1945.
15. W. E. Hearn, "Fast Slewing Monolithic Operational Amplifier," *IEEE J. Solid State Ckts.,* **SC–6**(1) (Feb., 1971): 20–24.
16. J. Giles, "Fairchild Semiconductor Linear Integrated Circuits Applications Handbook," Fairchild Semiconductor *App. Brief,* 1967.

6

Monolithic Power Circuits: Regulators and Power Amplifiers

The key component in a power integrated circuit is the output power device; therefore, many similarities exist between the design of power integrated circuits and discrete power transistors. In the case of monolithic circuits, the availability of a large number of active components on the chip at little or no extra cost allows one to design a number of overload protection schemes such as current-limiting current-foldback, and thermal-shutdown which are difficult to implement in discrete design. On the other hand, the presence of low-level signal processing circuitry in very close proximity to a power device poses a number of design problems due to thermal drifts and nonuniform temperature distribution on the chip, and requires special layout and packaging considerations. High power devices also exhibit a number of unique failure modes due to "electromigration" and "secondary breakdown" effects which will be described briefly in this chapter.

The two major classes of monolithic power circuits are the voltage regulators and power amplifiers. In this chapter basic guidelines will be outlined concerning the design of these classes of monolithic power circuits. Since the voltage regulators comprise a class of circuits not covered earlier in the text, some additional space will be devoted to basic design considerations for monolithic regulators, both for medium power as well as for high power applications.

In monolithic circuits, power handling and dissipation characteristics of the chip are to a large extent governed by the circuit package. Therefore, a portion of this chapter is also devoted to basic packaging considerations for high power integrated circuits.

6.1 DESIGN CONSIDERATIONS FOR MONOLITHIC VOLTAGE REGULATORS

The function of a voltage regulator is to maintain a constant output voltage level, irrespective of the changes of the input voltage or the output current. Voltage regulator circuits suitable for monolithic integration are "series" types connected in series between the load and the unregulated supply line. To be useful, a regulator circuit must exhibit very low "line" or "load" regulation and a high degree of "ripple-rejection." These basic performance parameters are defined as follows:

Line regulation: Percentage change of the output voltage for a specified change in input voltage.
Load regulation: Percentage change of the output voltage for a specified change in the load current.
Ripple-rejection: Ratio of the input peak-to-peak ripple to the output ripple voltage.

In addition, the output voltage level is required to have a very low temperature coefficient, and the output terminal be protected from burning out under accidental short-circuit conditions.

An integrated voltage regulator circuit in general consists of three main parts: the reference voltage element, the error amplifier, and the series pass element. Figure 6.1 shows the block diagram of a typical series regulator circuit comprised of these three basic blocks. The internal voltage reference generates a voltage level V_R which is independent of temperature or the supply voltage. Sampling resistors R_1 and R_2 provide a corrective feedback point for the error amplifier. Assuming that the error amplifier gain A is sufficiently high, the voltage drop across the pass element will vary with the variations of the unregulated input voltage to maintain output voltage V_2 constant and equal to:

$$V_2 = \frac{V_R A}{1 + \gamma A} \approx \frac{V_R}{\gamma} \tag{6.1}$$

where γ is the feedback factor determined by the sampling resistors, i.e.,

$$\gamma = \frac{R_2}{R_1 + R_2} \leq 1 \tag{6.2}$$

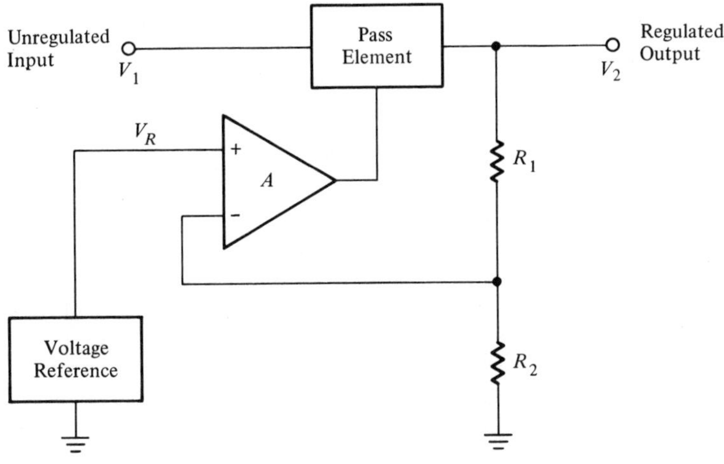

Figure 6.1 Block diagram of a series regulator.

The line regulation is directly related to the sensitivity of the voltage reference to the changes in the line voltage. For example, if the reference voltage changes by an amount ΔV_R for a specified change of the input line voltage, then the line regulation is

$$\text{Line regulation} = (\Delta V_R)\frac{A}{1 + \gamma A} \approx \frac{\Delta V_R}{\gamma} \tag{6.3}$$

Similarly, the load regulation, for a change ΔI_L of the load current, can be written as

$$\text{Load regulation} = \frac{(\Delta I_L)R_0}{1 + \gamma A} \tag{6.4}$$

where R_0 is the low frequency dynamic resistance of the pass element.

As indicated by the block diagram (Fig. 6.1), the range of regulated output voltage is $V_R \leqq V_2 < V_1$. In general, the upper bound on V_2 is always less than V_1, due to the finite voltage drop across the pass element. To obtain a wide dynamic range for the regulated output voltage, a low-value voltage reference is necessary. However, to obtain a high output voltage from a low V_R requires a low value of γ, i.e., a reduction of the regulator feedback loop gain. This in turn can result in a degradation of the line or the load regulation characteristics, as implied by Eqs. (6.3) and (6.4).

Figures 4.10 and 4.11 showed some practical circuit configurations which can be used to generate stable reference voltages. The thermal drift characteristics of these reference sources were described in Chapter 4, and

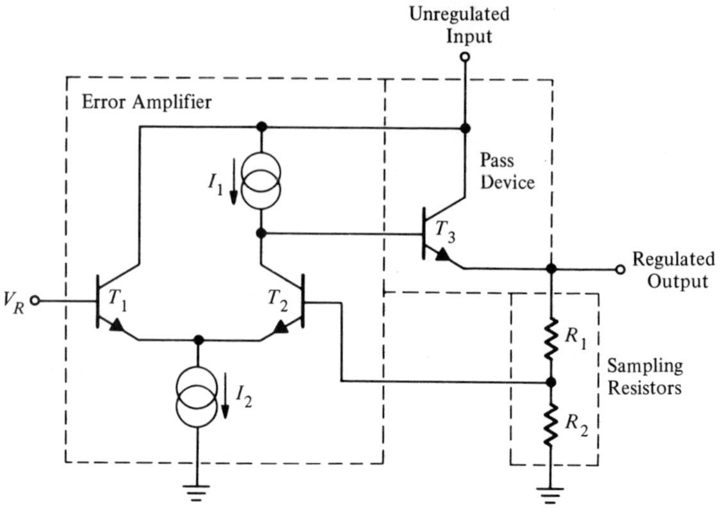

Figure 6.2 Simplified circuit diagram for a practical regulator.

will not be repeated here. Typical values of V_R obtainable from these circuits are in the range of 1.2 to 2.5 V.

Figure 6.2 shows a simplified circuit diagram for a practical series voltage regulator. For most applications the open-loop voltage gain A for the error amplifier is of the order of 60 to 70 dB, which can be obtained from a single high gain differential stage using active (current source) loads. The pass element is normally a power transistor connected in common-collector configuration. This corresponds to transistor T_3 of Fig. 6.2. In most cases a Darlington connection of two transistors is used as the pass element, in order to provide adequate buffering of the error amplifier from the load. In certain cases, the high gain error amplifier may have to be frequency-compensated to ensure stability under all operating conditions. For regulator circuits using a single gain stage, this can be accomplished by providing a compensation capacitance on the circuit chip,[1] similar to the case of internally compensated operational amplifiers (see Section 5.6). However, an excessive amount of frequency compensation may deteriorate the high frequency output impedance level for the regulator by causing a fall-off of the error amplifier gain at high frequencies. From Eq. (6.4), an effective output impedance $Z_o\ (j\omega)$ can be defined as

$$Z_o(j\omega) = \frac{R_0}{1 + \gamma A(j\omega)} \tag{6.5}$$

Thus, as the error amplifier response $A(j\omega)$ rolls off with frequency, the output impedance level tends to increase.

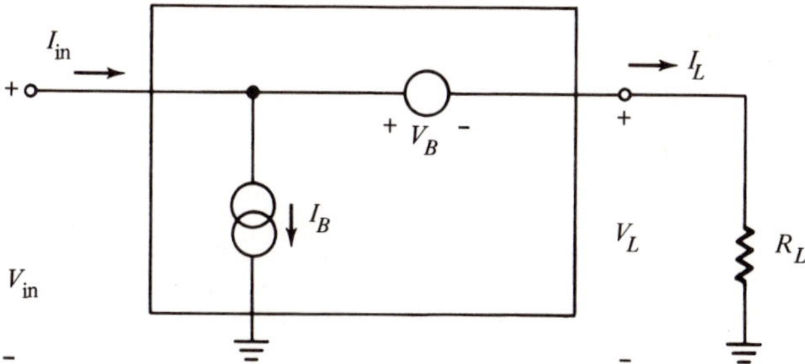

Figure 6.3 A simplified model for calculating efficiency of a series regulator.

Efficiency Considerations

In the case of a series-type voltage regulator, a simplified bias model shown in Fig. 6.3 can be used to calculate the efficiency.[2] In this model, current source I_B represents the current consumed in the regulator circuitry. Similarly, voltage source V_B represents the necessary voltage drop within the regulator to produce the desired load voltage from a given input voltage. For proper biasing of the regulator circuit, V_B is restricted to be greater than a minimum value, $(V_B)_{\min}$. For typical monolithic regulators, $(V_B)_{\min}$ is in the range of 1.5 to 3 V. The regulator efficiency, η, is defined as

$$\eta = (P_o/P_{\text{in}}) \tag{6.6}$$

where P_o and P_{in} are the output and the input power levels given as

$$P_o = I_L V_L \tag{6.7}$$

$$P_{\text{in}} = V_{\text{in}}(I_B + I_L) \tag{6.8}$$

Then, η can be expressed as

$$\eta = \frac{1}{(1 + V_B/V_L)(1 + I_B/I_L)} \tag{6.9}$$

As given by Eq. (6.9), regulator efficiency depends directly on the ratio of load voltage and load current to the bias voltage and bias current. Since the bias current I_B is more or less fixed by the regulator design, the maximum efficiency, $\eta_{\max}$, is obtained when the regulator is delivering its maximum rated current with the minimum input-output voltage differential. If the input contains large ac components, or large minimum-maximum fluctuations, the maximum efficiency is reduced, since the average level of the input voltage would have to be increased to ensure that the instantaneous value of V_B is greater than $(V_B)_{\min}$.

Overload Protection

Since a monolithic regulator is intended to handle sizeable currents under varying supply voltage or load current requirements, it is necessary to provide necessary overload protection in the circuit design to avoid destructive burn-out. This can be done by limiting the output current to a safe value under short-circuit conditions, using the output current-limiting technique described in connection with the operational amplifier design (see Section 5.4 and Fig. 5.16). In certain cases, a sustained value of excessive output current can cause damage to the regulator, even with short-circuit current-limiting. To avoid this, sometimes a "current-foldback" type limiting can be used at the output which not only limits the output current, but gradually reduces it to zero if the short-circuit condition prevails.

In addition to the output short-circiut protection, a "thermal-shutdown" mechanism can also be incorporated into the design which completely shuts off the regulator when the junction temperature on the chip reaches a critical level (typically 175 to 200°C) to prevent destructive burn-out. Design considerations for thermal shutdown and fault protection circuitry will be described further in Section 6.4.

6.2 PRACTICAL REGULATOR CIRCUITS

An ideal regulator is only a three-terminal device, consisting of unregulated supply, regulated output, and a ground terminal. However, in the design of a general purpose regulator circuit, additional control terminals may be needed to modify the regulator characteristics externally, to fulfill a speficic need, and to adjust current-limiting properties. Therefore, a general purpose integrated regulator circuit may have a functional block diagram similar to that shown in Fig. 6.4(*a*). Normally the reference is connected to the noninverting input of the error amplifier, and the feedback from the regulated output terminal is applied to the inverting input of the amplifier, through a resistive divider. The short-circuit protection is provided by a set of current-sense and current-limit terminals which can be used as a latching switch to limit the output current to a predetermined value, under sustained short-circuit conditions. The collector of the pass transistor can be brought out separately and used as the unregulated input if a separate positive supply is available to operate the control circuitry. Even though most integrated regulators are designed to regulate a positive voltage supply, a regulator circuit can be modified for use with negative voltages by adding an external *pnp* pass transistor to the output.

Figure 6.4(*b*) shows the actual circuit schematic for a general-purpose

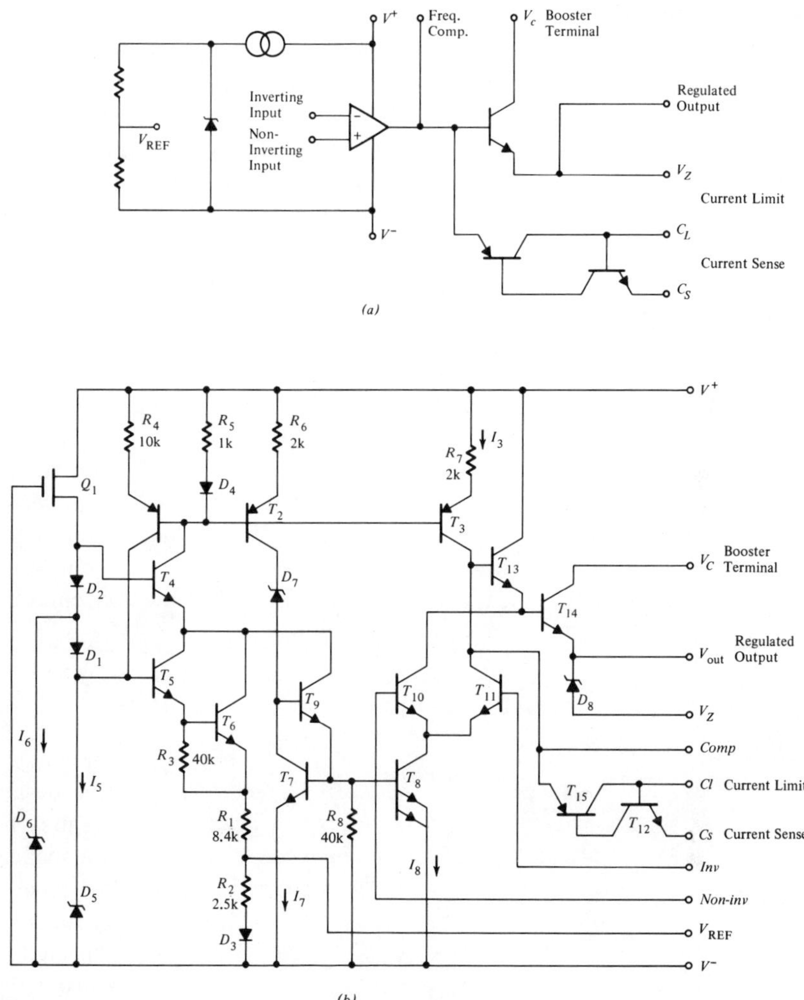

Figure 6.4 A monolithic medium power regulator: (*a*) functional block diagram; (*b*) actual circuit schematic (Signetics SE550).

medium power monolithic regulator (Signetics SE550). The circuit uses a temperature compensated voltage reference circuit similar to that shown in Fig. 4.10. In this case, the avalanche diode D_5 serves as the uncompensated voltage source (similar to D_B of Fig. 4.10), with the *pnp* current source in turn biased by a second constant-current stage formed by the *p*-channel MOS transistor Q_1. Darlington-connected transistors T_5 and T_6, along with D_3, R_1, and R_2 form the temperature compensated voltage reference, with a dc level 1.7 V above the negative supply. The *pnp* current source T_2 sets the bias currents through the error amplifier current source (T_8) and the active load T_3. Note that, since T_8 is a double emitter transistor (i.e., having twice the emitter area of T_1), bias current levels are set as $I_8 = 2\ I_7$. Transistors T_{13} and T_{14} form the output driver and the pass transistor for the regulator. Note that the collector of the pass transistor is left uncommitted. It can be connected to V^+ to form a Darlington-

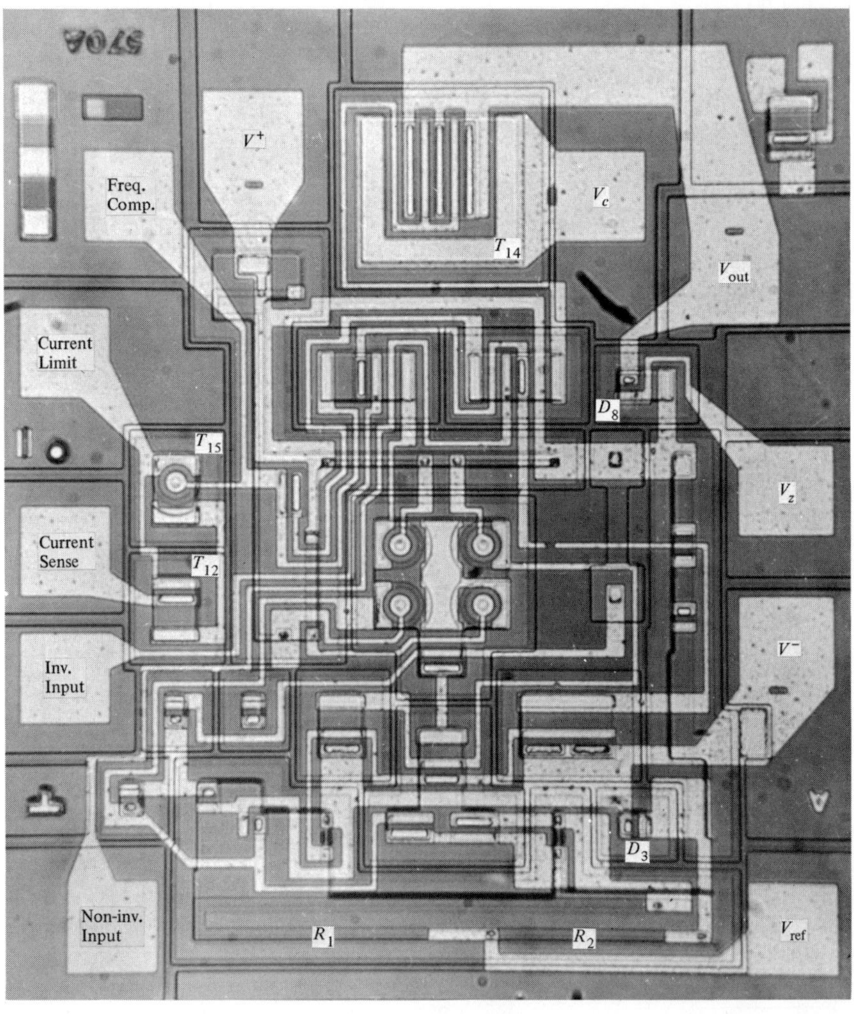

Figure 6.5 Circuit layout for the medium-power regulator. (*Photo: Signetics.*)

connected output stage, or it can be energized from the unregulated supply line if a separate supply voltage is available for the control circuitry. The circuit can be converted to a negative voltage regulator by connecting an external *pnp* pass transistor to the circuit. Avalanche diode D_8 is included in the circuit to provide the necessary dc bias for this purpose.

T_{12} and T_{15} form the output short-circuit protection by providing current-limiting and current-foldback characteristics. The operation of this type of a current-limiting stage is discussed further in Section 6.4.

Figure 6.5 shows the topological layout for the regulator circuit of Fig. 6.4. The external connections and some of the key circuit elements are also identified in the Figure. Typical electrical characteristics of the monolithic regulator are listed in Table 6.1. Since this particular regulator circuit is designed to operate at medium power levels (i.e., power dissipation <1 W), it does not contain thermal shutdown protection which is normally used in high-power regulator circuits.

A particular application for monolithic regulators is in digital systems, where they can be used to simulate a well-regulated power supply on a logic card. For this application a high current and fixed low voltage (≈ 5 V) is required. Figure 6.6 shows the simplified schematic for such a regulator circuit, designed to produce a 5 V constant output from an unregulated supply line.[3] This circuit uses a modified version of the temperature-compensated voltage reference shown in Fig. 4.11, where a temperature-independent voltage level is obtained by compensating the negative coefficient of V_{BE} with the positive temperature dependence of the base-emitter bias difference ΔV_{BE}. In Fig. 6.6 transistors T_5 and T_7 correspond to T_1 and T_2 of Fig. 4.11. T_7 amplifies the V_{BE} difference between T_5 and

TABLE 6.1 TYPICAL ELECTRICAL CHARACTERISTICS FOR THE VOLTAGE REGULATOR CIRCUIT OF FIGURE 6.4

Line regulation*	$= 0.2\%$ of V_{out}
Load regulation**	$= 0.03\%$ of V_{out}
Ripple rejection	$= 75$ dB
Temperature coef. of V_{out}	$= 0.002\%/°C$
Stand-by current drain	$= 1.3$ mA
Input voltage range	$= 8.5$ V to 50 V
Output voltage range	$= 2.0$ V to 40 V
Minimum $(V_{in} - V_{out})$ differential	$= 3.0$ V
Short-circuit current limit ($R_{sc} = 10\Omega$)	$= 60$ mA

*Measured for step change of V_{in} from 8.5 V to 50 V.
**Measured for step change of I_L from 1 mA to 50 mA.

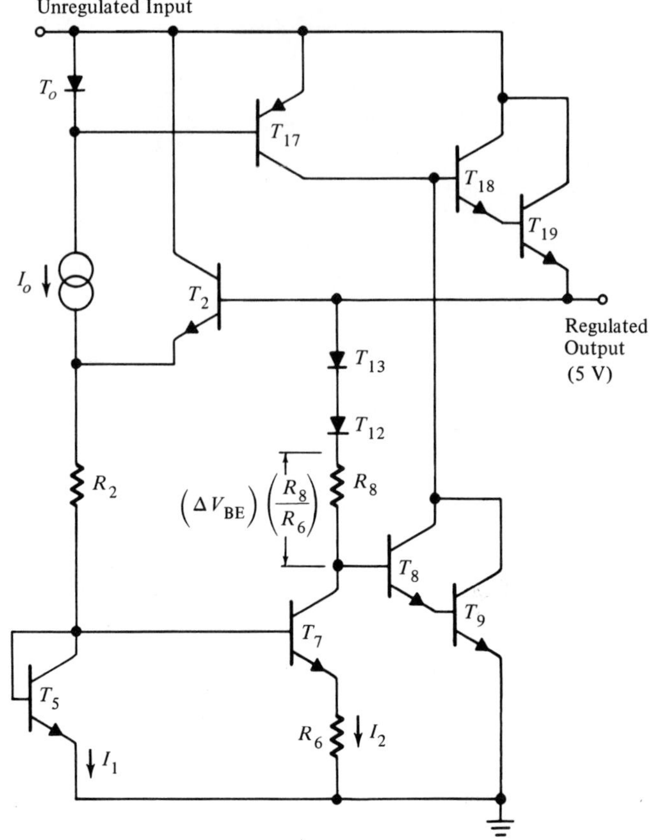

Figure 6.6 Simplified circuit diagram for high-current regulator.[3]

T_7, and produces a voltage across R_8 equal to

$$V_8 = I_2R_8 = (\Delta V_{BE})(R_8/R_6) \tag{6.10}$$

The Darlington-connected gain stage, T_8 and T_9, clamp the collector voltage of T_7 to 2 V_{BE} above ground. Thus counting the V_{BE} drops across T_{12} and T_{13}, the regulated output level can be calculated as

$$V_{\text{out}} = 4\ V_{BE} + (\Delta V_{BE})\ \frac{R_8}{R_6} \tag{6.11}$$

where

$$\Delta V_{BE} = (V_{BE})_5 - (V_{BE})_7 = V_T \ln\ (I_1/I_2) \tag{6.12}$$

Thus, the temperature coefficient of the output voltage can be reduced to zero by choosing the resistor and the current ratios such that

$$\frac{\partial V_{BE}}{\partial T} = \frac{R_8 V_T}{4\, R_6 T} \ln (I_1/I_2) \tag{6.13}$$

In the simplified circuit schematic of Fig. 6.6, a single-ended gain stage is formed by the Darlington-connected (T_8, T_9) pair, and the lateral *pnp* T_{17}. The error amplifier feedback loop is closed through emitter-follower buffer stage T_2. Current source I_o sets the bias level for T_{17}, by means of diode-connected transistor T_o.

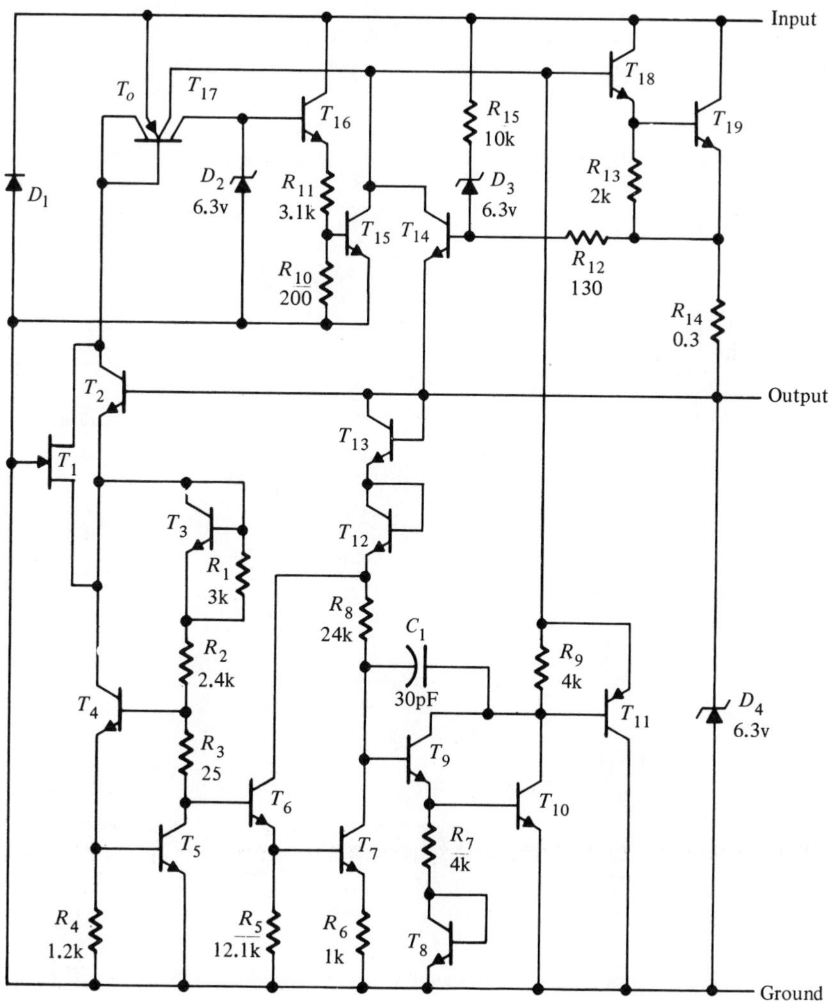

Figure 6.7 Complete schematic of the 5-V fixed-voltage regulator[3] (National Semiconductor LM109).

Figure 6.7 shows the complete circuit diagram for the regulator, including the overload protection circuitry. In the actual circuit, current source I_o is formed by a *n*-channel JFET, T_1. T_o and T_{17} are formed as a single multiple-collector device (see Fig. 4.19). A 30 pF monolithic capacitor is used to provide frequency compensation for the error amplifier. The substrate *pnp*, T_{11}, provides additional gain for the error amplifier. Avalanche diodes D_1 and D_4 provide overvoltage protection at the input and output terminals against voltage transients.

Figure 6.8 shows the photomicrograph of the high-current regulator chip. The circuit is designed to handle load currents as high as 1.5 A. Since all of this current has to be handled by the pass transistor (T_{19} of Fig. 6.7),

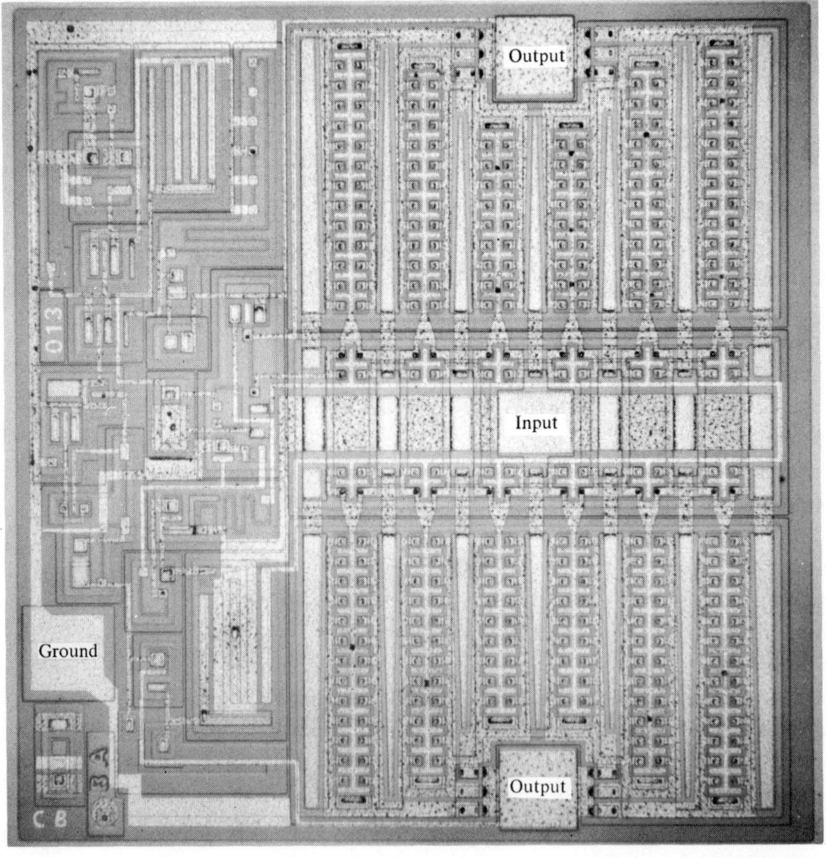

Figure 6.8 Topological layout of 5-V Fixed Regulator (National Semiconductor LM109).

TABLE 6.2 TYPICAL ELECTRICAL CHARACTERISTICS FOR 5-VOLT REGULATOR CIRCUIT OF FIGURE 6.7

Line regulation*	= 0.1% of V_{out}
Load regulation**	= 1% of V_{out}
Ripple rejection	= 80 dB
Temperature coefficient of V_{out}	= 0.01%/°C
Stand-by current drain	= 5.2 mA
Input voltage range	= 7 V to 35 V
Output voltage	= 5.05 V (fixed)
Maximum output current	= 1.5 A

*Measured for step change of V_{in} from 7 V to 25 V.
**Measured for a step change of I_L from 5 mA to 1.5 A.

this device actually occupies more of the chip area than the rest of the control circuitry combined. The design and layout rules for a transistor of these dimensions and power handling capability are discussed further in Section 6.5.

The high current regulator circuit of Fig. 6.7 and 6.8 is designed to operate over an input voltage range of 7 to 35 V, with an output current range of 5 mA to 1.5 A. The typical circuits listed in Table 6.2 give electrical characteristics for the 5-V regulator. Since the regulator is designed to handle high power levels, it also contains a built-in thermal overload protection mechanism, formed by D_2, T_{15}, and T_{16} of Fig. 6.7, which shuts off the regulator when the internal junction temperatures exceed the maximum safe value for device operation (typically 175 $\pm$ 10°C). The principle of operation of this thermal shutdown circuitry will be discussed in Section 6.4.

Since the fixed regulator circuit does not require any external components to set its operating levels or for frequency compensation, it needs only three external terminals. Therefore, it can be readily assembled in a relatively inexpensive three-lead power transistor package which has very low thermal resistance—such as the TO-3 type header, which has a thermal resistance of approximately 1.5°C/W.

The two monolithic regulator design examples reviewed in this section were chosen to illustrate the basic design techniques and the state-of-the-art performance characteristics associated with integrated voltage regulators. However, the examples also serve to illustrate a basic trade-off or design choice forced on a circuit designer: either to design for maximum versatility for a wide range of applications (general purpose design) or to optimize the design for a specific application (special purpose design). The circuit of Fig. 6.4 and 6.5 is designed for maximum flexibility, and therefore requires a number of external connections to be suitable for any one application. The 5-V fixed voltage regulator of Figs. 6.7 and 6.8 on the other hand is designed for digital system applications, with all the control and protection circuitry self-contained on the chip, and only the three terminals, input, output, and ground, brought out from the monolithic chip.

6.3 POWER AMPLIFIERS

Power amplifiers are designed to deliver a large amount of undistorted (sinusoidal) power into a low impedance load. Many of the basic design requirements discussed in connection with high power regulator circuits also apply to the design of monolithic power amplifiers. This is particularly true for the fault protection circuitry necessary to prevent the accidental burn-out of the circuit due to thermal or electrical overloads. A power amplifier is required to deliver the rated amount of power with a low total harmonic distortion (THD) at the output, and with a minimum amount of standby power. An additional desirable requirement is good ripple rejection, which enables the circuit to operate from a relatively poorly regulated dc supply.

Linearity (low distortion) requirements can be met by using local or overall feedback around the gain stages. The low standby power requirement can be satisfied by using class-B or class-AB type output power stages. The major circuit design problems encountered in the fabrication of monolithic power amplifiers arise from process limitations. A high performance class-B output stage requires complementary *npn* and *pnp* devices of comparable performance characteristics and current handling capability. However, both the lateral and substrate *pnp* devices available with the *npn* processing technology have poor frequency characteristics, and their current gains fall rapidly at high current levels. This latter effect is mainly due to the lightly doped base and emitter regions of these devices.

Figure 6.9 shows three basic class-B output stages suitable for power amplifiers. The circuit of 6.9(*a*) is the conventional class-B stage used in discrete design where high current *pnp* and *npn* devices are available. Circuits of Fig. 6.9(*b*) and (*c*) do not require high current complementary devices and can be fabricated in monolithic form using planar *npn* bipolar technology. The circuit of Fig. 6.9(*b*) is the "totem-pole" configuration discussed in Chapter 5 (see Section 5.4). The problem with this stage is that the voltage gain is asymetric with a poor negative swing.

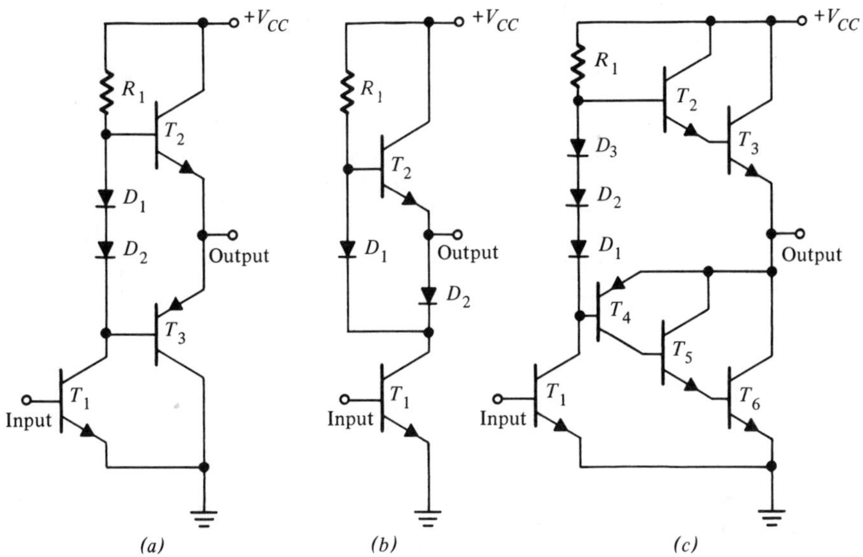

Figure 6.9 Power output stages: (*a*) discrete design with high current complementary devices; (*b*) all *np* class-B stage; (*c*) Class-B stage using low current *pnp*'s. Circuits (*b*) and (*c*) are suitable for integration.

The circuit of Fig. 6.9(*c*) is a modified version of the complementary class-B output stage. The *npn* power transistor of the discrete design is formed by Darlington-connected transistors T_2 and T_3. Similarly, the high-current *pnp* is simulated by using a lateral type, T_4, in connection with a Darlington-connected *npn* pair, T_5 and T_6. This connection is similar to that described in Chapter 4 (see Section 4.5), where the resultant device has the polarity of the *pnp* with the current handling capability of the Darlington-connected *npn* transistors. Since this circuit retains the basic advantages of the conventional complementary stage of Fig. 6.9(*a*), it is the one most often used in monolithic power amplifiers. The current handling and high frequency capability of this stage can be further increased by using a "drift-aided" lateral *pnp* (see Fig. 2.15) in place of T_4.

Figure 6.10 shows the circuit diagram for a high performance monolithic power amplifier[4] which includes short-circuit and thermal-overload protection circuitry. In this particular design, a two-stage amplifier is used, the first stage serving as a preamplifier for the output power stage. The basic sections of the overall amplifier are also identified in the diagram. The circuit is designed as a low distortion audio power amplifier, operating from a +40 V power supply and delivering 15 W sinusoidal output power into an 8 Ω load, with a THD less than 0.5 percent. Its quiescent current drain is 11 mA.

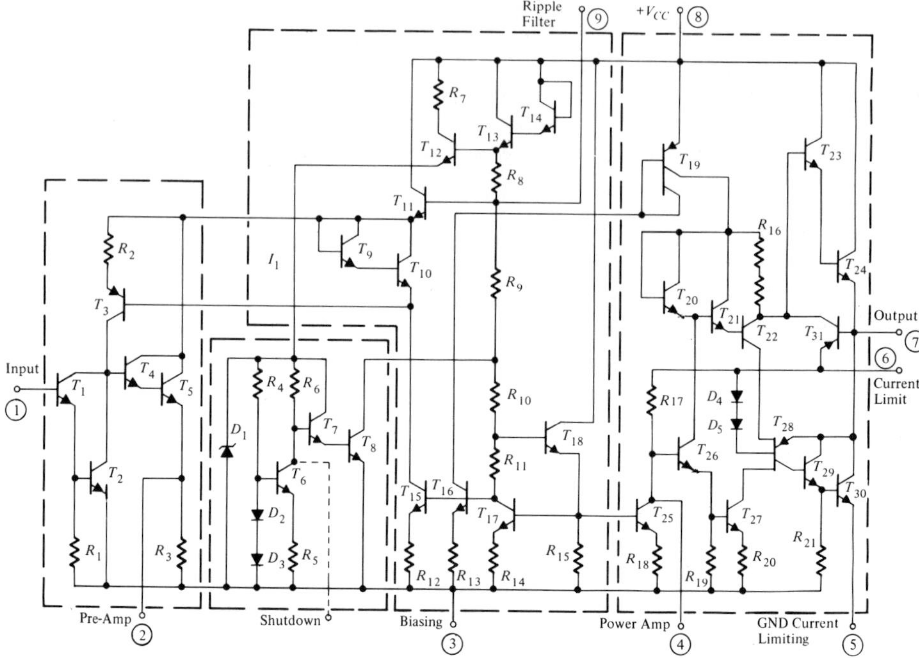

Figure 6.10 Circuit diagram of a 15W monolithic power amplifier.[4]

In the circuit of Fig. 6.10, the biasing network is designed to set the output dc level at $(V_{CC}/2)$ for maximum linearity and dynamic range, independent of the supply voltage and temperature, and depending solely on the resistor ratios and matching of the transistor characteristics. A detailed analysis of this bias circuit is covered elsewhere[4] and will not be repeated here. The output power stage uses the basic configuration given in Fig. 6.9(*c*), with the lateral *pnp* (T_4 of Fig. 6.9(*c*)) replaced by a drift-aided *pnp*, T_{28}, forming biasing diodes D_1, D_2 and D_3 of Fig. 6.9(*c*). Darlington-connected output transistor pairs (T_{23}, T_{24}) and (T_{29}, T_{30}) serve as the power devices. Output ground current protection is provided by connecting a small current-limiting resistor in series with the emitter of T_{30} (terminal 5 in Fig. 6.10). The output current-limiting can be provided by connecting a sensing resistor (R_{sc} of Fig. 6.11) between terminals 6 and 7. When the voltage drop across this sensing resistor exceeds the transistor V_{BE} drop, T_{31} turns on and limits the output current by cutting off the base drive for the Darlington pair (T_{23}, T_{24}). For large negative current swings, if the voltage drop across the sensing resistor exceeds one V_{BE} drop in the opposite direction, diodes D_4 and D_5, which are normally off, become con-

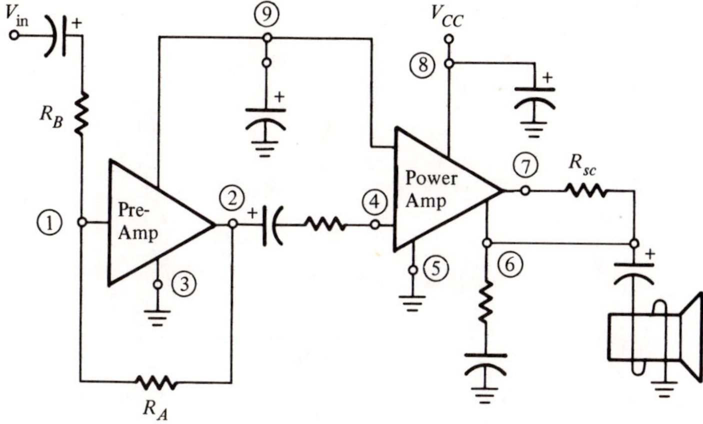

Figure 6.11 Typical connection diagram for 15-W amplifier.[4]

ducting and shunt the base drive to the drift-aided *pnp*, T_{28}. The ripple rejection of the amplifier can be improved by putting a bypass capacitor to terminal 9, which makes the bias voltage for the preamplifier stage immune to transient line voltage fluctuations.

Figure 6.11 shows a typical circuit connection for the 15 W amplifier. Note that resistors R_A and R_B set the preamplifier voltage gain as

$$A_v = -R_A/R_B \tag{6.14}$$

and the resistor R_{sc} connected between terminals 6 and 7 serves as the sensing resistor for output current-limiting.

The 15 W power amplifier of Fig. 6.10 is integrated on a 74 × 75 mil chip shown in Fig. 6.12. Note that the two output *npn* transistors are approximately 30 by 35 mils each and occupy about 40 percent of the total chip area. Particular design and layout considerations for these devices, which are designed to carry 3 A of current, will be discussed in Section 6.5.

6.4 FAULT PROTECTION

In order to prevent power circuits from burning out under accidental overload conditions, a number of fault protection precautions are necessary. In the case of monolithic power circuits, three separate kinds of fault protec-

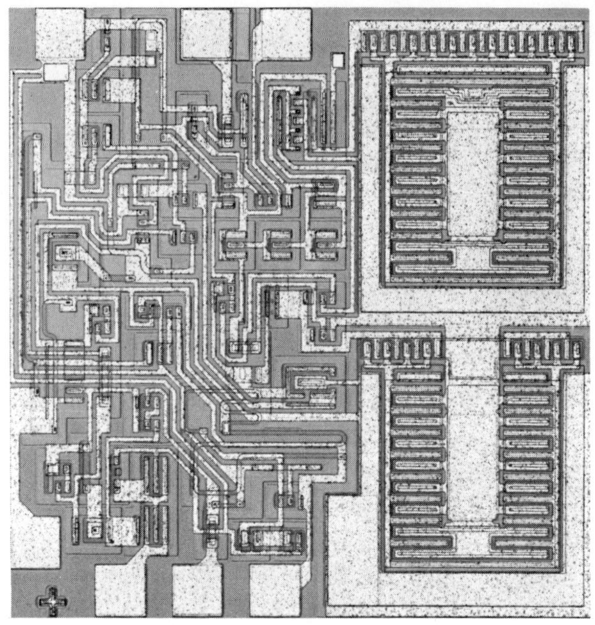

Figure 6.12 Circuit layout for the 15-W monolithic power amplifier. (*Photo: Motorola.*)

tion must be considered: (1) short-circuit protection; (2) overvoltage protection; (3) thermal-overload protection. By proper circuit design, all of these fault protection considerations can be incorporated into the monolithic chip.

Short-circuit protection is normally incorporated into the design by providing either active or passive current-limiting at the output. Passive limiting can be done by incorporating a small series resistance into the output circuit to limit the short-circuit output current to a value below the current handling capability of the power devices. Active current-limiting can be achieved by using a small voltage drop across a sensing resistor to turn on a normally off limiting transistor which, in turn, cuts off the base drive to the power device. An example of this is the external resistor R_{sc}, connected across terminals 6 and 7 of Fig. 6.11, which causes transistor T_{31} of Fig. 6.10 to conduct when the voltage across it reaches to a one V_{BE} drop.

In high power regulator-type circuits, where the exact nature of the load is not defined, an accidental short-circuit or current-overload condition may exist for a prolonged duration before it is detected. This may cause excessive heating and unnecessary power dissipation in the system. To overcome this problem, most monolithic regulators also incorporate a so-called "current-foldback" mechanism which operates as a latching switch, and reduces the output current to a fraction of its rated value if the current-overload condition persists. In most designs, the active devices necessary for current-limit and current-foldback protection are incorporated into the monolithic design, with sensing resistors left external to the circuit to allow the user to determine the limiting and foldback threshold according to his application.

Figure 6.13 shows the current-limit and foldback protection circuitry associated with the medium power regulator circuit of Fig. 6.4. Normally off transistors T_{12} and T_{15} form a latching switch. When the load current increases to a sufficiently high level, the voltage drop across the external current-sense resistor R_{sc} increases enough to cause T_{12} to turn on, which in turn causes T_{15} to conduct and reduce the base drive to pass transistors T_{13} and T_{14}. Once T_{12} and T_{15} begin conducting, they will latch up in that state and will remain conducting until I_L is reduced to a safe value. The external resistor R_{fb} is connected between the output and the base of the T_{12}, T_{15} combination and causes the output short-circuit current to exhibit a "foldback" characteristic where the steady state short-circuit current sustained by the regulator is less than the initial surge current necessary to actuate the protection circuitry.

Overvoltage protection is also utilized when the power circuit is expected to drive inductive loads such as servomotors. This can be incorporated into the design by using a normally off avalanche diode to clamp the output voltage to a maximum allowable level. This can be either a

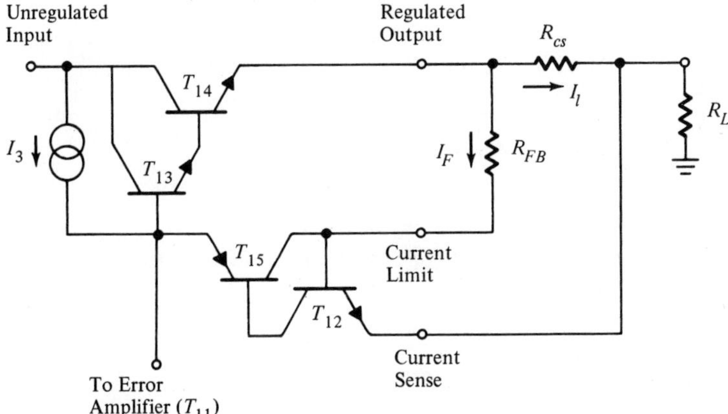

Figure 6.13 Short-circuit protection for the regulator circuit of Figure 6.4.

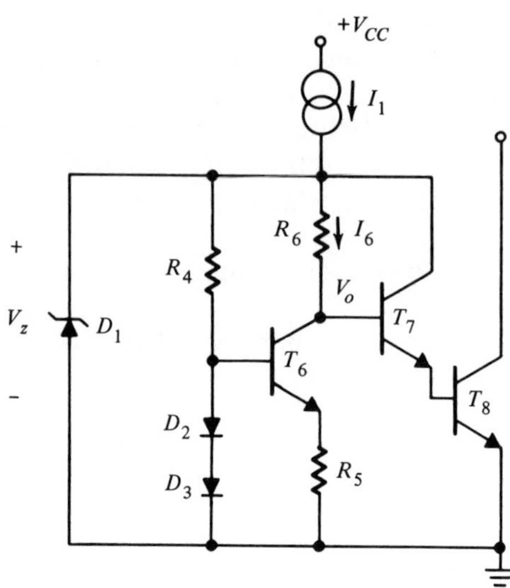

Figure 6.14 Thermal shutdown circuit for the power amplifier of Figure 6.10.

base-emitter or base-collector breakdown diode, or an active avalanche diode using a circuit configuration similar to that shown in Fig. 4.9(*b*). For example, the base-emitter breakdown diode D_4 of Fig. 6.7, connected between the output and the ground terminal is included in the circuit to prevent output voltage surges from exceeding 6.3 V.

Thermal overload protection is necessary to prevent the unit from sustaining permanent damage due to prolonged operation at elevated temperatures, in turn due to either self- or ambient heating. This can be incorporated into the design as a "thermal shutdown" mechanism which automatically shuts the power circuit off when junction temperatures on the chip exceed a safe operating value (typically 175°C). Figure 6.14 shows the diagram of the thermal shutdown control incorporated into the power amplifier circuit of Fig. 6.10. Transistor T_8 is normally off. When the temperature dependent voltage level V_o, at the collector of T_6, gets to be sufficiently high (i.e., $V_o = 2\ V_{BE}$), transistors T_7 and T_8 are turned on. When T_8 is conducting, it cuts off the power amplifier by shunting off the available bias current.

Temperature dependent bias voltage V_o is derived as follows: with reference to Fig. 6.14, assuming all forward diode voltages are identical, and equal to V_{BE}, collector current I_6 is approximately equal to (V_{BE}/R_5). Then V_o can be written as

$$V_o = V_z - V_{BE}(R_6/R_5) \tag{6.15}$$

where the voltage level V_z, at which transistor T_8 would begin to conduct can be calculated from (6.15) by setting $V_o = 2\ V_{BE}$. Then,

$$V_z(T_s) = V_{BE}(T_s)(2 + R_6/R_5) \tag{6.16}$$

The base-emitter breakdown voltage V_z and the diode turn-on voltage V_{BE} vary linearly with temperature, with respective temperature coefficients of α_z and α_{BE}. Then, the values of V_z and V_{BE} at shutdown temperature can be related to their room temperature (25°C) values V_{zo} and V_{BEO} as

$$V_z(T_s) = V_{zo}[1 + \alpha_z(T_s - T_o)] \tag{6.17}$$

and

$$V_{BE}(T_s) = V_{BEO}[1 + \alpha_{BE}(T_s - T_o)] \tag{6.18}$$

For a given fabrication process V_{zo}, V_{BEO}, α_z and α_{BE} are well characterized and highly repeatable. Therefore, a desired shutdown temperature T_s can be realized from Eq. (6.16) for a proper choice of the resistor ratio (R_6/R_5). From Eqs. (6.16) through (6.18), the appropriate choice of this resistor ratio for a given choice of T_s can be expressed as

$$\frac{R_6}{R_5} + 2 = \frac{V_{zo}}{V_{BEO}}\left[\frac{1 + \alpha_z(T_s - T_o)}{1 + \alpha_{BE}(T_s - T_o)}\right] \tag{6.19}$$

In most applications, T_s is chosen to be +175°C. In a well controlled fabrication process, this design value of T_s can be maintained to within ±10°C over successive production runs.

6.5 FAILURE MECHANISMS IN POWER DEVICES

There are two separate failure modes which are unique to monolithic devices and circuits designed to operate at high power levels. These are the so-called "secondary breakdown" and the "electromigration" effects. Secondary breakdown is a thermal instability effect encountered in bipolar power transistors and causes the transistor to switch to a low voltage, high current mode which can result in permanent deterioration of device characteristics or total burn-out. Electromigration is a mechanism which results in the physical movement of metal atoms in the interconnection layers, under high current density operation, and can eventually cause an open circuit in the interconnecting layer.

Secondary Breakdown

This phenomenon comes about because of nonuniform current distribution within the transistor junctions at high power levels, which cause nonuniform heating and formation of localized "hot spots" in junctions. The injection efficiency of the emitter increases rapidly with increasing temperature. Therefore, formation of localized hot spots leads to further nonuniformity and thus additional localized heating. Hence, the localized power density and temperatures can be very large despite any external stabilizing circuit.[5] If the junction temperature, under a given current and voltage level, reaches a "triggering temperature," the power device goes into secondary breakdown and becomes a virtual short circuit.

Secondary breakdown is a complex function of both the collector current I_c and the collector-emitter voltage V_{CE}. The collector current necessary to trigger secondary breakdown is much lower at high voltages. For example, a power transistor that can handle 3 A at 30 V can only handle 1 A at 50 V. Thus, for a given power transistor design, one has to define a "safe operating area" limit[6] on the device I-V characteristics and ensure that the current and voltage levels within the device are always maintained within this limit. In class-B power amplifiers, a useful rule-of-thumb is to choose a power transistor design that can handle the maximum short-circuit current simultaneously with one half of the supply voltage, without going into secondary breakdown.[4] Current distribution in the emitter region of the power transistor can also be improved by careful layout considerations and by using "ballast" resistors. This point will be examined further in the next section of this chapter.

Electromigration

This is a mass transport effect observed in metals under high current densities. It causes the metal atoms to migrate away from a high current density point in the metal, and can cause a conductor, such as an aluminimum interconnection on the chip, to become an open circuit at the point of highest current density. Electromigration is a slow process which speeds up as the current density or the temperature are increased, or as the lateral dimensions of the conductor normal to the current path are diminished. It starts as formation of voids on a conducting metal strip and eventually leads to a complete open circuit.

An approximate expression for the median-time-to-failure (MTF) of a conducting metal film can be given as[7]

$$\mathrm{MTF} = \frac{KWd}{J^2}\, e^{\phi/kT} \tag{6.20}$$

where K is a constant of proportionality
J = current density
W = conductor width

d = conductor thickness
ϕ = activation energy, depending on type and grain size of metal
k = Boltzmann constant
T = temperature in °K

In the case of aluminum, the activation energy for electromigration effects has been observed to be in the range of 0.48 eV (for small grain structures) to 1.2 eV (for large, well ordered grain structures). The electromigration effect is a slow "wear-out" type failure effect, similar in many ways to the "creep" failure in metals due to sustained mechanical stress. Because of the exponential nature of Eq. (6.20), it becomes significant especially at high temperatures. It should also be noted that, since J is inversely proportional to the cross-sectional area, at any given current level, electromigration effects increase as the third power of the cross-sectional area.

In monolithic power circuits, in order to obtain an acceptable degree of reliability (i.e., MTF $\approx 10^5$ h or approximately 10 yr), current densities in the interconnecting metal layers should be maintained at less than (5)(10^5) A/cm^2. For example, a 6000 Å thick aluminum interconnection layer should not be forced to carry more than 80 mA of current per mil width of the conductor thickness, if the medium time to failure due to electromigration is expected to be greater than 10 yr.

6.6 LAYOUT CONSIDERATIONS FOR MONOLITHIC POWER CIRCUITS

In a monolithic power circuit, the output power transistors are the key to the circuit performance. Therefore, the design and layout of the power transistors require additional care and attention. In most cases, more than half of the chip area is devoted to the power transistors (see chip photographs of Fig. 6.8 and 6.12), with the low power control or signal processing circuitry taking up a minor portion of the silicon area.

At high current levels, almost all of the current injection occurs along the edge of the emitter. To increase the emitter periphery, and to minimize the de-biasing effects due to base-spreading resistance, one normally uses an interdigitated structure for the emitter and the base regions. In the layout of the emitter stripes, voltage drops along the metal interconnection bars also need o be considered. Due to the exponential current-voltage dependence across the base-emitter junction, the injected current density in the emitter will decrease by a factor of 2 for every 18 mV drop along the metal conductor interconnecting the emitter stripes. With a typical aluminum interconnect sheet resistivity of about 30 mΩ per square, care has to be taken to keep the entire emitter surface injecting uniformly by using large overmetal strips over the emitters. To keep the current density reasonably uniform, the forward voltage across the base-emitter junction should not vary by more than 10 mV over the entire transistor.

To prevent the power transistor from being triggered into secondary breakdown, it is necessary to use small "current equilizing" or "ballast" resistors in series with each one of the individual emitter areas such that no localized heating or hot spots are formed due to unequal current concentration along the active emitter-base junction. Because of the distributed nature of the base-emitter junction, such a ballast resistor can be added into the circuit as a distributed resistor, made up of the bulk resistance of the n^+ diffused emitter region. This can be done by leaving a substantial distance (typically 15 to 30 μ) between the metal emitter contact and the active emitter periphery. This is shown schematically in Fig. 6.15. If any one point along the emitter periphery tends to conduct heavily and cause localized heating, the voltage drop across the ballast resistor will tend to de-bias that particular emitter region and equalize the current distribution.

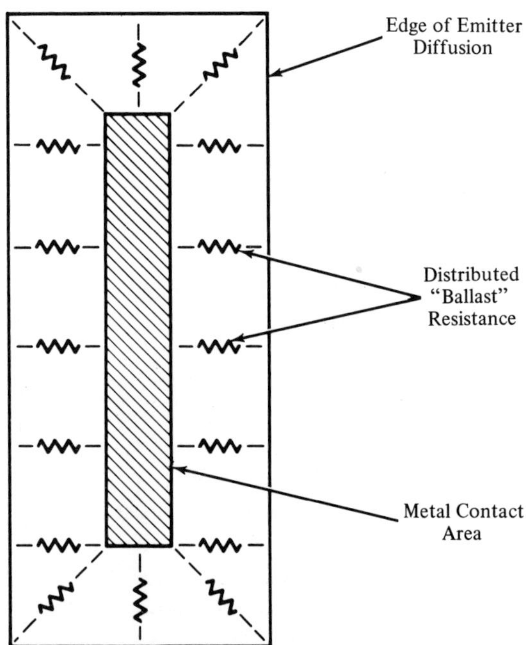

Figure 6.15 Distributed emitter ballasting using emitter bulk resistance.

In a monolithic power circuit, the presence of a power device in close proximity to the rest of the low level signal processing and control circuitry can require additional layout precautions. The power device serves as a concentrated heat source and can cause severe thermal gradients within the chip. This problem is particularly severe in the case of monolithic regulator circuits which must also contain a temperature independent voltage reference on the chip. Therefore, during the layout of the circuit, additional care must be taken to locate the control circuitry, and particularly the internal voltage reference, so as to minimize the effects of the temperature gradients within the chip.[8]

Figure 6.16 shows a practical block diagram for the physical layout of a high power monolithic regulator circuit. Normally, the distributed power source, the pass transistor, is located along one edge of the chip. The localized heating of the pass transistor generates a set of isothermal contours over the remainder of the chip, and these are more or less parallel to the longitudinal edge of the power device, as shown in the Figure.[9] The thermal drifts of monolithic components can be minimized by laying out the control circuitry so that the components to be matched are along the same isothermal contour—or equally distant—from the power dissipation source on the chip. Since the reference network is most affected by thermal gradients within the chip, it should be located as far away as possible from the pass transistor, as shown in the Figure. Similarly, the thermal shutdown

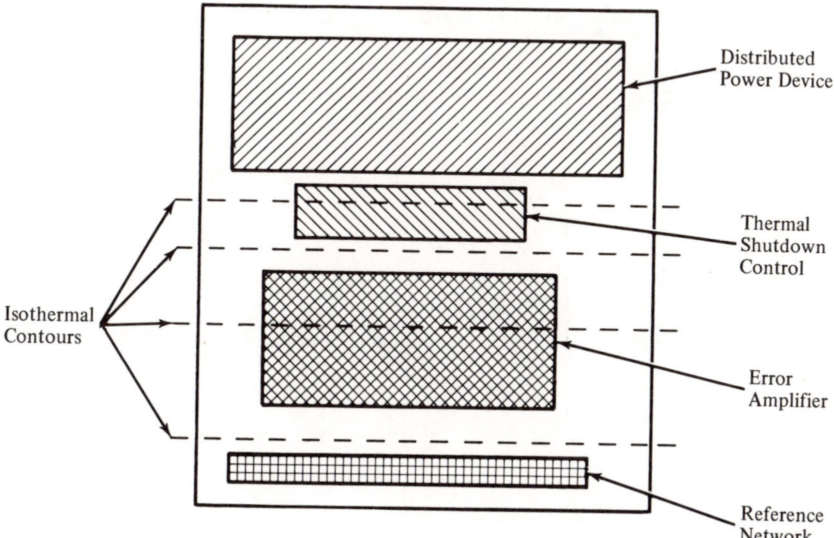

Figure 6.16 Layout considerations for a high power regulator circuit.

circuitry is required to sense the junction temperature of the power device; therefore it must be located near the output transistor.

6.7 THERMAL DESIGN AND PACKAGING

Heat dissipation is the most fundamental limitation on the power handling capability of a monolithic chip. Therefore, in a successful power circuit design, thermal considerations are every bit as important as electrical design. For safe and reliable operation of silicon devices, the maximum junction temperature has to be restricted to approximately 175°C. The conduction of heat from the chip to the package and to the surrounding medium presents a complex problem, depending on the chip layout and size, package type and dimensions, as well as the ambient temperature. However, it is possible to use a simple linearized model for heat conduction to estimate some of the basic parameters of thermal design. The heat flow Q through a thermal barrier having a net temperature difference ΔT across it can be related to the so-called "thermal resistance" R_1 for the barrier as

$$Q = (\Delta T)/R \tag{6.21}$$

There are two successive stages of heat conduction between the monolithic chip and the surroundings: heat flow from semiconductor junction to case and from case to ambient. In terms of its electrical analog, this effect can be described by a thermal resistance network[10] shown in Fig. 6.17, where T_j, T_c, and T_a are the junction, case, and ambient temperatures, and

R_{jc} = junction-to-case thermal resistance
R_{ca} = case-to-ambient thermal resistance

The thermal power dissipation properties of the circuit and the package can be approximated from the linearized model of Fig. 6.17 as

$$Q = (T_j - T_c)/R_{jc} + (T_c - T_a)/R_{ca} \tag{6.22}$$

The thermal resistance R_{jc} is determined by the chip layout and size and the means of attaching the chip to the case. The case to ambient resistance R_{ca} is partly under the control of the user, depending on the heat-sinking

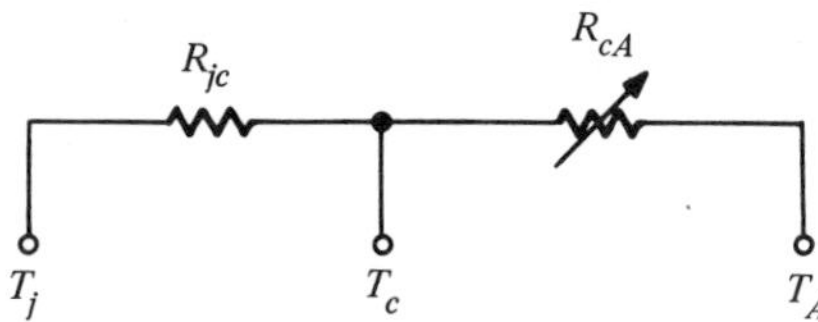

Figure 6.17 Electrical analog of thermal heat flow problem.

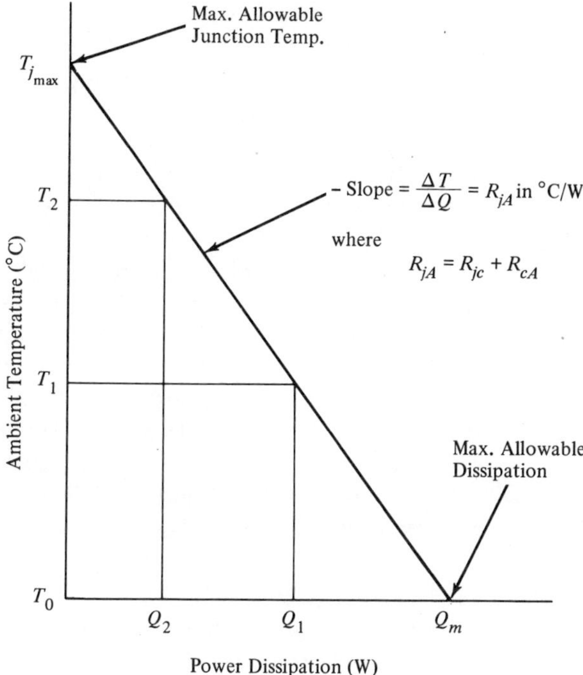

Figure 6.18 Derating curve for power devices

arrangements provided by the user. For this reason, R_{ca} is shown as a variable element in the electrical analog of Fig. 6.18. For a given design, exact values of the thermal resistances are normally estimated from empirical studies or from past experience.

In most cases, the power circuit is available as a self-contained monolithic package. In such a case, it is convenient to characterize the packaged unit directly in terms of junction-to-ambient heat dissipation characteristics. This can be done by defining a junction-to-ambient thermal resistance R_{ja}, where

$$R_{ja} = R_{jc} + R_{ca} \tag{6.23}$$

and calculate the power dissipation capability from

$$Q = (T_j - T_a)/R_{ja} \tag{6.24}$$

The maximum allowable junction temperature, T_{jm}, sets the maximum allowable dissipation Q for a given ambient temperature and thermal resistance. As indicated by Eq. (6.24), the power handling capability of the circuit decreases linearly with increasing ambient temperature, since T_{jm} is fixed (typically $T_{jm} \approx 175°C$). Thus, the power rating of the circuit has to be reduced as the ambient temperature is raised. This can be accomplished using a "derating curve,"[10] as shown in Fig. 6.18. In the Figure T_o normally corresponds to the room temperature (+25°C) where the device is rated to dissipate a maximum steady state power of Q_m watts. If the device is operated at ambient temperatures T_1 or T_2, then its power-handling capability has to be derated to Q_1 and Q_2 watts, respectively.

Packaging is the single most serious problem in the design of low cost power integrated circuits. Many of the conventional I-C packages, such as TO-5 and the plastic dual-in-line types, have too high a thermal resistance (typically $R_{jc} \approx 15°C/W$) to be suitable at power levels above 1 W.

Power packages can be divided into two groups:[8] plastic and metal (hermetic). Plastic packages for power circuits do not necessarily enjoy the cost advantages of small size plastic transistor or I-C packages for low-power circuits. Power dissipation of greater than several watts requires the use of a "bolt-on" type package which has provisions for direct attachment to a heat sink. The poor dimensional stability of the plastic package requires a reinforced mounting hole or an extended metal tab which bolts to a heat sink. In some cases, plastic and metal combination packages can be used, where the plastic package contains a metal plate along the base of the package. The chip is mounted directly on the inside face of this metal plate for low resistivity thermal conduction. With this type of structure, the thermal resistance of the package can be reduced to as low as 2°C/W. However, this approach requires a large area metal-to-plastic seal which has poor reliability characteristics.

The metal packages are more expensive but offer a higher degree of reliability and hermeticity. The metal packages, such as the three lead TO-3 package normally used for power transistors, can also be used for power I-Cs. An example of this is the 5 V fixed regulator circuit described in this chapter (see Section 6.2), which requires only three external terminals. Therefore, it can be readily packaged in a relatively low cost TO-3 metal package with a thermal resistance, R_{jc} of less than 3°C/W. For most I-C applications below the 5W range, the 9-pin steel TO-66 header ($R_{jc} \approx 3°C/W$) is a convenient package to use. For higher power applications, thermal resistance of steel packages is too high. Instead, more expensive aluminum or copper metal packages become necessary.

REFERENCES

1. M. M. Scott, "μA723 Precision Voltage Regulator Application Note," 1968.

2. T. M. Frederiksen, "A Monolithic High Power Series Voltage Regulator," *IEEE J. Solid State Ckts.* **SC–3** (1968): 380–387.
3. R. J. Widlar, "New Developments in IC Voltage Regulators," *IEEE J. Solid State Ckts.,* **SC–6** (1971): 2–7.
4. E. L. Long and T. M. Frederiksen, "High-Gain 15-Watt Monolithic Power Amplifier with Internal Fault Protection," *IEEE J. Solid State Ckts.,* **SC–6** (1971): 35–44.
5. H. A. Schafft and J. C. French, "A Survey of Second Breakdown," *IEEE Trans. Electron Dev.,* **ED–13** (Aug./Sept., 1966): 613–618.
6. F. Bergmann and D. Gerstner, "Some New Aspects of Thermal Instability of the Current Distribution in Power Transistors," *IEEE Trans. Electron Dev.,* **SC–13** (Aug./Sept., 1966).
7. J. R. Black, "Electromigration—A Brief Survey of Some Recent Results," *IEEE Trans. Electron Dev.,* **ED–16** (April, 1969): 338–347.
8. T. M. Frederiksen, "The Limits of Power ICs," *Electronic Products* (Aug., 1970): 22–28.
9. W. F. Davis, "Thermal Considerations in the Design of a 500 mA Monolithic Negative Voltage Regulator," *NEREM Record* (1969): 84–85.
10. M. Fogiel, *Microelectronics,* Research and Education Association, New York, 1968, pp. 394–402.

7 Analog Multipliers and Modulators

7.1 PROPERTIES OF AN ANALOG MULTIPLIER

In a variety of circuit or system applications, it is necessary to obtain an analog output signal which is a linear product of two input signals. The circuit block which can perform this function is the analog multiplier. In such a circuit block, the output voltage Z must be proportional to the product of the two inputs, X and Y, i.e.,

$$Z = K_1XY \tag{7.1}$$

where K_1 is the gain constant of the multiplier. In most applications it is also required that the output Z must conserve the polarity relationships between the two inputs such that each of the inputs can either be positive or negative; and the output would be of the polarity implied by Eq. (7.1). A multiplier which has this property is called a "four-quadrant" multiplier.

Since an analog multiplier deals with two separate input variables, X and Y, for a given output Z, its operating characteristics cannot be as readily defined as a single input system, such as an operational amplifier. Instead, a number of separate gain and offset parameters need to be defined to describe the performance characteristics of a nonideal multiplier. In a practical multiplier circuit, the output Z is related to any one of the inputs X and Y by an approximate expression of the form:

$$Z = K_1XY + K_xX + K_yY + K_o \tag{7.2}$$

where the constant K_1 is referred to as the multiplier gain constant; and the constants K_x, K_y, and K_o are known as the X, Y, and null-offset constants of the system. Thus, in a high accuracy multiplier system, at least four separate adjustments are needed to set the multiplier gain K_1, and to null out the remaining three offset constants. In most multiplier applications, the scale factor or the multiplier gain K_1 is set such that

$$Z = \frac{XY}{10} \tag{7.3}$$

Some of the most common terms used in describing the performance of a four-quadrant multiplier circuit are the "accuracy," "linearity" and the "bandwidth" of the multiplier. Each of these terms can be briefly defined as follows:[1]

Accuracy is defined as the maximum deviation of the actual output level from the ideal one, given by Eq. (7.3), for any choice of X or Y values within the dynamic range of the multiplier. It is normally specified as a percentage of full-scale output. For example, 1 percent full-scale accuracy means that for a four-quadrant multiplier with a ± 10 V output the actual output would be within ± 100 mV of the ideal level.

Linearity is most commonly defined as the maximum percentage deviation from "best straight line" data at the output, corresponding to equal magnitude inputs at the X and Y terminals. Since the maximum deviation from linearity occurs at the extreme ends of the multiplier dynamic range, it is also defined as a percentage of full-scale output. Thus, if a multiplier has 10 V inputs and an output given by Eq. (7.3), 1 percent linearity would mean a maximum deviation of $\pm$ 100 mV from a "best fit" straight line.

The high frequency capability of an analog multiplier is described in terms of its "bandwidth." Depending on the particular application of a multiplier, different definitions of "bandwidth" may be used. The "small signal 3-dB bandwidth" is defined as the frequency at which the output is 3 dB down from its low-frequency value, for a constant input level. The "1-percent absolute error bandwidth" is defined as the frequency where the output, $Z(j\omega)$, is down by 1 percent from its low frequency value. An alternate bandwidth criterion is the "1-percent vector error bandwidth," corresponding to the frequency where the output phase is shifted by 0.57° from its low frequency value, resulting in a 1-percent vector difference between $Z(0)$ and $Z(j\omega)$. The vector-error bandwidth is always much smaller than the corresponding absolute error bandwidth.

Temperature stability of a multiplier is also essential for most circuit applications. The stability is normally measured in terms of the temperature drift of the null-offset term K_o (in mV/°C) and the scale factor K_1 (in ppm/°C).

7.2 APPLICATIONS OF AN ANALOG MULTIPLIER

Similar to an operational amplifier, the analog multiplier forms a versatile building block for performing a number of mathematical operations such as multiplying, dividing, squaring, and square-root extraction. In most of these applications, it is used in conjunction with an operational amplifier to complement its functional capability. Figure 7.1 outlines some of the applications of an analog multiplier in performing basic mathematical operations. The multiplication and the squaring functions shown in the figure are self-explanatory. The division operation is performed by placing the multiplier in feedback around an operational amplifier, such that the multiplier output is summed at the operational input and forced to equal

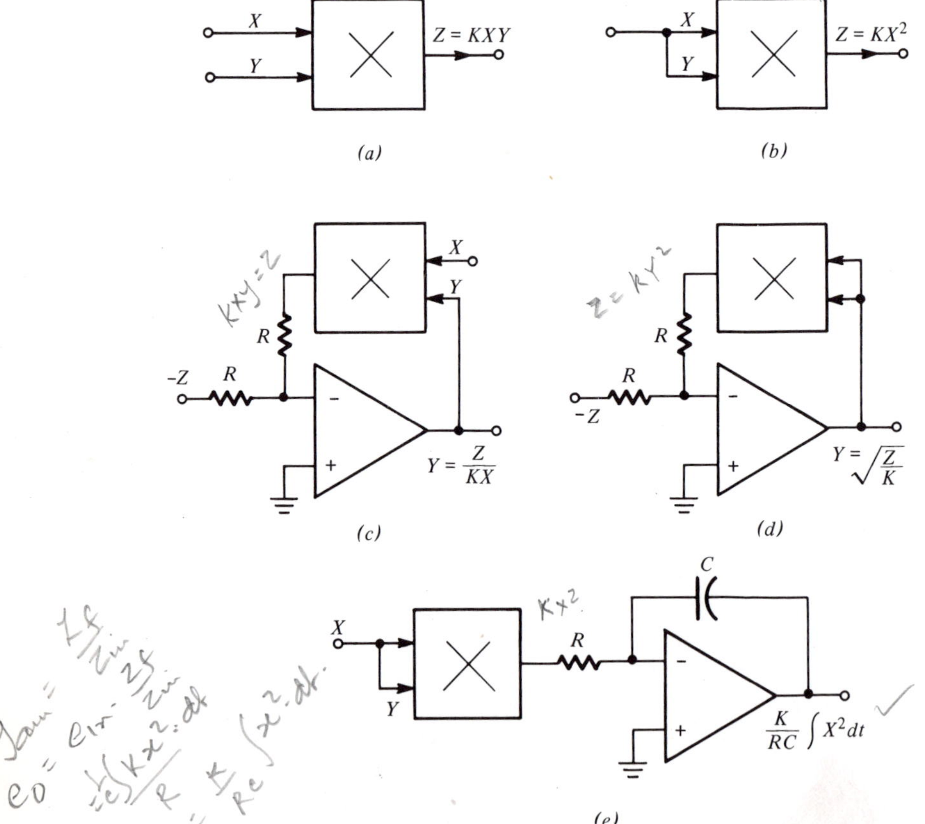

Figure 7.1 Some basic applications of an analog multiplier: (*a*) multiplication, (*b*) squaring, (*c*) division, (*d*) square root, (*e*) mean square.

the input signal Z. Then, the operational amplifier output Y necessary to satisfy this condition is

$$Y = \frac{Z}{K_1 X} \tag{7.4}$$

Similarly, by shorting the X and Y inputs, the same configuration can be used to obtain a square root of the input, i.e.,

$$Y = \sqrt{Z/K_1} \tag{7.5}$$

When using the multiplier in a feedback configuration around an operational amplifier,, such as in Fig. 7.1(*c*) and (*d*), care should be taken to ensure that $Y > 0$, to avoid an unstable positive-feedback condition across the operational amplifier. Also, it should be noted that the accuracy of the output in the circuit of Fig. 7.1(*c*) is inversely proportional to X. Thus, as X decreases, the divider error increases rapidly, which severely limits the dynamic range of the system.

In addition to performing multiplication or division, an analog multiplier can also be used as a modulator, frequency converter, phase detector, or a voltage-controlled attenuator. In many of these applications linearity or the four-quadrant capability are not required; therefore the necessary circuit configuration can be greatly simplified.

7.3 VARIABLE-TRANSCONDUCTANCE MULTIPLIER

There are a number of diverse circuit techniques which can be used to obtain an output proportional to the product of the two input signals.[2] The multiplication technique which is most readily suited to the monolithic circuits is the so-called "variable-transconductance" method. This method makes use of the dependence of the transistor transconductance on the emitter current bias. The basic principle of the transconductance multiplier can be readily demonstrated by the differential gain stage shown in Fig. 7.2. For small values of the input voltages V_1, such that $V_1 \ll V_T$, the differential output voltage V_o can be related to the input as

$$V_o = g_m R_L V_1 \tag{7.6}$$

where V_T is the thermal voltage, and g_m is the transconductance of the stage:

$$g_m = I_E/V_T \tag{7.7}$$

The transconductance can be varied by applying a second voltage V_2 as shown in the Figure. Then, if $I_E R_E \gg V_{BE}$, the bias voltage V_2 is related

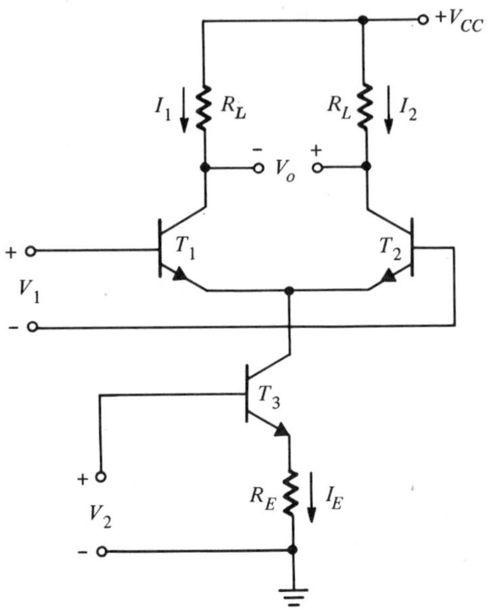

Figure 7.2 Basic differential stage as a transconductance multiplier.

to I_E as

$$V_2 \approx I_E R_E \tag{7.8}$$

From Eq. (7.7) and (7.8), the overall voltage transfer expression can be written as

$$V_o = g_m R_L V_1 = (V_1 V_2) \frac{R_L}{V_T R_E} \tag{7.9}$$

which is of the form shown in Eq. (7.1).

In the simple transconductance multiplier circuit, the total current I_E varies as a function of the modulating voltage V_2. This causes a large common-mode voltage swing in the circuit, which is objectionable if a single-ended output or dc coupling is required. This common-mode shift can be eliminated by using two differential stages in parallel, and cross-coupling their outputs,[3] as shown in Fig. 7.3. The two inputs V_1 and V_2 determine the partitioning of the total current I_T among different branches of the circuit. However, since I_T remains constant and since the devices

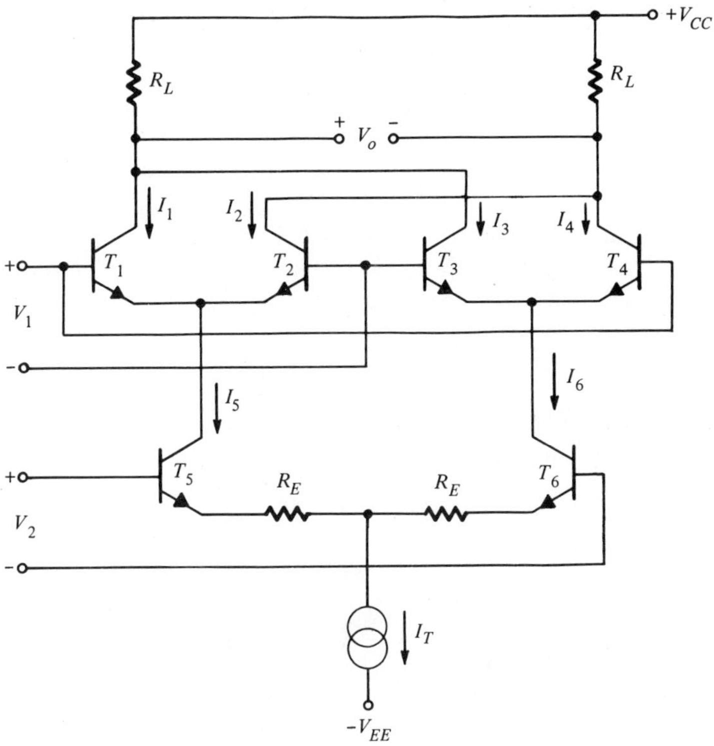

Figure 7.3 Cross-coupled differential stages as a variable-transconductance multiplier.

forming the circuit are symmetrically cross-coupled, the large common-mode shifts at the output are no longer present.

Since the circuit configuration of Fig. 7.3 is significantly more complex than that of Fig. 7.2, the operation of the circuit as a variable transconductance multiplier is not as readily apparent. However, in terms of the current partitioning within various branches of the circuit, the multiplication effect can be explained as follows: Assuming that all the devices in the circuit are well matched and that $\beta_0 \gg 1$ for all devices involved, the currents within the circuit are related as

$$\begin{aligned} I_1 + I_2 &= I_5 \\ I_3 + I_4 &= I_6 \\ I_5 + I_6 &= I_T \end{aligned} \tag{7.10}$$

Assuming that the input voltage V_1 is sufficiently small, such that $|V_1| \ll$

V_T, the current unbalance in the differential pairs (T_1, T_2) and (T_3, T_4) can be expressed as

$$(I_1 - I_2) = (g_m)_{12}V_1 \tag{7.11}$$

and

$$(I_3 - I_4) = -(g_m)_{34}V_1 \tag{7.12}$$

where $(g_m)_{12}$ and $(g_m)_{34}$ are the transconductances of the transistor pairs (T_1, T_2) and (T_3, T_4) respectively. Since there is no emitter degeneration resistance associated with these devices, the corresponding values of transconductance are directly proportional to the bias currents I_5 and I_6 through each stage, i.e.,

$$(g_m)_{12} = V_T/I_5 \quad \text{and} \quad (g_m)_{34} = V_T/I_6 \tag{7.13}$$

The total differential output voltage V_o can be expressed as:

$$V_o = R_L[(I_1 - I_2) + (I_3 - I_4)] \tag{7.14}$$

Then, from Eqs. (7.11) through (7.13), the output voltage of (7.14) can be expressed as:

$$V_o = V_1R_L[(g_m)_{12} - (g_m)_{34}] = \frac{V_1R_L}{V_T}(I_5 - I_6) \tag{7.15}$$

If the emitter series resistor R_E for T_5 and T_6 is chosen sufficiently large, such that

$$I_5R_E \gg V_T \qquad \text{and} \qquad I_6R_E \gg V_T \tag{7.16}$$

then the difference of currents I_5 and I_6 is directly proportional to V_2:

$$(I_5 - I_6) = V_2/R_E \tag{7.17}$$

Then, substituting Eq. (7.17) into (7.15), one obtains

$$V_o = \left(\frac{R_L}{R_EV_T}\right)V_1V_2 \tag{7.18}$$

which is the same as Eq. (7.9).

The circuit configurations of Figs. 7.2 and 7.3 demonstrate the basic principle of variable-transconductance multiplication. Unfortunately, the dynamic range of both of the circuits is severely limited for linear multiplier applications. This is because the results given in Eq. (7.9) or (7.18) are valid only for restricted values of V_1, i.e., $|V_1| \ll V_T$. Thus, for linear operation of the circuit, the values of V_1 are restricted to be no more than several millivolts. If V_1 is comparable to or larger than the thermal voltage V_T, transistor pairs (T_1, T_3) and (T_2, T_4) function as synchronous switches which turn on and off, depending on the polarity of V_1. Then the circuit functions as a "balanced modulator" rather than a linear multiplier. This application of the circuit will be discussed further in a later section.

7.4 MONOLITHIC FOUR-QUADRANT MULTIPLIERS

In trying to increase the dynamic range of the multiplier circuits described in the previous section, one faces a problem: The linear multiplication effect is obtained because the transconductance of transistors T_1 and T_2 in Fig. 7.2 (or T_1, T_2 and T_3, T_4 pairs in Fig. 7.3) is proportional to their emitter current. This is only true if the transistors are operated with no emitter degeneration, and under small signal conditions, i.e., $|V_1| \ll V_T$. With reference to Fig. 7.2, one can see that under large-signal conditions the voltage-to-current transfer characteristics of the differential pair T_1 and T_2 are no longer linear. Instead, collector currents I_1 and I_2 are related to the applied voltage V_1 as

$$I_1/I_2 = e^{V_1/V_T} \tag{7.19}$$

In order to obtain linear multiplication over a wide dynamic range, it is necessary to reduce, somehow, the exponential current-voltage transfer characteristics of Eq. (7.19) to a linear onee. This would be possible if one were to process the actual input, V_1, in a nonlinear manner, with a predetermined and well controlled nonlinearity, before applying it to the bases of T_1 and T_2. In other words, suppose that input signal V_1 were applied to the differential pair. Then, the actual differential voltage (V_1) appearing at the bases of T_1 and T_2 would be logarithmically related to V_1, and the exponential relationship of Eq. (7.19) would be linearized. This is the basic design philosophy which is used in designing monolithic four-quadrant multipliers.[4]

Figure 7.4 shows a simplified circuit scheme explaining how one can obtain a logarithmic input voltage (V_1), in order to linearize the transfer characteristics of a grounded-emitter differential stage. In this case T_1 and T_2 are biased through diode-connected transistors T_A and T_B, in a manner described in Section 4.1. The diode-connected transistors are then driven by controlled current sources I_A and I_B. The net bias voltage V_1 is then derived from the difference between currents I_A and I_B, and can be expressed as

$$V_1 = V_T \ln [I_B/I_A] \tag{7.20}$$

Substituting Eq. (7.20) in (7.19), one can show that currents I_1 and I_2 of the differential stage are directly related to the bias currents as

$$I_1/I_2 = I_B/I_A \tag{7.21}$$

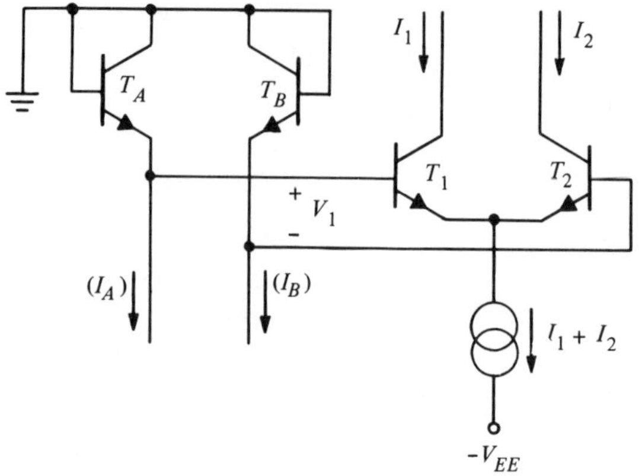

Figure 7.4 Generation of a logarithmic bias voltage input for a differential stage.

irrespective of the nonlinear transfer characteristics of the base-emitter diodes. Similarly, the currents in respective branches of the circuit are related to the total bias current levels as

$$\frac{I_A}{I_A + I_B} = \frac{I_2}{I_1 + I_2} = \frac{1}{1 + e^{V_1/V_T}} \tag{7.22}$$

and

$$\frac{I_B}{I_A + I_B} = \frac{I_1}{I_1 + I_2} = \frac{e^{V_1/V_T}}{1 + e^{V_1/V_T}} \tag{7.23}$$

It should be noted that in deriving Eqs. (7.20) through (7.23), no assumptions were made to restrict the amplitudes of the device currents. Therefore, these sets of equations are valid over a broad range of current levels as long as the device characteristics are well matched and V_{BE} drops continue to obey the simple diode equation (see Eq. (2.1). Since the close matching is an inherent property of monolithic devices, this latter requirement is readily met in integrated circuits.

A practical four-quadrant multiplier circuit which uses the input preconditioning method described above is shown in Fig. 7.5.[5] The left-hand portion of the circuit processes the X input signal V_x, to generate an intermediate voltage V_1 across diodes D_1 and D_2. As will be shown by the basic circuit equations, the nonlinearity introduced into the X input in generating V_1 from V_x is exactly the inverse of the nonlinearity associated with the base-emitter junctions of the quad-connected transistors (T_5, T_6) and (T_7, T_8). Thus, the output voltage V_o is proportional to the linear product of the two input voltages.

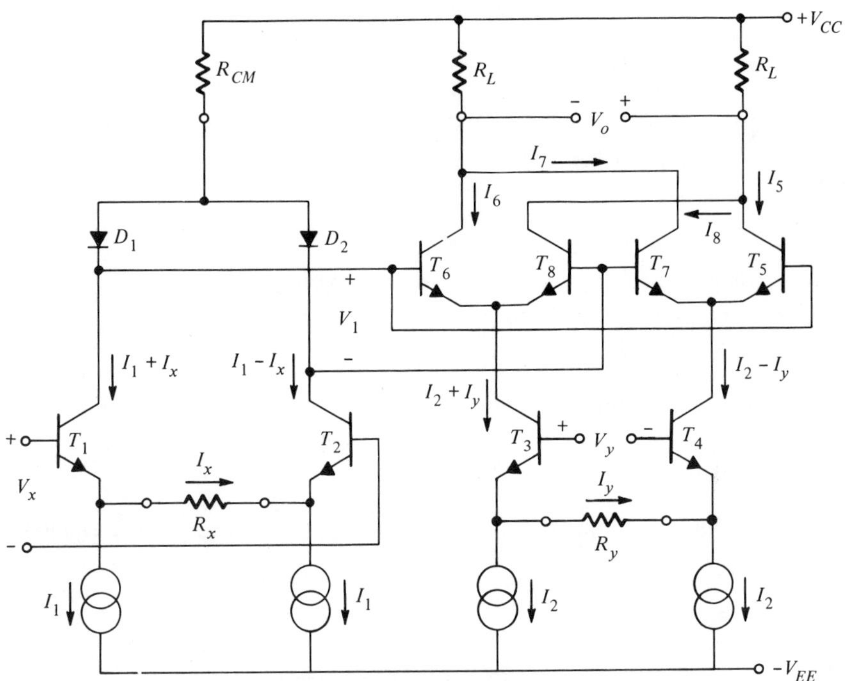

Figure 7.5 Circuit diagram of a four-quadrant multiplier.

The respective emitter degeneration resistors R_x and R_y provide a linear conversion of the input voltages to differential currents I_x and I_y, where

$$I_x = V_x/R_x \qquad \text{and} \qquad I_y = V_y/R_y \tag{7.24}$$

To ensure linear voltage to current conversion at the input terminals, R_x and R_y are chosen to be

$$R_x \gg V_T/I_1 \qquad \text{and} \qquad R_y \gg V_T/I_2 \tag{7.25}$$

With reference to the circuit diagram of Fig. 7.5, the output voltage V_o can be expressed as

$$V_o = R_L[(I_6 + I_7) - (I_5 + I_8)] \tag{7.26}$$

Following the derivations leading to Eqs. (7.22) and 7.23, each of these currents can be related to the currents in input transistors T_1 through T_4 as

$$\frac{I_6}{I_3 + I_y} = \frac{I_5}{I_3 - I_y} = \frac{I_1 + I_x}{2\,I_1} \tag{7.27}$$

and

$$\frac{I_8}{I_3 + I_y} = \frac{I_7}{I_3 - I_y} = \frac{I_1 - I_x}{2\ I_1} \tag{7.28}$$

Substituting Eqs. (7.27) and (7.28) in Eq. (7.26) to eliminate currents I_6 through I_8, and rearranging terms, one obtains

$$V_o = \frac{2\ R_L}{I_1}\ (I_x I_y) \tag{7.29}$$

Since I_x and I_y are linearly related to input voltages V_x and V_y (see Eq. 7.24), V_o can also be written

$$V_o = KV_xV_y \tag{7.30}$$

where the scale factor K_1 is given by

$$K_1 = \frac{2\ R_L}{I_1 R_x R_y} \tag{7.31}$$

For most multiplier applications, scale factor K is chosen to be equal to

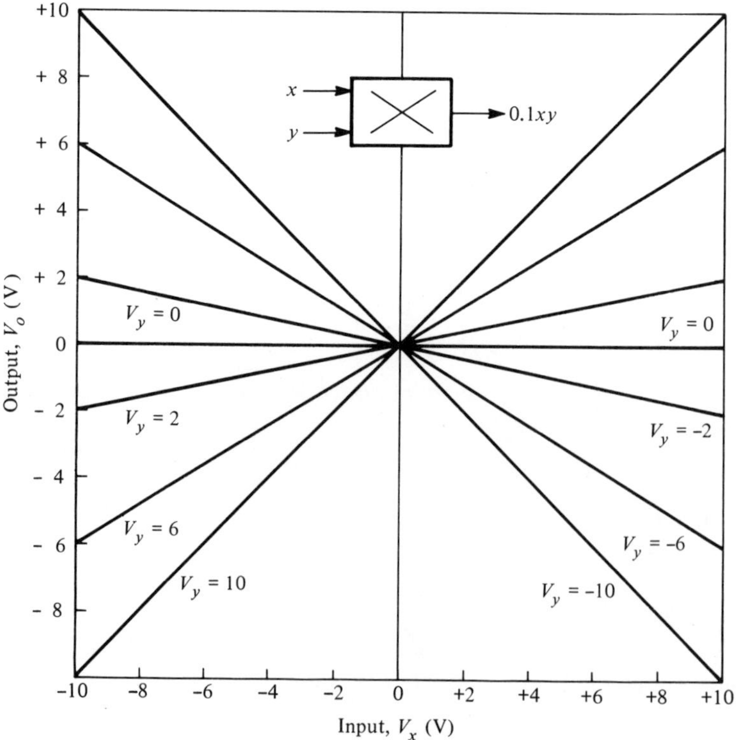

Figure 7.6 Four-quadrant transfer characteristics for the multiplier circuit of Figure 7.5.

0.1 (see Eq. 7.3). This choice of the scale factor is generally preferred because it makes the output of the functional multiplier block compatible with other types of analog multipliers.[2] Figure 7.6 shows the typical four-quadrant transfer characteristics of the transconductance multiplier circuit of Fig. 7.5, with $K_1 = 0.1$.

The four-quadrant multiplier circuit of Fig. 7.5 has been fabricated in a monolithic form. The photomicrograph of the integrated circuit chip is shown in Fig. 7.7. For maximum versatility, resistors R_L, R_{CM}, R_x, and R_y have been left external to the circuit to enable the user to adjust the scale factor and the common-mode ranges of the circuit independently.

Typical electrical characteristics of the monolithic multiplier circuit of Fig. 7.5 are listed below for four-quadrant operation of the circuit with $V_{cc} = +32$ V, $V_{EE} = -15$ V, $I_1 = 1$ mA, $R_L = 11\,\text{k}\Omega$ $R_x = R_y = 15\,\text{k}\Omega$

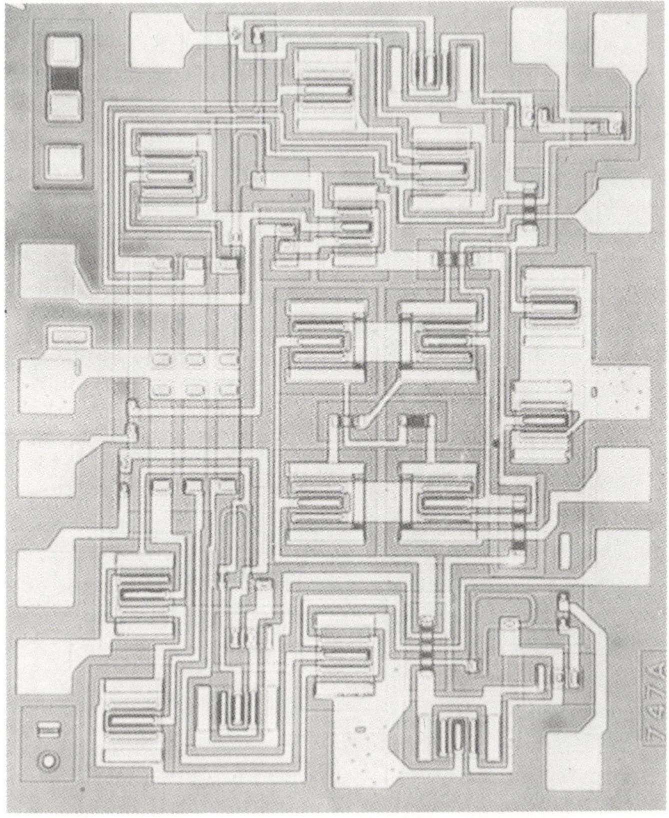

Figure 7.7 Topological layout of the multiplier circuit of Figure 7.5 (resistors R_L, R_{CM}, R_x and R_y are external to the chip). (*Photo: Signetics.*)

Linearity:
Output Error in % of Full Scale

(a) X input ($V_y = 10$ V) $-10 < V_x < 10$	$\pm 0.5\%$
(b) Y input ($V_x = 10$ V) $-10 < V_y < +10$	$\pm 1\%$

Bandwidth:

(a) 3-db bandwidth	3 MHz
(b) 3° relative phase-shift bandwidth	750 kHz
(c) 1% absolute phase-shift bandwidth	30 kHz

Common-mode Rejection: 60 dB

The four-quadrant multiplier principle described in this section is a good example of how one can efficiently utilize the nearly perfect matching and tracking properties of monolithic devices through imaginative circuit design. The linear operation of the circuit of Fig. 7.5 depends solely on the matching of the monolithic components and not on their absolute values, thus making it ideally suited for monolithic integration.

7.5 BALANCED MODULATORS

In a number of applications, the four-quadrant capability of a linear multiplier is not necessary. Typical examples of this class of circuit are the "modulator" or "mixer" circuits. In a modulator circuit, linear response is required with respect to only one of the inputs. This linear input is known as the "modulating input." The second input of the circuit, often referred to as the "carrier input," is driven by a constant amplitude ac signal.

A circuit configuration which is suitable for modulator applications is the cross-coupled differential stage shown in Fig. 7.3. In this case the carrier signal is applied to the nonlinear terminal V_1 and the modulating signal is applied to terminal V_2 shown in the Figure. As will be described shortly, because of the differential symmetry of the circuit, the component of the carrier signal appearing at the output is greatly suppressed. Consequently, the circuit is also called a "balanced modulator." Because of its balanced nature the common-mode bias levels within the circuit are not affected by the input signals.

If the two inputs of the circuit are driven by two sinusoidal input signals $v_1(t)$ and $v_2(t)$ given as

$$v_1(t) = E_1 \cos \omega_1 t \qquad \text{and} \qquad v_2(t) = E_2 \cos \omega_2 t \tag{7.32}$$

Following an analysis similar to that of the preceding section, one can show that the output voltage $v_0(t)$ is related to the two input voltages as[3]

$$v_o(t) = \frac{R_L}{R_E} v_2(t) \tanh \frac{E_1}{2 V_T} \tag{7.33}$$

where the hyperbolic tangent term comes about because of the nonlinear signal processing at the base-emitter junctions of (T_1, T_2) and (T_3, T_4) of Fig. 7.3. (See Eqs. 7.20 through 7.23). If $v_1(t)$ is a high level input signal such that $E_1 \gg V_T$, the hyperbolic tangent reduces to a switching waveform $S_1(t)$ shown in Fig. 7.8(*a*). Physically, this means that, for high level

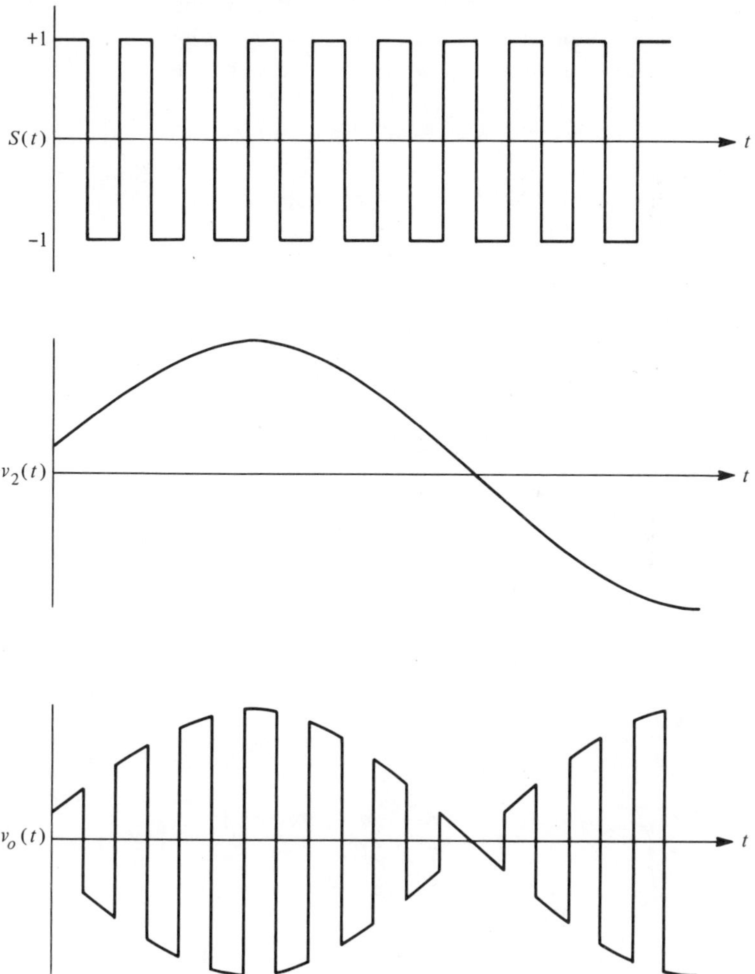

Figure 7.8 Modulated output generated by a switching function $S(t)$, and modulating input V_2 (t).

inputs, (T_1, T_2) and (T_3, T_4) of Fig. 7.3 function as two synchronous single-pole, double-throw switches, as shown schematically in Fig. 7.9. Thus, the modulating input, $v_2(t)$, is effectively "chopped" at the frequency of the carrier input $v_1(t)$ to produce an output waveform of the form,

$$v_o(t) = \frac{R_L}{R_E} v_2(t)S_1(t) \tag{7.33}$$

In this case, the output voltage corresponds to a symmetrical square wave whose amplitude is modulated by $v_1(t)$, as shown in Fig. 7.8(*c*).

In addition to the basic balanced modulator circuit of Fig. 7.3, a number of alternate monolithic circuit configurations is possible. Figure 7.10 shows another circuit topology which has the same basic properties of the all-bipolar modulator circuit.[6] In the circuit of Fig. 7.10, the junction gate FET, F_1, is switched "on" and "off," and serves as a voltage-controlled switch between the emitters of T_1 and T_2. In a balanced circuit, where T_1 and T_2 are well matched, there is negligible dc bias between the source

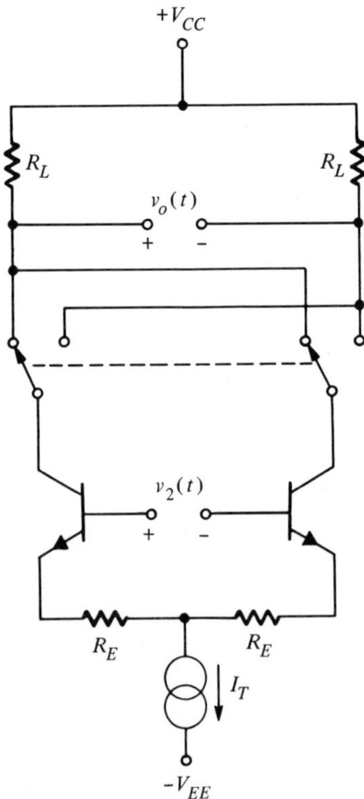

Figure 7.9 Approximate equivalent circuit for a balanced modulator.

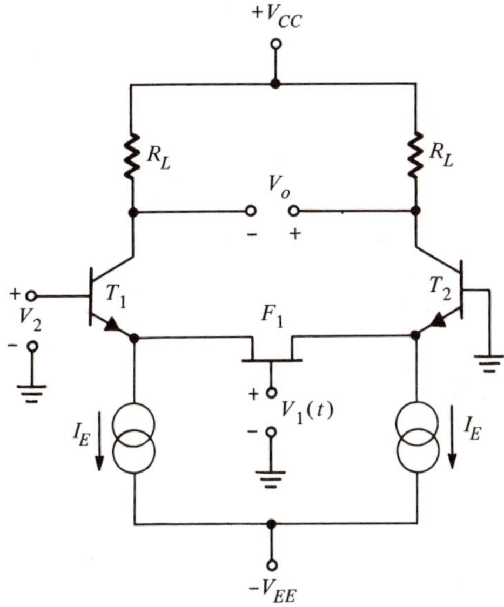

Figure 7.10 An alternate circuit configuration.

and the drain terminals of F_1; thus the chopping signal $v_1(t)$ does not affect dc bias levels within the circuit. If the FET gate is driven by a high level switching signal with an amplitude greater than the FET pinch-off voltage V_p, then the output voltage can be expressed as

$$v_o(t) = \frac{R_L}{R_{sd}} v_2(t) S_1(t) \tag{7.34}$$

which has the same form as Eq. (7.33). R_{sd} in Eq. (7.34) is the FET source-drain resistance under "on" condition and is assumed to be much larger than the dynamic emitter resistances of T_1 or T_2.

Compared to the all-bipolar modulator of Fig. 7.3, the balanced modulator circuit of Fig. 7.10 offers a simpler circuit topology for the same function. However, its main drawback is that it requires a higher level switching signal, since $V_p \gg V_T$.

A fundamental property of a balanced modulator circuit is its ability to suppress any one of the input signals at the output, when the other input is not present. This is known as the "carrier suppression" or the "null-suppression" property of the modulator. In an actual circuit, the inherent component mismatches, which are present even in monolithic circuits, unbalance the circuit, and somewhat degrade this property. For example, if the circuit of Fig. 7.3 had inherent device mismatches which could be represented as effective dc offsets E_{o1} and E_{o2} at the corresponding terminals, then for $E_{10} \ll V_T$ the output might be expressed as

$$v_o(t) = \frac{R_L}{R_E}\left[v_2(t)S_1(t) + \frac{S_1(t)E_{o2}}{I_T R_E} + \frac{V_2(t)E_{o1}}{V_T}\right] \tag{7.35}$$

where the last two terms of Eq. (7.35) represent the so-called carrier and the modulation "feed-through" due to the circuit unbalance. In a monolithic circuit where the matching and tracking of the components are far superior to their discrete counterparts, the carrier supression of a balanced modulator circuit such as that of Fig. 7.3 can be in excess of 50 dB.

7.6 APPLICATIONS OF A BALANCED MODULATOR

Balanced modulators find a wide range of applications in analog communications or frequency translation circuits. The most direct application of modulator-type circuits is in the generation of amplitude modulated (AM) signals. In addition to AM signal generation, they can be used for automatic gain control, frequency multiplication, phase detection and synchronous AM or FM demodulation, and frequency discrimination. As described in the previous section, in most of these applications, the carrier input is normally driven with a high level signal, such that the modulator circuit functions as a set of synchronous switches, and effectively "chops" the modulating signal. This results in an output of the form given by Eq. (7.33). More generally, letting subscripts $_m$ and $_c$ stand for the modulation and the carrier components, the output voltage can be expressed as

$$v_o(t) = K_1 v_m(t) S_c(t) \tag{7.36}$$

where $S_c(t)$ is the symmetrical square wave of unity amplitude shown in Fig. 7.8, having the same fundamental frequency ω_c as the carrier signal; and K_1 is the modulator gain. In terms of its Fourier spectrum, $S_c(t)$ can be expressed as an infinite sum of discrete frequencies at the integral multiples of the carrier frequency, ω_c, i.e.,

$$S_c(t) = \sum_{n=1}^{\infty} A_n \cos n\omega_c t \tag{7.37}$$

where the Fourier coefficients are given as

$$A_n = \frac{\sin (n\pi/2)}{(n\pi/2)} \tag{7.38}$$

In most applications, the higher order harmonics of $S_c(t)$ can be filtered out by means of a low-pass filter at the output of the modulator, as shown in Fig. 7.11. If the low-pass filter bandwidth is chosen to be approximately

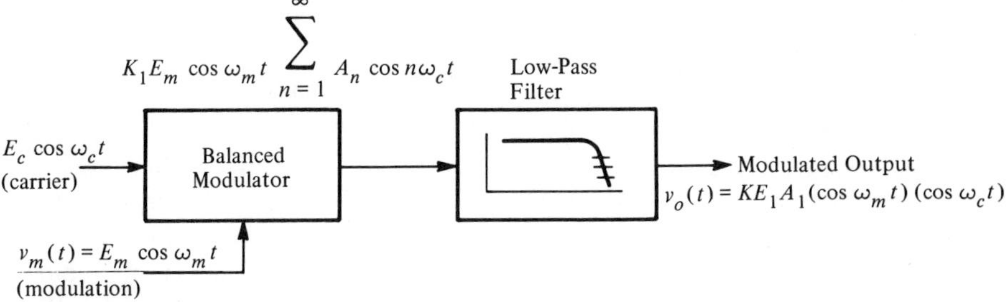

Figure 7.11 Use of a low-pass filter to eliminate higher order harmonics in modulator output.

equal to ω_c, the resulting output of the system is of the form,

$$v_o(t) = \frac{2\ K_1E_1}{\pi}(\cos \omega_m t)(\cos \omega_c t) \tag{7.40}$$

Using trigonometric identities, the output waveform of Eq. (7.40) can also be expressed as

$$v_o(t) = \frac{K_1E_1}{\pi}[\cos(\omega_c + \omega_m)t + \cos(\omega_c - \omega_m)t] \tag{7.41}$$

Equation 7.41 implies that, in this type of a modulator, the output does not contain any energy at the carrier frequency. Instead, the total energy of the output is concentrated at two discrete frequencies, $(\omega_c + \omega_m)$ and $(\omega_c - \omega_m)$, which are known as the "upper" and "lower" sidebands. Since there is no component of the output at the carrier frequency, this type of modulation is known as "suppressed carrier modulation," and it comes about from the basic null-suppression property of a balanced modulator, as discussed in the previous section.

If the modulating signal contains a dc component, i.e.,

$$v_m(t) = E_m(1 + m \cos \omega_m t) \tag{7.42}$$

then the output is of the form,

$$v_o(t) = E_o[\cos \omega_c t + m/2 \cos(\omega_c + \omega_m)t + m/2 \cos(\omega_c - \omega_m)t] \tag{7.43}$$

which corresponds to a conventional AM signal. It is made up of a carrier component at frequency ω_c, as well as the two symmetrical sideband components at frequencies $(\omega_c \pm \omega_m)$. The coefficient m is known as the modulation index, and is restricted to be $0 < m < 1$.

In addition to amplitude modulation, the balanced modulator circuits of Fig. 7.3 or 7.10 can be used for a number of other related applications.

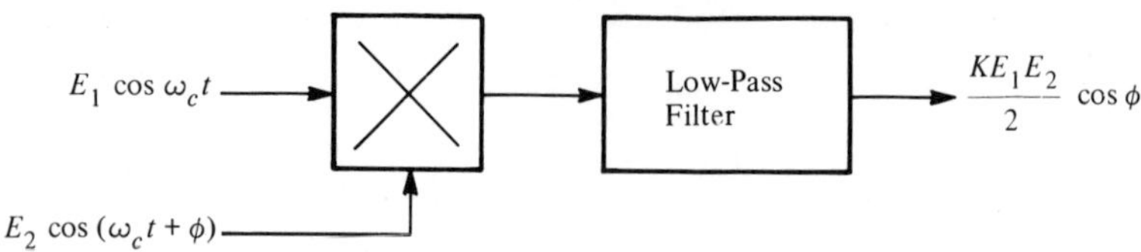

(a) Phase Detection

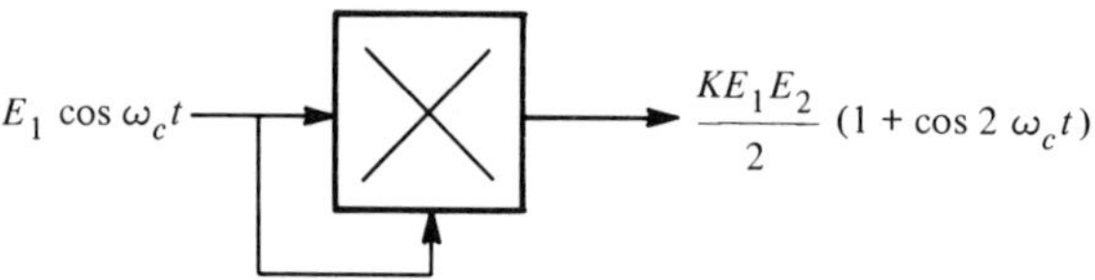

(b) Frequency Doubling

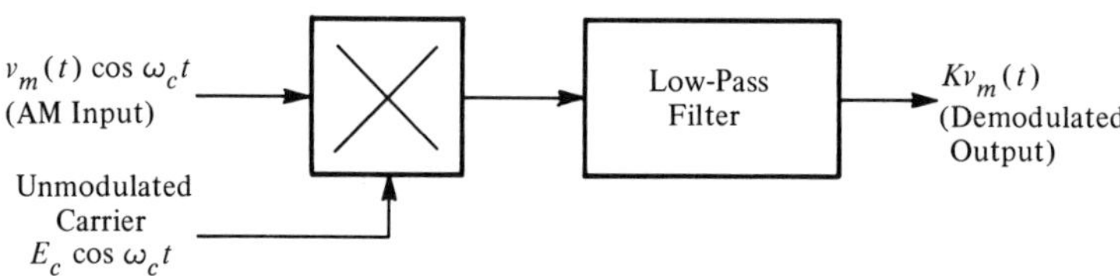

(c) Synchronous AM Detection

Figure 7.12 Some applications of a balanced modulator.

Some of these are shown in Fig. 7.12. If two input signals at the same frequency ω_c, but having a relative phase difference ϕ, are applied to the inputs, as shown in Fig. 7.12(*a*), the output of the modulator would be of the form;

$$v_o(t) = \frac{K_1E_1E_2}{2} [\cos (2\omega_c t + \phi) + \cos \phi] \qquad (7.44)$$

The ac component of Eq. 7.44 is at twice the carrier frequency, and can be readily filtered out by a low-pass filter. Then the remaining dc component of the output is proportional to the cosine of the phase angle:

$$v_o \Big|_{dc} = \frac{K_1E_1E_2}{2} \cos \phi \qquad (7.45)$$

Conversely, if the modulator output is ac-coupled, then the dc term in Eq. 7.44 is filtered out, resulting in an ac signal at twice the input frequency, and the circuit operates as a frequency doubler.

Figure 7.12(*c*) shows the use of a balanced modulator for synchronous AM detection. If an amplitude modulated signal is applied to the linear input of a balanced modulator, and an unmodulated carrier is applied to the switching input, neglecting the higher order harmonics of the switching waveform, one obtains an output of the form,

$$v_o(t) = K_1[V_m(t) \cos \omega_c t][\cos \omega_c t] \tag{7.46}$$

The output waveform of Eq. (7.46) can also be written as

$$v_o(t) = \frac{K_1 V_m(t)}{2} [1 + \cos 2\, \omega_c t] \tag{7.47}$$

If the high frequency component of the output at twice the carrier frequency is filtered out by a low-pass filter, the remaining portion of $v_o(t)$ is proportional to the demodulated output:

$$v_o(t) \Bigg|_{\substack{\text{L.P.} \\ \text{filtered}}} = (K_1/2)\, V_m(t) \tag{7.48}$$

In the case of conventional double-sideband AM signals, the unmodulated carrier reference necessary for synchronous detection can be obtained by passing the AM signal through a "limiter" circuit which removes the modulation and generates a switching waveform, $S_c(t)$, at the carrier frequency. Then, by multiplying this switching waveform with the modulated input, as shown in Fig. 7.12(*c*), one can extract the demodulated output.

The phase detection property of the balanced demodulator circuit given in Eq. 7.45 can also be used for frequency discriminator applications or for demodulating frequency modulated (FM) signals. For this application, the balanced modulator circuits of Figs. 7.3 or 7.10 are interconnected with a broad-band limiter and a band-pass phase-shift network as shown in Fig. 7.13.[3] The limiter circuit removes all the amplitude modulation from the input FM signal, and generates a high-level switching

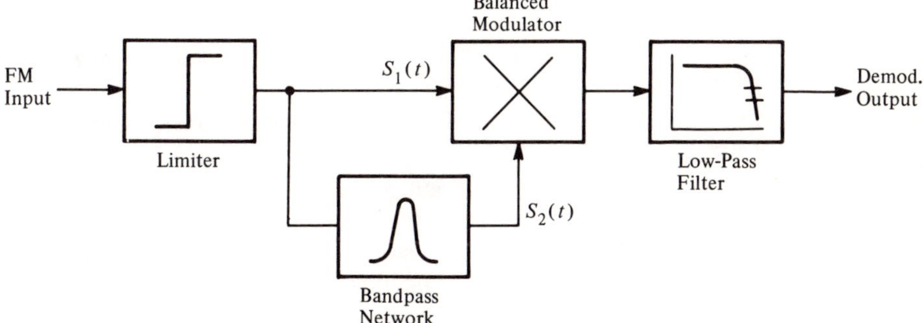

Figure 7.13 Use of a balanced modulator as FM detector and discriminator.

signal $S_1(t)$ at the same frequency as the input. The band-pass phase-shift network is tuned to the carrier frequency ω_c of the input FM signal. It produces a second signal $S_2(t)$ to drive the balanced modulator. If Q is the selectivity of the band-pass network, the phase shift ϕ introduced by this network for small frequency deviations $\Delta\omega$ in the vicinity of ω_c can be expressed as

$$\phi \approx -\pi/2 \pm 2\,Q\,\frac{\Delta\omega}{\omega_c} \tag{7.49}$$

for $(2Q\Delta\omega) \ll \omega_c$. Then the low-pass filtered output of the modulator is related to the frequency deviations of the input signal as

$$v_o(t)\Big|_{\substack{\text{L.P.}\\ \text{filtered}}} \approx K\,\frac{\Delta\omega}{\omega_c} = K\,\frac{\Delta f}{f_c} \tag{7.50}$$

where K is the combined gain of the filter and the detector sections, and $f = (\omega/2\pi)$. Since, in an FM signal, the frequency deviation of the carrier, $\Delta\omega$, represents the information, the output given by Eq. 7.50 represents the demodulated signal.

The balanced modulator circuit of Fig. 7.3 is used as a basic building block in monolithic systems designed for commercial radio or TV receiver applications. For example, the FM detection scheme of Fig. 7.13 is today widely used as a monolithic replacement for the detector section of commercial FM receivers or TV sound systems.[7] In addition, many of the monolithic circuits designed for FM stereo decoding,[8] color TV chroma demodulation,[9] and video detection[10] are all based on the balanced-modulator principle.

REFERENCES

1. R. Stata, "Unspooking Multiplier Specs" *Electronic Products* (May, 1970): 92–96.
2. "Special Survey: Packaged Analog Multipliers," *EEE Magazine,* (Nov. 1968): 80–94.
3. A. Bilotti, "Applications of a Monolithic Analog Multiplier," *IEEE J. Solid State Ckts.,* **SC–3**(4) (1968): 373–380.
4. B. Gilbert "A Precise Four-Quadrant Multiplier with Subnanosecond Response," *IEEE J. Solid State Ckts,* **SC–3**(4) (1968): 365–373.
5. E. Renschler, "Theory and Application of a Linear Four-Quadrant Monolithic Multiplier," *EEE Magazine,* **17**(5) (May, 1969).
6. A. B. Grebene and H. R. Camenzind, "Phase Locking as a New Approach for Tuned Integrated Circuits," Int. Solid State Ckts. Conf., *Digest Tech. Papers,* **12** (1969): 100–101.

7. A. Bilotti and R. S. Pepper, "A Monolithic Limiter and Balanced Discriminator for FM and TV Receivers," *IEEE Trans. Broadcast TV Receivers,* **BTR–13** (1967): 54–65.

8. R. H. Stockwell and H. R. Camenzind, "An Inductorless Monolithic Stereo Decoder," Int. Solid State Ckts. Conf., *Digest Tech. Papers,* **13** (1970): 106–107.

9. L. Blaser, D. Bray and J. Rennick, "A Monolithic Chroma Demodulator Integrated Circuit for Color TV Receivers," *IEEE Trans. Broadcast TV Receivers,* **BTR–14**(3) (1968): 52–57.

10. G. Lunn, "A Monolithic Wideband Synchronous Video Detector for Color TV," *IEEE Trans. Broadcast TV Receivers,* **BTR–15**(2) 11 (1969): 159–166.

8 Wideband Amplifiers

In the design of broad-band high frequency amplifiers, the inherent limitations of monolithic devices impose severe restrictions on possible design approaches. The most significant design restriction is the lack of integrated inductors which make conventional broad-banding techniques such as shunt-peaking not suitable for monolithic wideband amplifiers. Furthermore, large coupling capacitors or impedance matching transformers are not compatible with integrated circuits. In spite of these basic drawbacks, a large variety of small signal wideband amplifier configurations are available which are well suited to monolithic integration. In basic circuit topologies and broad-banding techniques will be examined to emphasize the differences in the design and analysis of discrete and integrated wideband amplifiers.

In monolithic amplifiers ac coupling between the stages is in general not possible. Therefore, the ac design and performance of the circuit cannot be considered as a separate problem from dc bias considerations. The availability of a large number of well matched active devices again provides a distinct advantage for the designer. For example, one can now use compound connection of two or more devices to replace a single transistor for improved high frequency performance. Similarly, since the devices are well matched, their parasitics can be neutralized by proper circuit layout or interconnections.

Among the monolithic devices, the *npn* bipolar transistor has the best high frequency capability. For this reason almost all wideband amplifiers use *npn* devices in the signal path; the lateral or substrate-*pnp* devices, when used, are confined to biasing applications. Feedback techniques offer by far the most powerful broad-banding method for monolithic circuits. In this chapter some of these basic feedback configurations will be reviewed in terms of their high frequency performance.

8.1 HIGH FREQUENCY DEVICE MODELS

An efficient analysis of a wideband amplifier requires a good knowledge of the monolithic device parameters and circuit parasitics. The starting point of the design is the small signal high frequency equivalent circuit for the active elements. In the case of the *npn*-bipolar, the most convenient model is the hybrid-π circuit. This model was described in Chapter 2 (see Section 2.1), and will be briefly reviewed in this section. Figure 8.1(*a*) shows a simplified version of the small signal hybrid π equivalent circuit for a bipolar transistor. In this model, whenever possible, the component values are expressed in terms of the readily measurable parameters of the device, such as low frequency current gain β_0, gain bandwith product ω_T, and transconductance g_m, where

$$g_m = I_E/V_T \tag{8.1}$$

and

$$\omega_T = 2\pi f_T \tag{8.2}$$

The remaining components correspond to the inherent device parasitics: r_b is the parasitic base-spreading resistance, C_c is the collector-base junction capacitance, r_{cs} is the collector series resistance, and C_{ss} is the collector-substrate capacitance. For a typical monolithic small signal transistor (see Fig. 2.4), some representative values of these device parameters are

$$\begin{aligned} \beta_0 &= 100 & r_{cs} &\approx 80\ \Omega \\ f_T &\approx 450\ \text{MHz}\ (I_E = 1\ \text{mA}) & C_c &= 0.6\ \text{pF} \\ r_b &\approx 50\ \Omega & C_{ss} &= 2\ \text{pF} \end{aligned}$$

where the capacitance values are assumed to be measured with 3 V reverse bias.

The simplified equivalent circuit of Fig. 8.1(*a*) ignores the lateral current flow in the transistor base as well as the excess phase shifts due to the distributed nature of the actual transistor parameters. Therefore it can be quite inaccurate at frequencies approaching the f_T of the transistor.

The presence of the base-collector capacitance C_c causes the model

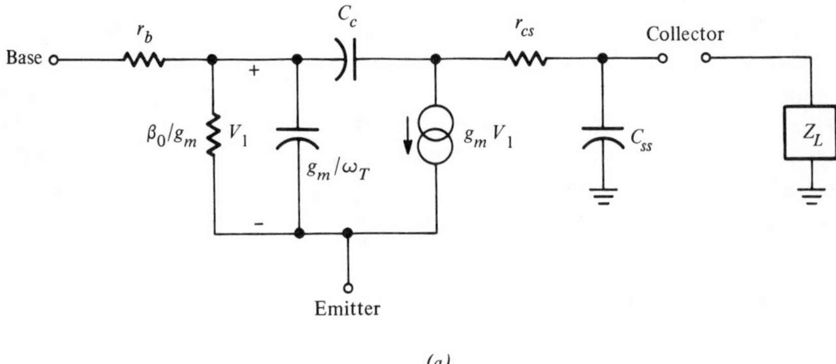

(a)

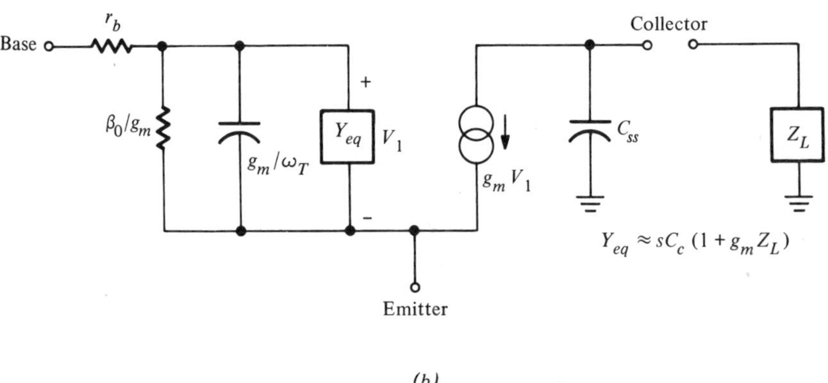

(b)

Figure 8.1 Simplified hybrid-π model for bipolar transistor (*a*), and its approximate unilateral equivalent (*b*).

to be bilateral and makes manual calculations quite cumbersome. For an "order-of-magnitude" estimation of the device performance, the model of Fig. 8.1(*a*) can be further reduced to its approximate unilateral equivalent shown in Fig. 8.1(*b*), where Y_{eq} is an approximate equivalent admittance to simulate the feedback provided by C_c. Y_{eq} is related to the output load impedance Z_L as

$$Y_{eq} = sC_c(1 + g_m Z_L) \tag{8.3}$$

where s is the complex frequency variable, and Z_L is assumed to include the substrate capacitance C_{ss}. If C_{ss} is negligible and the load is purely resistive (i.e., $Z_L = R_L$), then Y_{eq} reduces to

$$Y_{eq} = sC_c\,(1 + g_m R_L) \tag{8.4}$$

which is the well known "Miller effect" approximation.[1]

In many cases, the simplified hybrid-π circuits of Figure 8.1 do not provide a sufficiently accurate description of device performance at high frequencies. The unilateral model of Fig. 8.1(*b*) can introduce significant error even at frequencies as low as 5 to 10% of f_T, especially when capacitive loads are present. The simplified models of Fig. 8.1 do not account for the distributed nature of the device parasitics, such as r_b, r_{cs}, C_c, and C_{ss}. This omission can also lead to significant errors in the phase response of the transistor for frequencies approaching the f_T of the device.

Characterization and accurate modeling of active circuit components at high frequencies is a difficult problem. Recently a number of modeling and measurement techniques have been developed which characterize the performance of bipolar devices to a higher degree of accuracy then the simple models of Fig. 8.1.[2,3] However, these models of the devices require a large number of additional parameters to be defined and measured; therefore they are better suited to numerical, rather than analytical, computations. With the advances in the state-of-the-art and the availibility of computer-aided analysis techniques, the additional complexity of these models does not present a serious problem. The use of computer-aided analysis and optimization techniques for wideband amplifier design will be discussed briefly, in a later section of this chapter.

8.2 COMPOUND DEVICES

The gain-bandwidth product of an amplifier stage can be significantly improved by using multiple active devices as a direct replacement for a single transistor. Figure 8.2 shows two such compound device configurations which can be used to replace a common-emitter gain stage. Note that the circuit diagrams of Fig. 8.2 are for ac signals only; therefore the bias networks are not shown in the figure. It should be noted that the compound device connections shown in the figure are not new or novel in transistor circuit design.[4] However, their particular application and usefulness in analog integrated circuit design stems from the fact that the added active devices and their associated bias circuitry can be incorporated into the design with minimal increase of circuit cost or complexity.[5]

Figure 8.2(*a*) shows the ac circuit diagram of a common-collector/common-emitter (*cc-ce*) stage. In this compound device configuration, T_1 effectively buffers gain stage T_2 from the input. Therefore, capacitive loading at the input due to the Miller capacitance of T_2 is greatly reduced. This results in an improved gain-bandwidth product for the compound device over a single common-emitter stage. The improvement in frequency performance is particularly significant for large values of the source resistance R_S.[6]

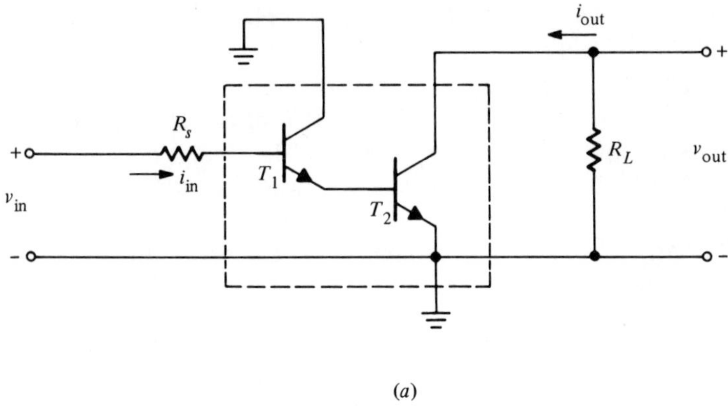

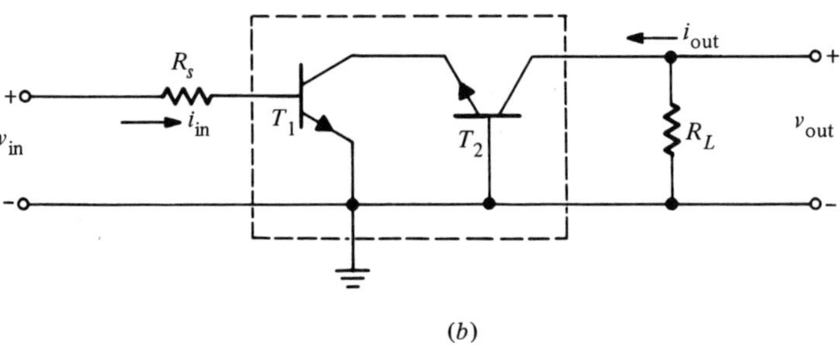

Figure 8.2 Ac circuit diagrams for compound device stages: (*a*) common-collector / common-emitter stage; (*b*) cascode or common-emitter / common-base stage.

The cascode, or common-emitter/common-base stage of Fig. 8.2(*b*) also forms a useful compound device topology suitable for wideband amplifier design. The improvement in high frequency performance is obtained by the impedance mismatch between transistors T_1 and T_2. The collector load of T_1 is formed by the input impedance of T_2. Since T_2 is operated in common-base configuration, its input impedance is very low. This in turn keeps the voltage gain across T_1 low, and minimizes the capacitive loading at the input due to the base-collector capacitance of T_1. The circuit has a low frequency current gain A_I given as

$$A_I = i_{out}/i_{in} = \beta_1 \alpha_2 \approx \beta_1 \tag{8.5}$$

where β_1 and α_2 are the common-emitter and the common-base current gains for T_1 and T_2 respectively. Since α_2 is very close to unity, the overall current gain of the stage is approximately equal to that of transistor T_1. Since output transistor T_2 buffers current gain stage T_1 from the load, the frequency broad-banding advantages of a cascode stage become significant as the value of the load resistance R_L is increased. For example, in the case of medium voltage gain amplifier stage design ($R_L > 2\ k\Omega$), the cascode-connected compound devices can offer up to an order of magnitude improvement in the 3 dB bandwidth over that obtained from a simple common-emitter stage.[6]

In both the *cc-ce* pair and the cascode configuration, the compound devices exhibit a higher degree of isolation between the input and output terminals of the stage than in the case of a simple common-emitter transistor. This isolation is obtained because the reverse transmission across the compound-device stage is greatly reduced by the impedance mismatches between T_1 and T_2. In other words, in each case, one of the devices (T_1 in the case of *cc-ce pair*; and T_2 for the cascode stage) performs the function of an impedance transformer. This effect makes the compound device configurations particularly attractive for the design of high frequency tuned amplifier stages where the parasitic cross-coupling between the input and the output tank circuits can make the alignment of the stage difficult.

In the compound device connections of Fig. 8.2, use of a second transistor adds an additional time constant or a pole to the overall circuit response. In the case of the *cc-ce* pair, the magnitude of this pole becomes comparable to the dominant pole of the stage as the source resistance R_S is increased. The same is true for the case of the cascode circuit, this time for increasing values of R_L. Since the *cc-ce* pair and the cascode circuit stages offer the most improvement in their 3 dB bandwidth for large values of R_S and R_L respectively, this implies that when the compound devices are used for gain-bandwidth optimization, they can cause the circuit to exhibit rapid high frequency roll-off and excess phase lag. Therefore, additional care must be taken in applying feedback around amplifier stages containing compound devices.

Emitter Coupled Stages

For high frequency integrated circuit applications the emitter-coupled or the "paraphase" configuration shown in Fig. 8.3 offers considerable advantage over a simple common-emitter stage. The emitter-coupled stage is a special case of the compound device connections, made up of a common-collector (T_1) and a common-base (T_2) transistor. It also corresponds to a special case of the basic differential amplifier stage described in Chapter 5, except in this case the circuit is driven single-endedly, and the collector load of the input stage T_1 is replaced by a short circuit. In the emitter-

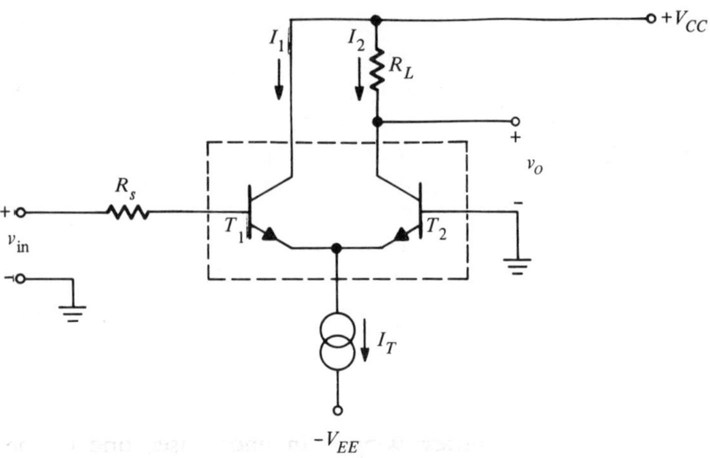

Figure 8.3 Emitter-coupled gain stage.

coupled gain stage common-collector input transistor T_1 minimizes the input capacitance and provides the necessary current amplification. The common-base transistor T_2 in turn converts this current gain into a voltage amplification across load resistance R_L. Similar to the *cc-ce* pair and the cascode stages, the emitter-coupled stage also eliminates the narrow-banding effect of the Miller-feedback capacitance C_c, and greatly reduces the reverse transmission from the output back to the input. However, unlike the compound device stages of Fig. 8.2, the emitter-coupled configuration is a noninverting gain stage and does not provide a phase reversal between the input and the output signals. Therefore, the common-collector/common-base compound device does not provide a direct replacement for a common-emitter gain stage, particularly when negative feedback is involved. The noninverting nature of the emitter-coupled stage also makes it difficult to use in negative feedback configurations.

For low frequency operation, the voltage gain A_v of an emitter-coupled stage is approximately equal to one half of that available from a similar common-emitter stage. For equal bias currents through T_1 and T_2, the low-frequency gain can be expressed as

$$A_v \approx \frac{\beta_0 R_L g_m}{g_m R_S + 2\,\beta_0} \tag{8.6}$$

where g_m is the transconductance of T_1 or T_2. Even though the voltage gain is approximately one half a common-emitter transistor, the 3 dB bandwidth of the emitter-coupled stage is significantly larger than twice the common-emitter bandwidth. Since the emitter-coupled configuration combines most of the advantages of the *cc-ce* pair and the cascode configurations, its bandwidth advantages are particularly significant for large values of R_S and R_L.[6]

The emitter-coupled gain stage offers a very high degree of bias stability, because of the common-mode feedback provided by the bias current source I_T. Therefore, it is well suited to direct coupling and can be used as a building block for multistage amplifier and limiter circuits.[7,8]

Figure 8.4 shows a practical wideband circuit configuration using compound devices.[8] This circuit also illustrates the possibility of interconnecting a number of the basic compound device configurations of Figs. 8.2 or 8.3 to form a more complex compound device. For example, in the circuit of Fig. 8.4, the input stage is formed by the compound connection of a *cc-ce* pair/cascode configuration with T_1, T_2, and T_3 operated in common-collector, common-emitter, and common-base configurations respectively. Then, this compound device is in turn emitter-coupled to a symmetrical

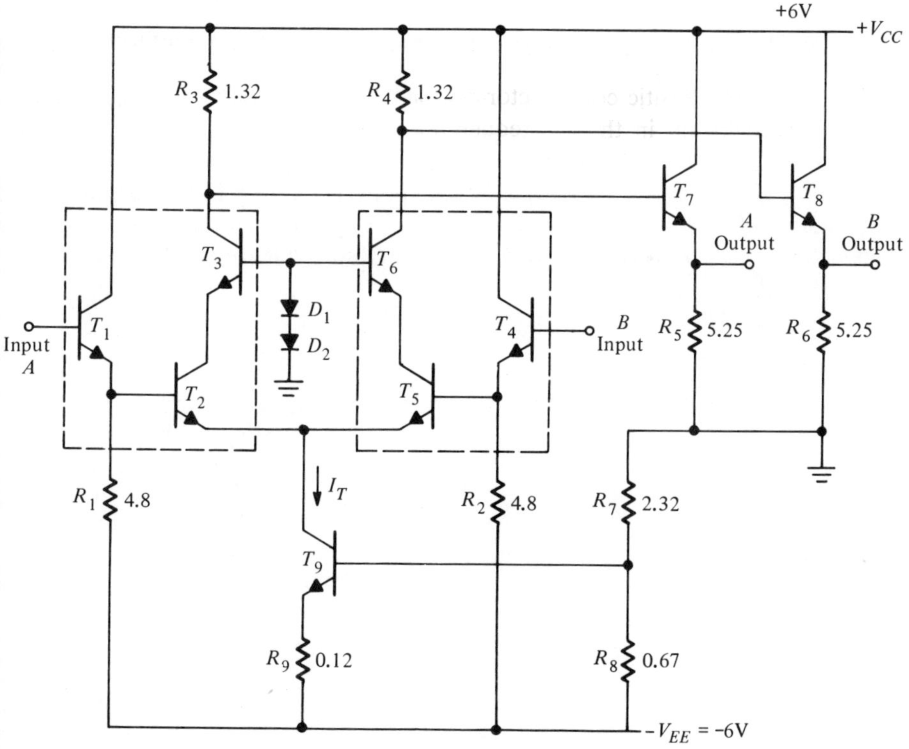

Figure 8.4 A wideband amplifier circuit using compound devices (RCA CA3040). (all resistor values in $k\Omega$).

configuration of T_4, T_5, and T_6 to form a differential input stage. These compound devices are identified within the dotted lines in the circuit schematic. Common-collector stages T_7 and T_8 are used to provide low impedance differential outputs for the circuit. The low frequency voltage gain of the circuit for single-ended input can be expressed as

$$A_v \approx g_m R_3 \tag{8.7}$$

where g_m is the transconductance of the compound input devices composed of (T_1, T_2, T_3) or (T_4, T_5, T_6), and is given as

$$g_m = (I_T/2\ V_T) \tag{8.8}$$

I_T is the common-mode bias current drawn by T_9.

Figure 8.5 shows the single-ended voltage gain for the wideband amplifier circuit of Fig. 8.4, with $R_S = 50\ \Omega$. Note that the 3 dB bandwidth is approximately 55 MHz, with a voltage gain of approximately 30 dB. However, due to the compound devices, the frequency response exhibits a rapid fall-off (typically −18 dB/octave) at high frequencies.

8.3 NEUTRALIZATION OF COLLECTOR-BASE CAPACITANCE

The parasitic collector-base feedback capacitance C_c is the most significant limitation in the frequency response of a common-emitter gain stage. In the case of a differential amplifier stage, this parasitic capacitance can be significantly reduced by utilizing a bridge-neutralization scheme as shown in Figure 8.6(*a*).[9] In a balanced differential stage, the ac voltages at the collectors of T_1 and T_2 have the same amplitude, but differ in phase by

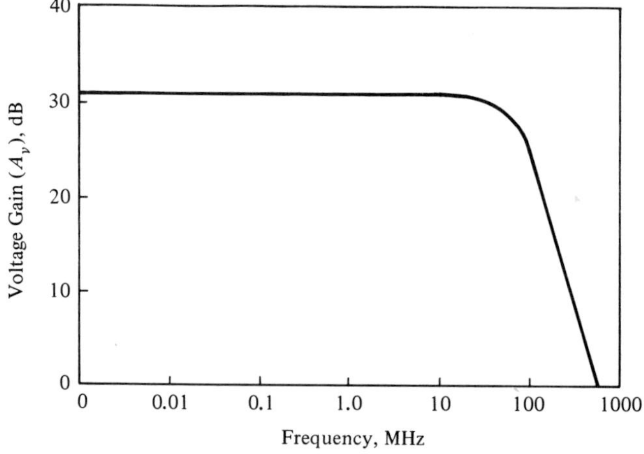

Figure 8.5 Voltage gain as a function of frequency for the circuit of Figure 8.4. (single-ended input and output with $R_s = 50\Omega$, $R_L = 1\ \text{k}\Omega$).

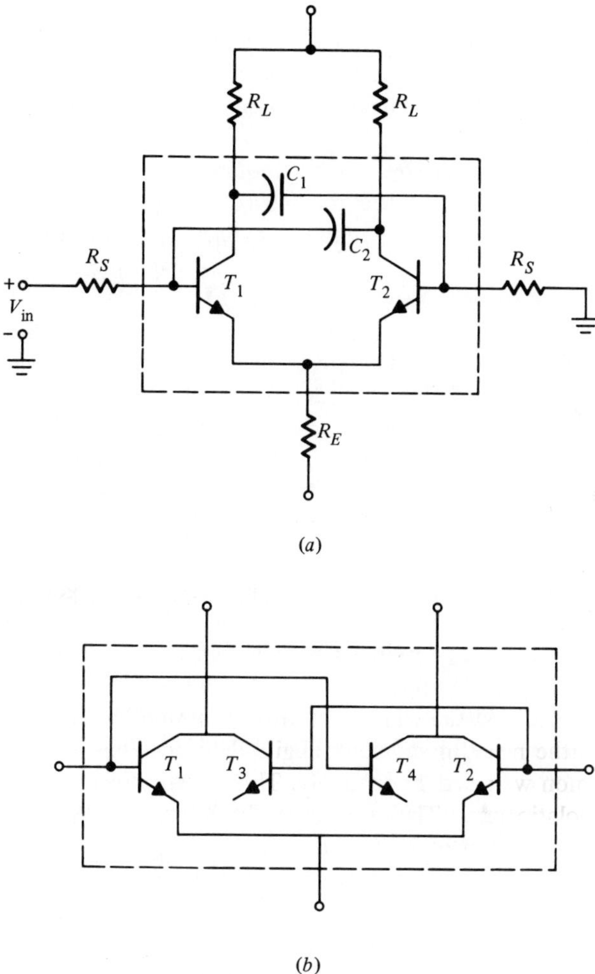

Figure 8.6 Neutralization of collector-base capacitance in a differential gain stage.

180°. Therefore, the feedback to the bases of T_1 and T_2 through C_2 and C_1 is out of phase with the internal feedback due to the C_c of each transistor. By choosing C_1 and C_2 to have the same value as the C_c, it is possible to neutralize the parasitic feedback through C_c.

In discrete (nonintegrated) circuit design, this neutralization technique has severe practical limitations because an efficient neutralization of C_c requires that feedback capacitors C_1 and C_2 be matched to C_c to better than ±10 percent. In practical devices, C_c is less than 1 pF, and its absolute value is predictable, at best, to ±20 percent. Furthermore, C_c is a function of the device operating point, frequency, and temperature. Discrete capacitors cannot match or track these variations of the base-collector capacitance and cannot be effectively used in the neutralization scheme of Fig. 8.6(*a*).

Since the C_c neutralization scheme requires matching and tracking of the components, rather than accurate absolute values, it is ideally suited to monolithic circuit design. Feedback capacitances C_1 and C_2 can be realized in integrated form as the collector-base capacitances of additional transistors T_3 and T_4, as shown in Fig. 8.6(*b*). All four transistors in the figure can be designed to have identical geometrics and are simultaneously fabricated on the same chip. Therefore, they exhibit nearly ideal matching and thermal tracking of their characteristics. The collector-base capacitances of T_3 and T_4 very nearly match the C_c of T_1 and T_2, and when cross-coupled as shown in Fig. 8.6(*b*), they provide an efficient neutralization of C_c associated with T_1 and T_2. The reverse transmission through C_c is most troublesome in the design of bandpass intermediate frequency (I-F) amplifiers where the reverse interaction between the input and output of the stage makes the "alignment" or tuning of multiple gain stages quite difficult. Using the neutralization scheme of Fig. 8.6 the reverse transmission through C_c can be reduced by as much as 95 percent over a wide range of operating conditions.[10]

It should be noted that, in the case of a monolithic circuit, the added cost or complexity due to the neutralizing devices is negligible. T_3 and T_4 have their collectors common with T_1 and T_2, respectively. Therefore, they do not require separate isolation islands. Thus, the increase of circuit chip size due to the addition of T_3 and T_4 is negligible.

The neutralization scheme of Fig. 8.6 has one serious drawback. It relies on the presence of symmetrical dc and asymmetrical ac signals in the circuit. This condition is realized only in differential gain stages. Therefore, the application of the C_c neutralization technique is limited to differential designs.

8.4 FEEDBACK CASCADES

In designing wideband amplifiers, one relies heavily on the use of feedback to optimize the high frequency performance of the amplifier, and to exchange bandwidth for gain. In an amplifier circuit involving multiple gain stages, one has the choice of applying the feedback locally (i.e., around each individual gain stage) or around the overall circuit. Local feedback is somewhat easier to apply from the overall frequency stability point of view since it allows one to design each stage somewhat independently of the others.

Local feedback is a direct carry-over from discrete design techniques. However, in monolithic circuits where no ac coupling is available, the number of identical feedback stages which can be directly cascaded without requiring extensive dc level shifting is limited to two or, at most, three gain stages. Most of the dc level-shift circuits discussed in Chapter 4 are not suitable for wideband amplifier design because of their frequency limitations. This is particularly true of the level-shift circuits of Figs. 4.14, 4.15, and 4.17. The only two possible level-shift stages which can offer some potential use in wideband amplifier design are those of Fig. 4.13(*b*) and (*c*). In the case of the avalanche diode level-shift scheme of Fig. 4.13, the noise generated in the avalanche diode is a serious detriment for wideband amplifier applications.

Figure 8.7 shows the two basic single-stage feedback configurations used in broadband amplifier design. The shunt feedback of Fig. 8.7(*a*) results in a low input impedance gain stage which is basically a current amplifier, and operates best when driven from a high source impedance. The series feedback configuration of 8.7(*b*) increases the input impedance of the stage. This type of a stage is best suited for voltage amplification, and it is normally operated from a low impedance source.

In cascading feedback stages, the designer has the freedom to choose any particular kind of feedback for any given stage. However, the respective driving source impedance requirements of shunt and series feedback stages are best satisfied if one alternates between the two types of feedback for each of the cascaded stages. In this manner the shunt-feedback stage can provide the necessary low-impedance drive for the series feedback

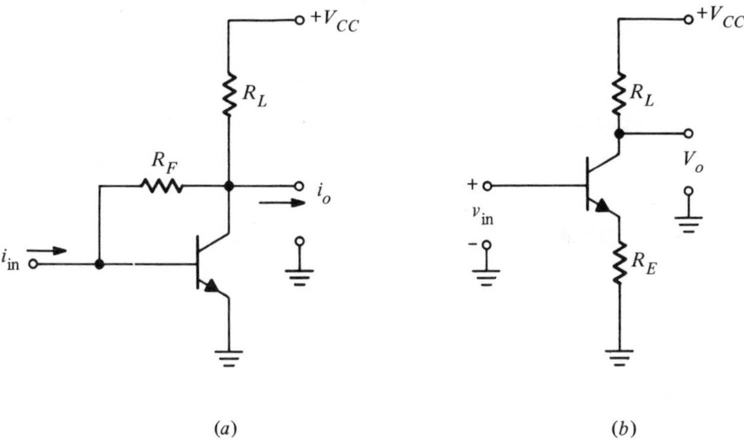

Figure 8.7 Single-transistor feedback stages: (*a*) shunt-feedback; (*b*) series-feedback.

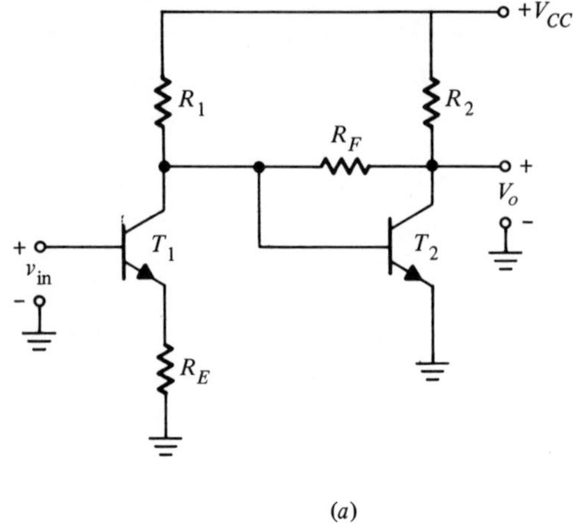

(a)

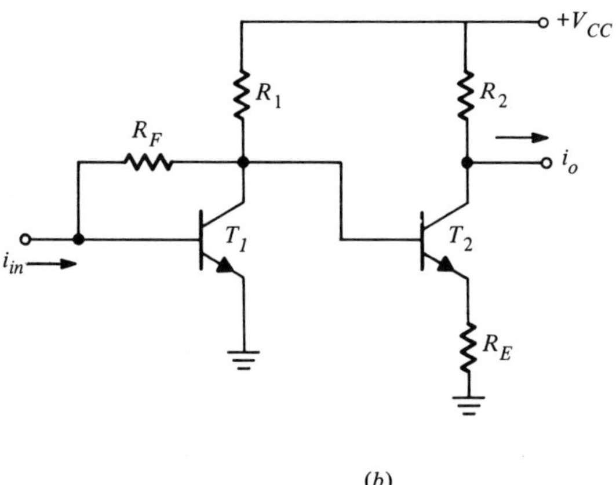

(b)

Figure 8.8 Feedback Cascades: (*a*) Series-Shunt Cascade; (*b*) Shunt-Series Cascade.

stage; and conversely the series feedback output can provide the high impedance drive for the shunt feedback stage succeeding it.

Figure 8.8 shows the two basic cascade pairs used in wideband amplifier design. The series-shunt cascade of Fig. 8.8(*a*) operates best as a voltage amplifier, since it offers a relatively high input impedance and output admittance. Its voltage gain can be approximated as

$$A_v = (v_o/v_{\text{in}}) \approx (R_F/R_E) \tag{8.9}$$

The shunt-series cascade configuration is useful as a wideband current amplifier, since it offers a low input impedance. Its low frequency current gain can be approximated as

$$A_I = (i_{out}/i_{in}) \approx (R_F/R_E) \tag{8.10}$$

Feedback cascades show somewhat poorer dc stability than other types of wideband amplifier configurations which use overall feedback.[6] In monolithic integrated circuits where a large number of active devices are readily available, this dc bias stability problem can be partly eliminated by using a differential circuit configuration. Since a differential stage inherently employs a high degree of common-mode feedback (see Chapter 5) it offers a higher degree of dc bias stability and also simplifies dc level-shift

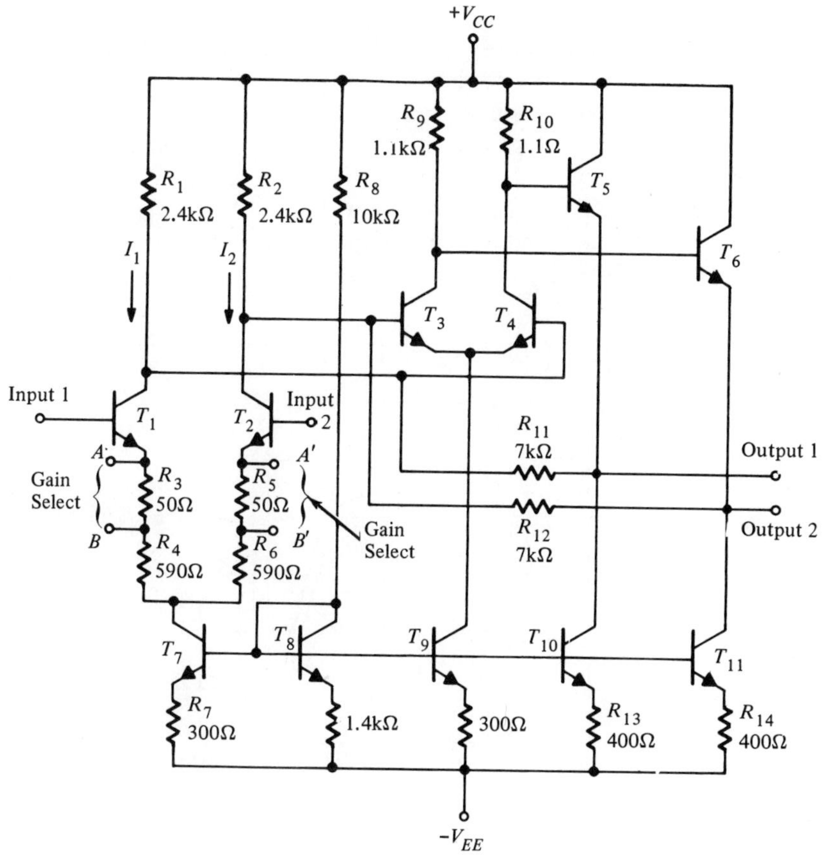

Figure 8.9 A differential series-shunt feedback cascade (Fairchild μA 733).

problems. Both of the feedback cascade configurations of Fig. 8.8 are readily adaptable to differential design. Figure 8.9 shows a differential series-shunt cascade circuit especially designed for wideband applications. The overall single-ended gain of the circuit can be approximated:

$$A_v = v_o/v_{in} \approx \frac{R_{11}}{R_3 + R_4 + r_1} \tag{8.11}$$

where r_1 is the dynamic emitter impedance of T_1, given as

$$r_1 = 1/g_{m1} = V_T/I_1 \tag{8.12}$$

Note that the gain of the circuit can be externally adjusted by changing the amount of series feedback in the first stage. This can be done by externally shorting the resistor taps (A, A') and (B, B') which cause (R_4) and $(R_3 + R_4)$ to drop out of the series feedback, and thus increase the gain.

Figure 8.10 shows the single-ended frequency response for the wideband amplifier stage of Fig. 8.9, for various values of gain, for ±6 V operation of the circuit. Note that the series-shunt cascade circuit offers a significant improvement in the 3 dB bandwidth for the comparable values of gain, over the compound device circuit of Fig. 8.4. The photomicrograph of the monolithic circuit chip is shown in Fig. 8.11.

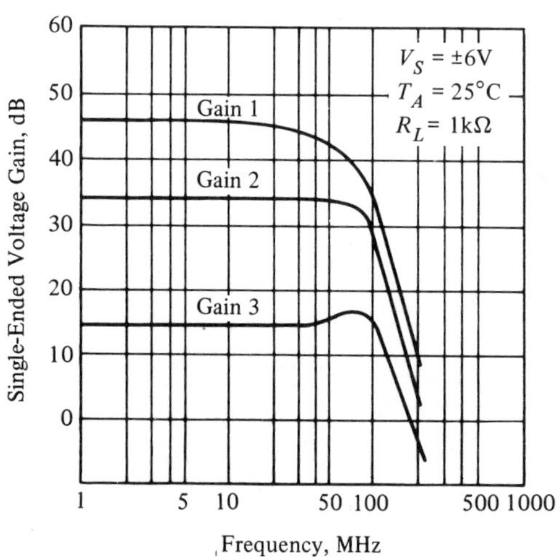

Figure 8.10 Frequency response of the series-shunt cascade amplifier of Figure 8.9: Gain 1: *A-A'* shorted; gain 2: *B-B'* shorted; gain 3: all taps open.

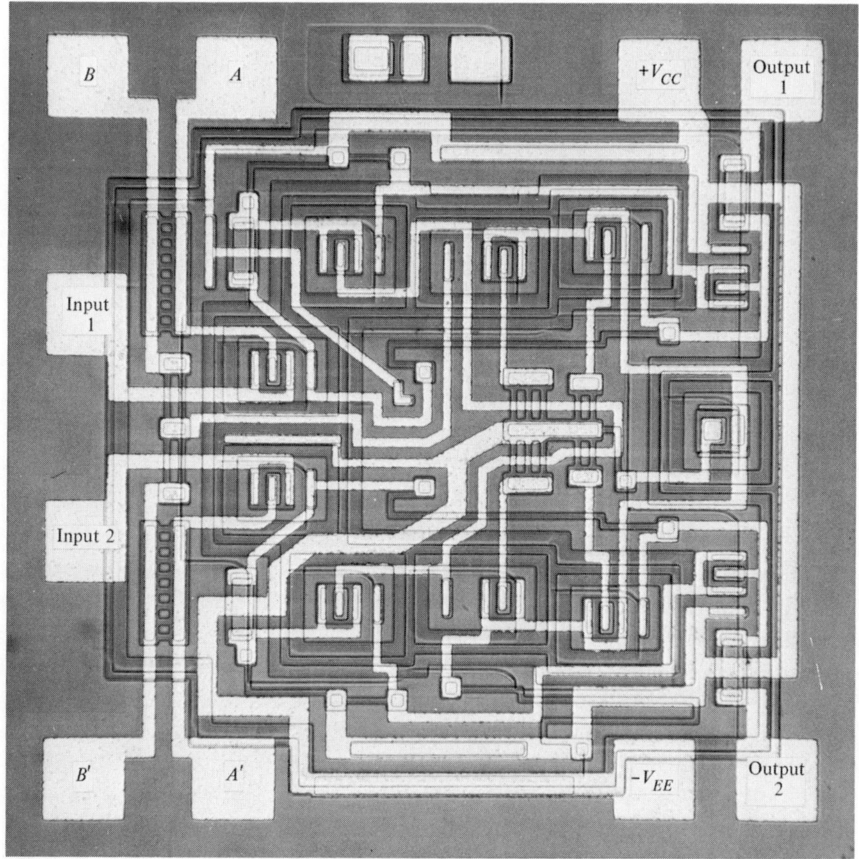

Figure 8.11 Circuit layout for the differential series-shunt cascade. (*Photo: Fairchild*).

8.5 FEEDBACK PAIRS AND TRIPLES

An alternate approach to the design of high frequency feedback amplifiers is the use of overall feedback. In this case one applies feedback around two or more gain stages to obtain the desired frequency capability. As compared with the local-feedback cascades described in the previous section, the overall feedback offers a higher degree of bias stability and desensitivity to individual gain tolerances. However, since the overall feedback approach involves a larger number of active devices within the feed-

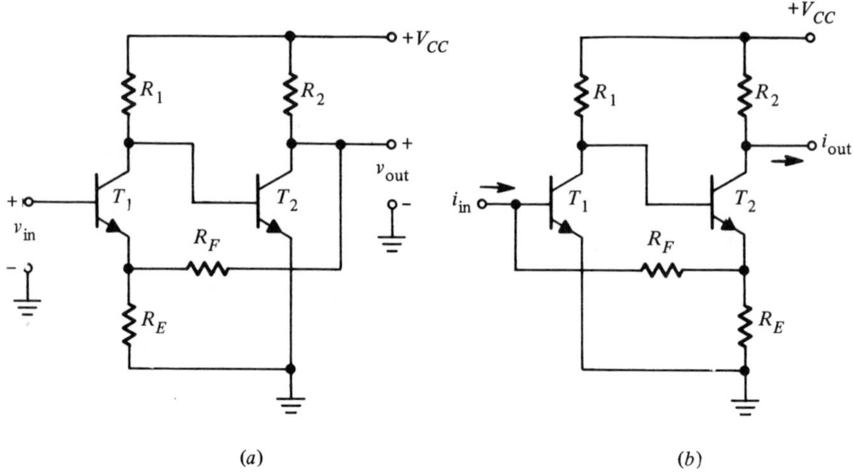

Figure 8.12 Basic feedback pairs: (*a*) series-shunt pair; (*b*) shunt-series pair.

back loop, it requires a more careful consideration of the nondominant poles and excess phase shifts associated with the active devices, to ensure stability. In general, stability considerations limit the total number of gain stages which can be enclosed in a single overall feedback loop. Figures 8.12 and 8.13 show some of the basic feedback configurations suitable for monolithic design. Depending on the number of active devices in the loop, these circuits can be classified as feedback "pairs" and "triples."

The series-shunt pair of Fig. 8.12(*a*) is a convenient circuit configuration for broadband voltage amplification. Because of the series feedback associated with the input stage, the circuit offers a high input impedance. Similarly, the shunt feedback associated with the output results in a low output impedance. Assuming that transistor collector resistors R_1 and R_2 are chosen to provide sufficiently large open-loop gain, the overall closed-loop voltage gain can be approximated as

$$A_v \approx (R_E + R_F)/R_E \tag{8.13}$$

The shunt-series feedback pair of Fig. 8.12(*b*) has a low input and high output impedance; therefore it is particularly useful as a current amplifier. Assuming that the overall open-loop gain of the stage is sufficiently high, and that the circuit is driven from a high impedance source, its overall current gain is determined by the feedback elements:

$$A_I = i_{out}/i_{in} \approx (R_E + R_F)/R_E \tag{8.14}$$

The feedback triples shown in Fig. 8.13 employ three gain stages

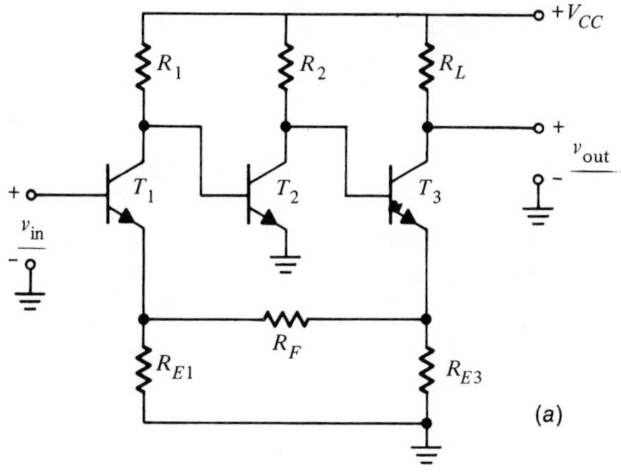

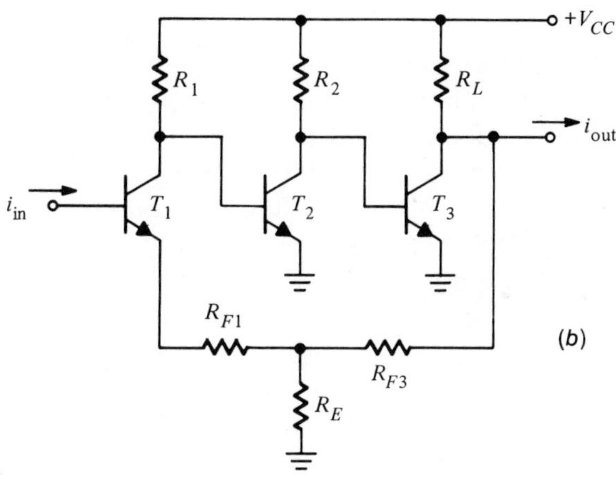

Figure 8.13 Feedback triples: (*a*) series-series triple; (*b*) shunt-shunt triple.

within the feedback loop. Therefore, they offer a higher open-loop gain and bias desensitivity than the feedback pairs. The series-series triple of Fig. 8.13(*a*) has a high input impedance; therefore it operates best as a voltage amplifier, driven with a low source impedance. Its mid-band voltage gain can be approximated as

$$A_v \approx -\frac{R_L(R_{E1} + R_{E3} + R_F)}{R_{E1}R_{E3}} \tag{8.15}$$

The shunt-shunt triple of Fig. 8.13(*b*) is most suited for current amplification. Its mid-band gain can be approximated as

$$A_I \approx -\frac{R_{F1}R_{F3} + R_E(R_{F1} + R_{F3})}{R_E R_L} \tag{8.16}$$

In the design of integrated wideband amplifiers, the series-shunt pair (Fig. 8.12(*a*)) and the series-series triple (Fig. 8.13(*a*)) are the most commonly utilized circuit configurations. One can show that the overall performance of the two circuits are comparable, although the series-shunt pair in general can offer a slightly higher 3 dB bandwidth for the same gain.[11] However, the higher dc loop gain obtainable in the series-series triple results in a higher degree of bias stability. Because of this advantage, the series-series triple is one of the few circuit configurations which can be integrated without using a differential configuration and common-mode feedback for stability.

Figure 8.14 shows a complete circuit schematic for an integrated wideband amplifier circuit using the series-series triple as the gain element.[12]

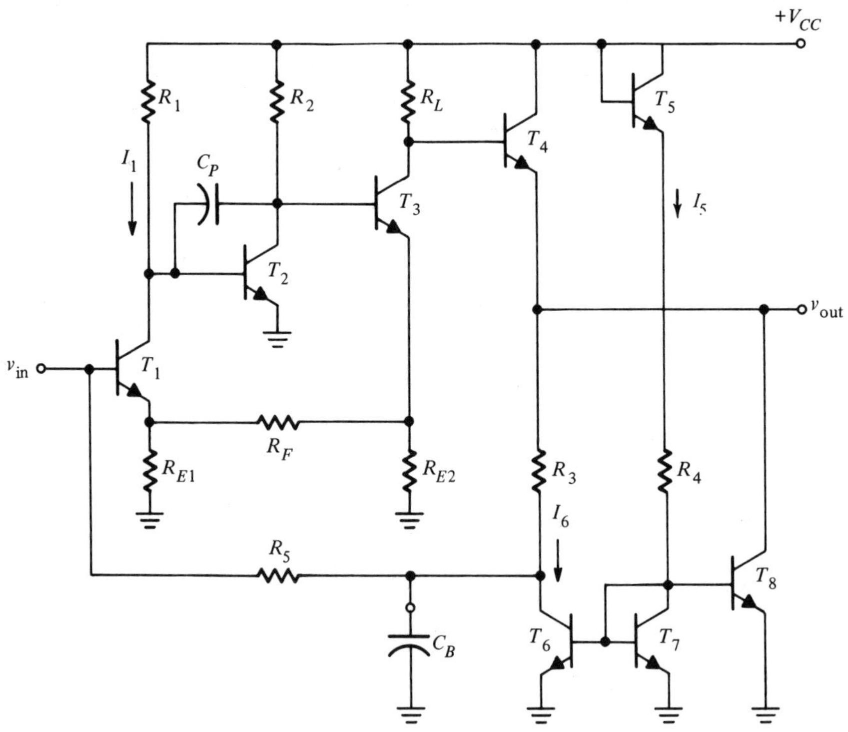

Figure 8.14 A wideband amplifier using the series-series triple (Motorola MC 1553).

Note that the circuit contains a large number of additional components and active devices in addition to the basic series-series triple. The triple is formed by transistors (T_1, T_2, T_3) and the feedback resistors R_{E1}, R_{E2}, and R_F. The emitter follower T_4 is used to provide a low-impedance output for the circuit. C_p is a small ($\approx$2 pF) internal capacitance providing additional broad-banding for the circuit. This particular method of broadbanding, called "pole-splitting," will be described further in the following section. In addition to the basic feedback loop, the circuit of Fig. 8.14 also contains an additional shunt dc feedback path formed by R_3, T_6, and R_5, which provide both a stabilized dc bias to input transistor T_1 and also keep the output dc level at approximately $V_{CC}/2$, relatively independent of component absolute value tolerances and power supply voltages. The external bypass capacitor C_B is used to eliminate ac coupling between the input and the output through this bias feedback path. If the circuit is driven from a low impedance source, C_B may not be necessary, since the ac signal in the bias feedback path would be attenuated by the resistor ratio (R_s/R_5), where R_s is the impedance of the input signal source.

The bias feedback scheme used in the series-series triple of Fig. 8.14 gives a good illustration of the efficient use of the two key advantages of monolithic integrated circuits: availability of a large number of active devices and the close matching and tracking of component values. If one makes the reasonable assumption that all transistor V_{BE} drops are well matched, and that the transistor current gain $\beta \gg 1$, then with reference to Fig. 8.14, the output dc level V_{ODC} can be written as

$$V_{ODC} = I_1 R_{E1} + V_{BE} + I_6 R_3 \tag{8.17}$$

Since T_6 and T_7 are well matched and $\beta \gg 1$, then currents I_5 and I_6 are approximately equal and given as

$$I_5 \approx I_6 = (V_{CC} - 2\,V_{BE})/R_4 \tag{8.14}$$

Similarly, the collector of T_1 is only a V_{BE} above ground; therefore, I_1 can be related to V_{CC} as

$$I_1 = (V_{CC} - V_{BE})/R_1 \tag{8.19}$$

Combining Eqs. (8.18) and (8.19), one can then rewrite V_{ODC} as

$$V_{ODC} \approx V_{CC}(R_{E1}/R_1 + R_3/R_4) + V_{BE}(1 - 2\,R_3/R_4 - R_{E1}/R_1) \tag{8.20}$$

By proper choice of resistor values, the second term of Eq. (8.20) can be made negligibly small, thus making V_{ODC} proportional to V_{CC} and set by the resistor *ratios* alone. In the actual design of the circuit of Fig. 8.13(*a*), the following resistor values are used in the bias network:

$$R_{E1} = R_{E2} = 0.1\ \text{k}\Omega \qquad R_3 = 3\ \text{k}\Omega$$
$$R_1 = 9\ \text{k}\Omega \qquad R_4 = 6\ \text{k}\Omega$$
$$R_2 = 5\ \text{k}\Omega \qquad R_5 = 12\ \text{k}\Omega$$
$$R_L = 0.6\ \text{k}\Omega$$

By substituting these values into Eq. (8.20), one can readily show that the output dc level is almost exactly equal to $V_{CC}/2$, thus ensuring a maximum output voltage swing for the given choice of supply voltage. Once the bias levels within the circuit are established, the closed-loop gain can be determined by the proper choice of the ac feedback resistor R_F, as given by Eq. (8.15). Note that, since both ends of R_F are at approximately the same dc potential, negligible bias current flows through R_F; therefore, dc bias conditions are not affected by the choice of R_F to obtain a predetermined closed-loop voltage gain.

The series-series triple circuit of Fig. 8.14 is designed to operate with a nominal current drain of 11 mA at $V_{CC} = 6$ V. When driven from a 50 Ω source, the small-signal 3 dB bandwidth of the amplifier approximately 45 MHz and 50 MHz, for voltage gains of 100 and 50 respectively.

As a consequence of overall feedback, the effect of device nonlinearities and the temperature dependence of the output dc level and the voltage gain are greatly reduced. Total harmonic distortion at the output is less than 0.2 percent; the gain variation with temperature is less than ±0.25 dB over a −55 to +125°C operating temperature range.

8.6 ROOT-LOCUS TECHNIQUES

In optimizing the frequency response of feedback amplifiers, root-locus methods provide one of the most versatile design techniques. Although root-locus methods have been initially developed for analysis of feedback and servo systems, they are readily adaptable to feedback amplifier design. In the design of wideband feedback amplifiers using root-locus methods, the first step is to break the overall circuit into two segments, $A(s)$ and $F(s)$, which designate the transfer functions of the forward and reverse signal flow paths respectively.[13] This is shown schematically in Fig. 8.15. Normally the forward signal path contains active devices, and therefore its signal transmission properties are nearly unilateral. The feedback path $F(s)$ is generally a passive, bilateral network. To be able to analyze the network with basic root-locus techniques, one normally makes the basic assumptions that the forward signal transmission is almost totally through $A(s)$, and the reverse transmission is almost totally through the feedback network $F(s)$. Then, one can readily express the transfer function $H(s)$ of the feedback network in terms of the transfer characteristics of the sub-

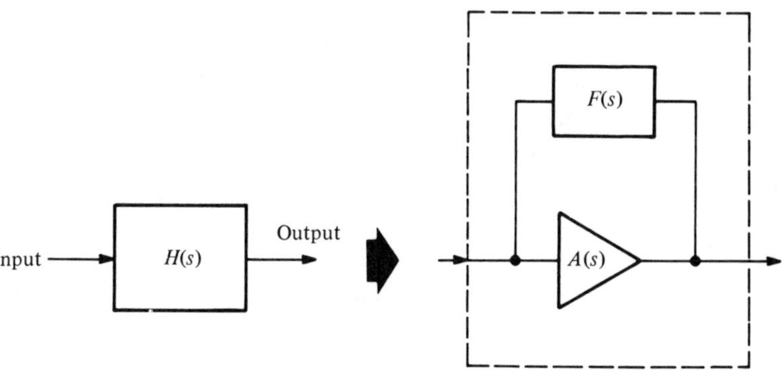

Figure 8.15 Decomposition of a feedback amplifier into forward and reverse transmission sections.

networks $A(s)$ and $F(s)$ as

$$H(s) = \frac{A(s)}{1 + A(s)F(s)} \tag{8.21}$$

where s is the complex frequency variable. The roots of the expression $[1 + A(s)F(s)]$ can be obtained graphically, using root-locus techniques, to determine the poles of the closed-loop transfer function.

Figure 8.16(a) shows the typical root-locus diagram for a three-stage feedback amplifier similar to the series-series triple discussed in the preceding section. The open-loop poles are shown with crosses; and the closed-loop poles are designated as boxes. Note that the negative real axis is shown to be discontinuous so that both the dominant and nondominant time constant or poles of the circuit can be included in the same Figure. The three dominant open-loop poles, p_1, p_2, and p_3 normally are due to the three respective gain stages in the forward path. The nondominant poles p_4, p_5, and p_6 are usually associated with the circuit parasitics and the distributed nature of the circuit elements. The closed-loop poles move along the root-loci defined by the solid line, as a function of the overall loop gain. To avoid peaking in the closed-loop response, the closed-loop poles are in general required to be located within the 45° radials from the origin. Thus, the maximum 3 dB bandwidth is obtained when the closed loop poles q_1, q_2, and q_3 are located as shown in Fig. 8.16(a). One can show that the resulting 3 dB bandwidth is approximately equal to the distance of q_1 from the origin. If the overall open-loop gain is increased, poles q_1 and q_2 would move further along the locus toward the $j\omega$ axis, which would result in undesirable peaking in the frequency response, and possible instability. It is possible to modify the location of the open-loop poles by introducing a small amount of capacitance into the proper nodes

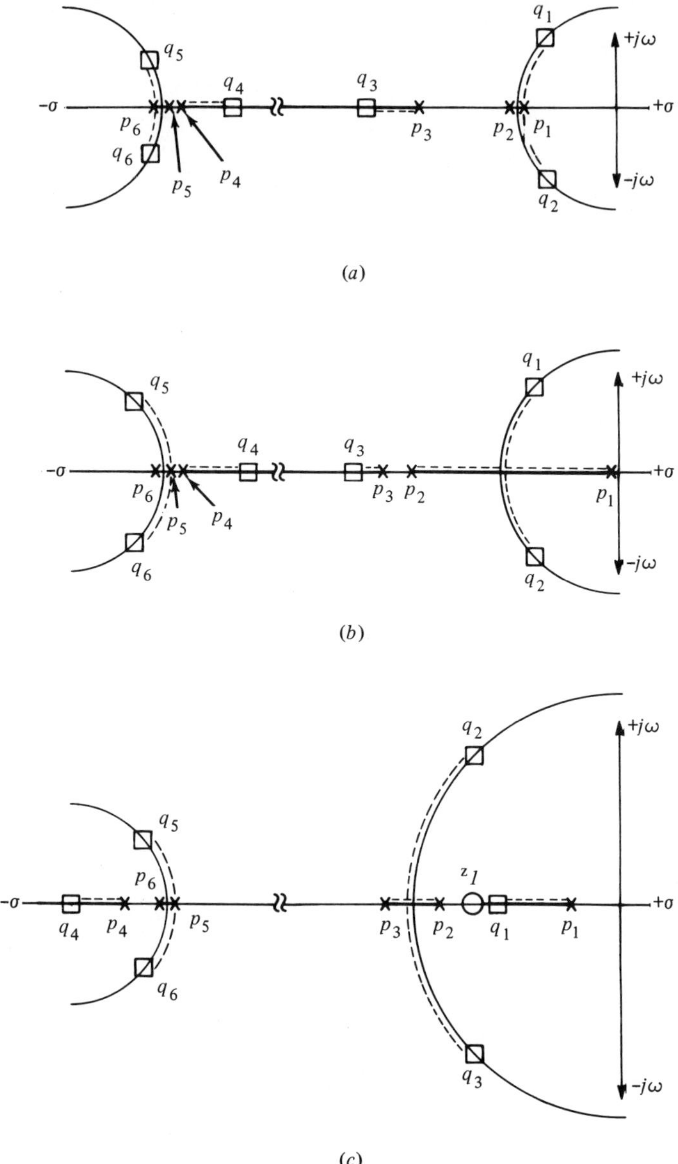

Figure 8.16 Frequency broad-banding using root-locus: (*a*) root-locus for three-stage feedback amplifier; (*b*) root-locus using pole-splitting; (*c*) root-locus using feedback zero.

of the circuit, either in the signal or the feedback path. This in turn can result in a modified set of locations for the open-loop poles and zeros, and lead to a more favorable set of closed-loop poles. Two such techniques which are readily compatible with monolithic circuits are the "pole-splitting" and the "feedback-zero" methods. These are briefly described below:

Pole-Splitting

In general it is not desirable to have open-loop poles p_1 and p_2 very close to each other. It can be shown that the root-loci of q_1 and q_2 can be displaced further away from the origin and the $j\omega$ axis if these two poles can be "split" apart.[12] In a practical design this can be achieved by narrow-banding one of the gain stages in the signal path by introducing some shunt capacitance, at a given circuit node. In the case of the series-series triple of Fig. 8.13, this corresponds to connecting a capacitor C_p across the collector base junction of T_2. This capacitor splits poles p_1 and p_2 by causing p_1 to move closer toward the origin and p_2 away from the origin. Then, when the overall feedback loop is closed, closed-loop poles q_1 and q_2 move on a new root-locus shown in Fig.16(*b*). Since the new locus is now further away from the origin, the 3 dB bandwidth of the closed-loop amplifier is significantly higher. In most cases, the value of the pole-splitting capacitor necessary for broad-banding is quite small, typically of the order of 1 to 10 pF; therefore it can be readily incorporated in the monolithic design without requiring an excessive amount of chip area. In the case of the circuit of Fig. 8.14, the value of C_p is approximately 2 pF, which takes up no more chip area than an ordinary small signal transistor.

Feedback-Zero

It can be shown that by introducing a transmission zero into the feedback path of a broad-band amplifier, the root-locus can be significantly altered.[12,13] This can be achieved by placing a small shunt capacitor across the feedback network. In the case of the feedback triple of Fig. 8.14, this feedback zero can be obtained by putting a small capacitor C_F across feedback resistor R_F, and this results in a real axis transmission zero z_1, given as

$$z_1 = -\frac{1}{R_F C_F} \tag{8.22}$$

If by proper choice of C_F, z_1 is located just to the right of p_2, between p_1 and p_2, then the root-locus pattern can be modified as shown in Fig. 8.16(*c*). Note that the poles of the complex pole-pair now break away from the real axis between p_2 and p_3, instead of p_1 and p_2; therefore closed-loop poles q_2 and q_3 on this portion of the poles are now located further away from the origin. In most design examples, the value of C_F necessary to generate a feedback zero is of the order of several picofarads.[12] Therefore, it can also be readily incorporated into the monolithic design without a significant increase in chip area or complexity.

8.7 CURRENT AMPLIFIERS: THE "GAIN-CELL"

In the design of very wide bandwidth monolithic circuits, current amplifiers have an intrinsic advantage over voltage amplifiers. Since most of the parasitics associated with the monolithic devices are capacitive, the amplifier bandwidths can be improved if most or all of the signal processing on the chip can be done in terms of current rather than voltage amplification, thus eliminating voltage swings across parasitic capacitances. For example, as a current amplifier, the transistor is useful for frequencies up to its cut-off frequency f_T. However, the useful range of most voltage amplifiers is significantly below that because of the excessive phase shifts associated with the voltage transfer across a transistor at high frequencies. Therefore, to utilize the maximum frequency capability of a bipolar transistor, it is necessary to utilize it as a current, rather than voltage, amplifier whenever possible. Even in the cases where voltage amplification is required, the input voltage signal can be converted to a current, amplified through several current gain stages and finally converted back to a voltage swing at the output.

Figure 8.17 shows a simple cross-coupled current gain stage, or a "gain-cell," which is well suited for current amplification over a frequency range comparable to the f_T of the individual transistor.[14] The linear operation of the circuit relies very strongly on the very close matching and tracking of monolithic device characteristics, particularly the base-emitter voltage drops.

The basic gain-cell structure shown in Fig. 8.17 is very closely related to the four-quadrant transconductance multiplier circuit discussed in Chapter 7 (see Figs. 7.4 and 7.5).[15] Assuming that the devices are well matched and the transistor current gain $\beta \gg 1$, the operation of the gain-cell can be explained as follows: The input drive is provided by differential current sources I_{B1} and I_{B2} which partition a total bias current I_B as

$$I_{B1} = xI_B \qquad \text{and} \qquad I_{B2} = (1 - x)I_B \tag{8.23}$$

Then, following the analysis presented in Section 7.4, the collector currents of the inner transistor pair, T_2 and T_3 are also partitioned proportionately as

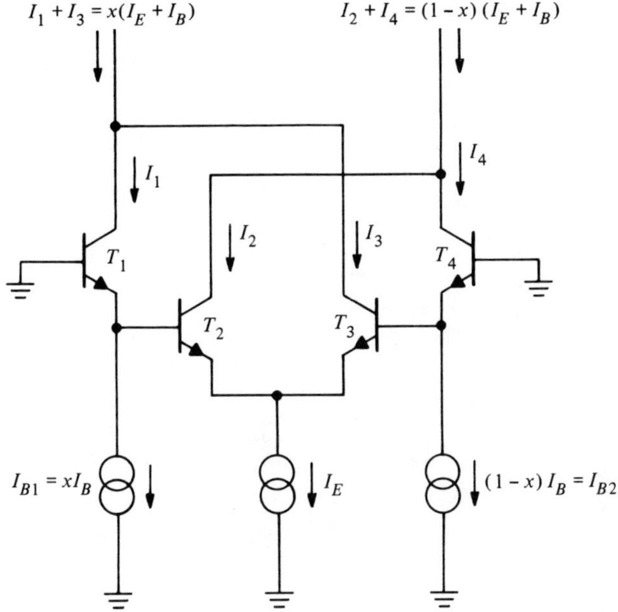

Figure 8.17 Circuit diagram of the "gain-cell" current amplifier stage.

$$I_2/I_3 = I_{B2}/I_{B1} \tag{8.24}$$

or, in other words,

$$I_2 = (1 - x)I_E \qquad \text{and} \qquad I_3 = xI_E \tag{8.25}$$

Since the collectors of T_2 and T_3 are cross-coupled to the collectors of T_1 and T_4, currents (I_1, I_3) and (I_2, I_4) are summed together to produce a set of output currents $(I_1 + I_3)$ and $(I_2 + I_4)$. The net current gain of the stage is then given as

$$A_I = i_{\text{out}}/i_{\text{in}} = \frac{I_1 + I_3}{I_{B1}} = \frac{I_2 + I_4}{I_{B2}} \tag{8.26}$$

As shown by Eq. (8.26), the overall current gain of the gain-cell is determined solely by the ratio of the bias currents associated with the inner and the outer transistor pairs of the cell.

Figure 8.18 shows a circuit schematic for a voltage amplifier stage which uses the current gain-cell as the intermediate amplifier stage. The emitter degeneration resistors provide a linear conversion of the input voltage to a current signal; similarly the matched load resistors R_L convert the output current signal into a differential voltage. The net voltage gain of the stage can be expressed as

$$A_v = (R_L/R_E)(1 + I_E/I_B) \tag{8.27}$$

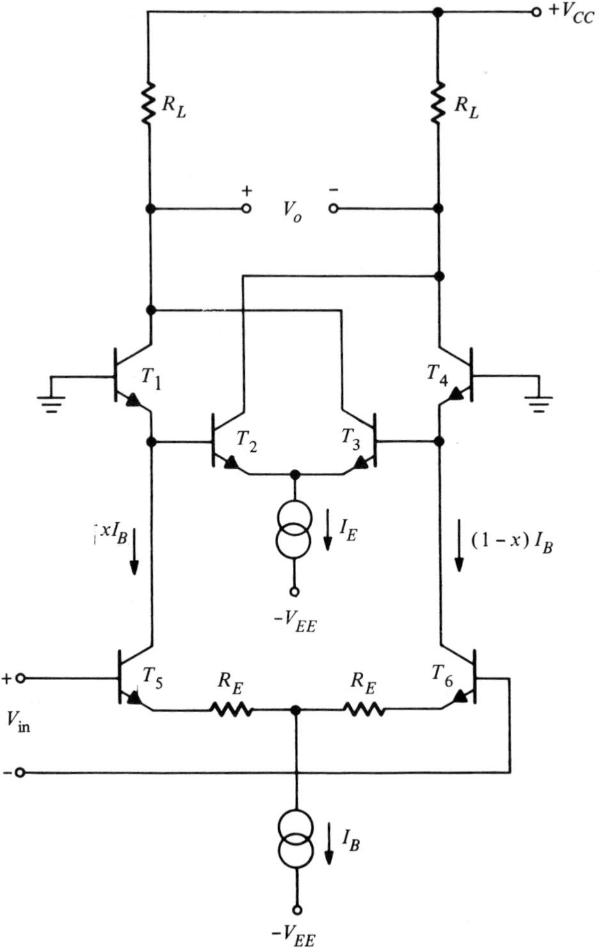

Figure 8.18 A wideband voltage amplifier using current gain-cell configuration.

When higher values of current amplification are desired, a number of identical gain-cells can be directly cascaded, as shown in Fig. 8.19. Note that in this case the total current gain A_I, for a cascade of n stages becomes

$$A_I = 1 + 1/I_B \sum_{j=1}^{n} I_{E_j} \tag{8.28}$$

Since the basic gain-cell of Fig. 8.17 is a current amplifier, the cascade of several gain-cells as shown in Fig. 8.19 also provides an efficient use

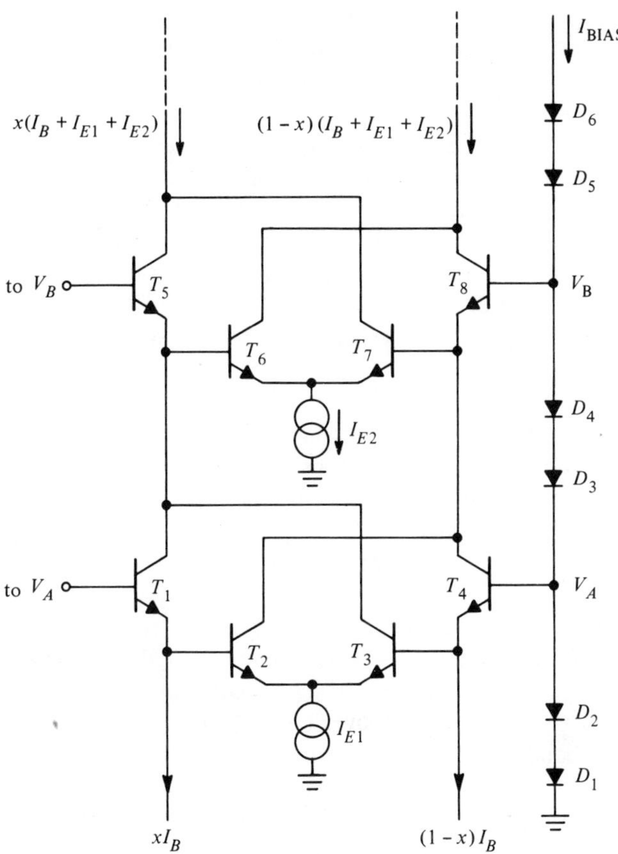

Figure 8.19 Cascaded gain-cells.

of the supply current since most branches of the circuit are connected in series with the power supply and use the same supply current. Since the voltage level shifts necessary to bias the successive gain stages is quite low, a diode string similar to that shown in Fig. 8.19 can be used to provide sufficient bias to all stages of the cascade, without the need for ac decoupling.

It can be shown that the high frequency capability of the gain-cell amplifier is directly proportional to the current gain-bandwidth product f_T of the transistors used.[14] For a given value of f_T, the frequency response of the gain-cell can be approximated by a single dominant pole p_1 where

$$p_1 \approx -\frac{2\pi f_T}{A_I} \tag{8.29}$$

where A_I is the current gain of the cell, given by Eq. 8.26. For a given choice of transistor geometry, f_T is maximum at a given emitter-current level. Therefore, in designing a gain-cell amplifier, care should be taken to scale the device geometrics such that the f_T of each of the devices is nearly optimum at their bias level of operation. This implies that when cascading a number of gain-cells, as shown in Fig. 8.19, the emitter areas for the upper transistor pair (T_5, T_8) should be made larger than those for the lower pairs (T_1, T_4) by an amount depending on the current I_{E1}.

If a number of identical gain-cell stages are cascaded to obtain a desired value of overall current gain, the total 3 dB bandwidth of the cascaded stages is reduced by the bandwidth shrinkage factor γ[16] where

$$\gamma = (\sqrt[n]{2} - 1)^{\frac{1}{2}} \tag{8.30}$$

where n is the number of identical gain stages forming the cascade. Thus, if n identical gain-cells, each with a current gain A_I, were cascaded, the 3 dB bandwidth of the combination would be approximately equal to:

$$f_{3\text{dB}} \approx \frac{f_T}{A_I} \gamma \tag{8.31}$$

A monolithic version of the cascaded gain-cell structure of Fig. 8.19 has been fabricated on a 50 mil × 60 mil monolithic die.[14] The circuit uses a cascade of 3 gain-cells made up of high frequency integrated transistors with f_T of approximately 1200 MHz. The following values of overall current gain and bandwidth and rise-time were reported:[14]

Current gain	Rise-time (nanosec)	3dB Bandwidth (MHz)
10	0.63	500
20	0.81	420
50	2.0	170
100	4.0	92

The gain-cell concept is a good example of one of the many circuit techniques ideally suited to the design of integrated circuits. It again demonstrates that, with some imagination, the circuit designer can overcome the inherent limitations of integrated circuits, such as the lack of inductors and the limited choice of active devices, and end up with a design which relies on the advantages of monolithic structures such as the control of device geometries and close matching and tracking of the components.

8.8 ELECTRONIC GAIN CONTROL

In a number of applications utilizing wideband amplifiers, it is often desirable to control the gain of the amplifier, electronically, without effecting any other performance parameter. This type of electronic gain control is particularly useful in communication circuits such as R-F and I-F amplifiers to improve the signal handling capability or the dynamic range of the amplifier. The electronic gain control capability allows the amplifier gain to be controlled by an automatic gain control (AGC) loop.

In conventional (nonintegrated) solid state amplifiers electronic gain control is usually achieved by shifting the dc operating point of one or more transistor stages[17] or by employing diode attenuators between stages. In monolithic integrated circuits, where ac coupling between successive stages is not available, these conventional gain control methods lead to undesirable dc level shifts within the circuit and are generally not suitable. However, by using the matching properties of the integrated devices, some alternate gain control mechanisms can be devised for monolithic wideband amplifiers.

A class of wideband amplifier circuits which are readily suited to monolithic integration are those derived from the "balanced modulator" circuits described in Chapter 7 (see Sections 7.5 and 7.6). Figure 8.20 shows the circuit diagram of a wideband amplifier stage with electronic gain control capability.[18] This circuit is basically a modified version of the balanced modulator circuit shown in Fig. 7.10. The circuit uses the voltage-

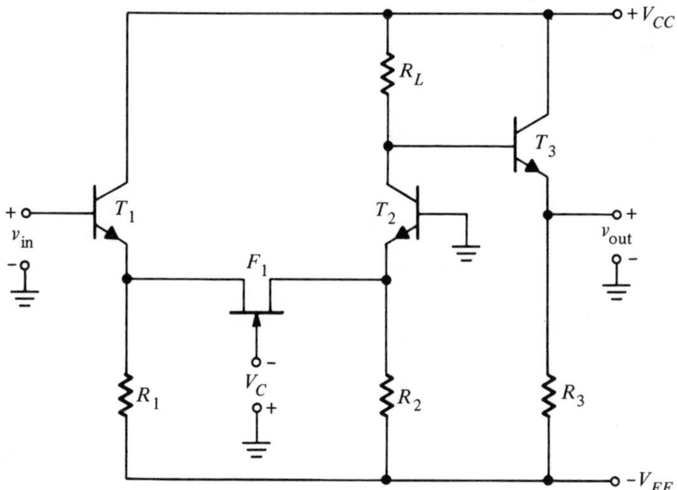

Figure 8.20 A broadband amplifier stage with electronic gain control.[18]

dependent source-drain resistance R_{sd} of a field-effect transistor F_1 as the voltage-controlled coupling element between bipolar stages T_1 and T_2. The n-channel FET can be readily fabricated, simultaneously with the bipolar transistors, using a structure similar to that shown in Fig. 2.22. Since both the source and the drain of F_1 are at very nearly the same dc potential, negligible dc current flows through the channel. Therefore, dc bias conditions within the circuit are not affected by the gain-control voltage V_c applied to the gate of F_1. The ac gain of the circuit is very nearly equal to the ratio of the load resistor R_L to the FET source-drain resistance R_{sd}. R_{sd} is a function of the control voltage V_c. For an FET with a uniformly doped channel (see Fig. 2.22), it can be approximated as

$$R_{sd} = \frac{R_{sdo}}{\sqrt{1 - V_c/V_p}} \tag{8.32}$$

where R_{sdo} is the value of source-drain resistance with zero control bias, and V_p is the FET pinch-off voltage. Thus, the overall voltage gain of the stage can be expressed as

$$A_v \approx (R_L/R_{sdo}) \sqrt{1 - V_c/V_p} \tag{8.33}$$

The circuit configuration of Fig. 8.20 is suitable as a wideband automatic gain control (AGC) block, offering a 3 dB bandwidth of approximately 50 MHz with a voltage gain of 20 dB, and provides an AGC range of −40 dB over this frequency range.

The gain-cell circuit of the preceding section is also readily suitable for electronic gain control. With reference to the voltage amplifier of Fig. 8.18, the overall voltage gain is given by Eq. (8.27) as

$$A_v = (R_L/R_E)(1 + I_E/I_B) \tag{8.34}$$

Thus, by controlling the bias currents I_E or I_B, the voltage gain can be varied electronically.

The doubly balanced modulator circuit of Fig. 7.3 is also suitable as a wideband AGC circuit since the output voltage is determined by the two inputs V_1 and V_2. Assuming that the input signal V_{in}, is applied to the bases of T_5 and T_6, and the second input, V_1, is replaced by a differential control bias V_c, the voltage gain of the stage can be written from Eq. (7.33) as

$$A_v = (R_L/R_E) \tanh (V_c/2 V_T) \tag{8.35}$$

Thus, the voltage gain of the stage is equal to zero at ($V_c = 0$), and approaches (R_L/R_E) as the control voltage is increased such that $|V_c| \geqq 2 V_T$. If a linear relationship is required between the control voltage and the stage gain, then the four-quadrant multiplier circuit of Fig. 7.5 can be used as gain-controlled amplifier stage, with either the X or Y input as the control terminal. The basic problem associated with the doubly balanced modulator circuit of Fig. 7.3 for electronic gain control applications is that the zero gain condition obtained at $V_c = 0$ depends very critically on the matching of the device characteristics. This has been discussed earlier in connection with the "null-suppression" characteristics of the doubly balanced modulator (see Eq. 7.35), and it severely limits the AGC range at high frequencies.

Figure 8.21 shows a modified version of the balanced modulator circuit which overcomes this difficulty.[19] In this case, the circuit is no longer

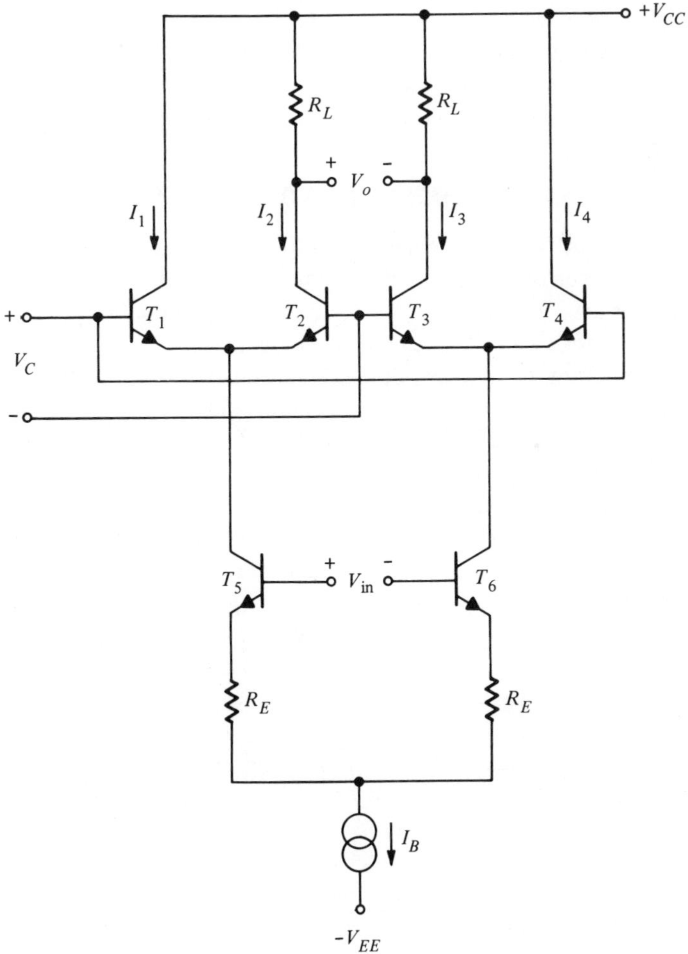

Figure 8.21 A wideband amplifier with large AGC range.[19]

doubly balanced since the collectors of T_1 and T_4 are directly connected to V_{CC}. The gain control voltage V_c applied to differential pairs (T_1, T_2) and (T_3, T_4) controls the partitioning of bias currents I_1 through I_4 between the respective transistors. From the analysis of Chapter 7 (see Eqs. 7.22 and 7.23), the net voltage gain of the stage can be written as

$$A_v = R_L/R_E \left(\frac{1}{1 + e^{V_c/V_T}}\right) \tag{8.36}$$

Note that the dc currents I_2 and I_3 through the load both vary as a function of the gain control voltage. However, as long as the devices are well matched, the variations of I_2 and I_3 would track very closely, and the differential output would not be affected.

In the circuit of Fig. 8.21, the exponential dependence of gain on the control voltage is particularly useful, since it allows one to control the gain over a wide dynamic range with a relatively small control voltage.

8.9 COMPUTER-AIDED DESIGN AND OPTIMIZATION

Design and optimization of wideband amplifiers is a difficult problem. In optimizing the performance of high frequency circuits, one often pushes the active devices to the limits of their frequency capability—where the simple device models of Fig. 8.1 are no longer accurate. Therefore, using manual calculation techniques, it is often difficult to predict the actual circuit performance beyond a first-order approximation.

Even with the basic hybrid-π model of Figure 8.1(*a*) analysis of a multistage feedback amplifier circuit becomes a tedious task. If more complete and sophisticated models[2,3] were utilized, a closed-form analytical description of circuit performance becomes very cumbersome. If one goes beyond the simple analysis and tries to optimize the performance of a given circuit configuration, one is faced with a large number of independent design variables. Under these circumstances, computer-aided analysis and optimization techniques provide a powerful design tool for the engineer.

Over the recent years, computer-aided design has become a subject of much interest. A large number of analysis and optimization programs have been developed.[20-22] Some of these programs have been aimed specifically at the design of monolithic broad-band amplifiers.[21]

Optimization of wideband amplifier performance, even with the aid of a computer, amounts to a tedious problem. The circuit performance is a complex function of a large number of variables such as dc bias levels, device layout and geometry, passive component values and the circuit and package parasitics. Therefore, the first step in any optimization attempt is the definition of a specific performance criterion to be maximized. This can be done by defining a "performance index" or a "figure-of-merit" for circuit operation which is the actual quantity to be optimized, within realizable bounds. The actual optimization process then becomes a numerical search procedure to obtain the proper combination of values of design variables which give a maximum or a minimum value of the performance index. The choice of a maximum or a minimum depend on the actual definitions or the choice of the performance index.

A commonly used performance index for wideband amplifiers is the "least-squared error" criterion for a specified low frequency gain and small signal bandwidth.[21] For a wideband amplifier this performance index E can be specified as an analytical function of the form,

$$E = \sum_{i=1}^{m} W(\omega_i)[|A_V(\omega_i) - A_o|]^2 \tag{8.37}$$

where ω_i are a set of specified frequency points with $i = 1, 2, \ldots, m_1$, and $W(\omega_i)$ are a set of scalar nonnegative weighting parameters. The weighting parameters permit the designer to emphasize response at various regions of the amplifiers frequency band of operation. For example, to ensure adequate realization of a specific dc gain value, $\omega_1 = 0$ may be given a significantly larger weight than other points near the response band edge.

Once the performance index is defined, a computer program can be written to perform the numerical optimization of the performance index. For this purpose, a number of numerical search and optimization algorithms have been developed.These are well covered in the references.[20-22]

REFERENCES

1. J. Millman and C. C. Halkias, "Electronic Devices and Circuits," McGraw-Hill, New York, 1967, pp. 348–350.
2. H. K. Gummel and H. C. Poon, "A Compact Bipolar Transistor Model," Int. Solid-State Ckts. Conf., *Digest Tech. Papers,* **13** (Feb., 1970): 78–79.
3. J. Mar, "A Time-Domain Method of Measuring Transistor Parameters," Int. Solid-State Ckts. Conf., *Digest Tech. Papers,* **13** (Feb., 1970): 80–81.
4. R. D. Thornton, L. L. Searle, D. O. Pederson, R. B. Adler and E. S. Angelo Jr., "Multistage Transistor Circuits," *SEEC,* Wiley, New York, 1965, Vol 5, pp. 172–182.
5. H. R. Camenzind and A. B. Grebene, "An Outline of Design Techniques for Linear Integrated Circuits," *IEEE J. Solid State Ckts.,* **SC–4**(3) (June, 1969): 110–122.

6. B. A. Wooley, "Monolithic Wideband Amplifiers," Memorandum ERL–M243, Electronics Research Laboratory, University of California, Berkeley, California, March, 1968.
7. A. Bilotti and R. S. Pepper, "A Monolithic Limiter and Balanced Discriminator for FM and TV Receivers," *Proc. Natl. Electronics Conf.,* **13** (Oct., 1967): 489–494.
8. "RCA Linear Integrated Circuits," RCA Technical Services IC–42, pp. 235–273; RCA Corporation Solid State Division, Somerville, N.J., 1970.
9. G. W. Haines, "C_c Compensated Transistors," 1966 IEEE Electron Devices Mtg., Washington, D.C., Oct., 1966.
10. J. A. Mataya, G W. Haines and S. B. Marshall, "I-F Amplifier Using C_c Compensated Transistors," *IEEE J. Solid State Ckts.,* **SC–3**(4) (Dec., 1968): 401–407.
11. B. A. Wooley, "Automated Design of DC-Coupled Monolithic Broad-Band Amplifiers," *IEEE J. Solid State Ckts.,* **SC–6**(1) Feb., 1971): 24–34.
12. J. E. Solomon and G. R. Wilson, "A Highly Desensitized Wide-Band Monolithic Amplifier," *IEEE J. Solid State Ckts.,* **SC–1**(1) (Sept., 1966): 19–28.
13. M. S. Ghausi and D. O. Pederson, "A New Design Approach for Feedback Amplifiers," *IRE Trans. Ct. Theory,* **CT–8** (Sept., 1961): 274–284.
14. B. Gilbert, "New Wideband Amplifier Technique," *IEEE J. Solid State Ckts.,* **SC–3**(4) (Dec., 1968): 353–365.
15. B. Gilbert, "A Precise Four-Quadrant Multiplier with Subnanosecond Response," *IEEE J. Solid State Ckts.,* **SC–3**(4) (Dec., 1968): 365–373.
16. E. J. Angelo, Jr., "Electronic Circuits," McGraw-Hill, New York, 1958, pp. 489–510.
17. W. F. Chow and A. P. Stern, "Automatic Gain Control of Transistor Amplifiers," *Proc. IRE,* Sept., 1955, 119–127.
18. A. B. Grebene and R. S. Pepper, "A Wide-Band AGC Block Suitable for Integrated Realization," Int. Solid State Ckts. Conf., *Digest Tech. Papers,* **8** (Feb., 1965): 96–97.
19. W. R. Davis and J. E. Solomon, "A High Performance Monolithic I-F Amplifier Incorporating Eletcronic Gain Control," *IEEE J. Solid State Ckts.,* **SC–3**(4) (Dec., 1968): 408–416.
20. G. C. Temes and D. A. Callahan, "Computer-aided Network Optimization—The State-of-the-Art," *Proc. IEEE,* **55** (Nov., 1970): 1832–1863.

21. B. A. Wooley, "Automated Design of DC-Coupled Monolithtic Broadband Amplifiers," *IEEE J. Solid State Ckts.,* **SC–6** (Feb., 1971): 24–34.

22. R. A. Rohrer, L. W. Nagel and R. Meyer, "CANCER—Computer Analysis of Nonlinear Circuits Excluding Radiation," *Int. Solid State Ckts., Conf., Digest of Tech. Papers,* **14** (Feb., 1971): 124–125.

9

Frequency Selective Circuits

In the design of frequency selective integrated circuits, the lack of integrated inductors is a significant disadvantage. In many applications, this drawback can be overcome by utilizing active filter techniques. In designing active filters one uses a combination of resistors, capacitors, and gain blocks in a feedback loop to obtain the desired frequency selectivity without the need for inductors. Since the basic blocks for an active filter, namely the resistors, capacitors, and amplifiers, are available in integrated form, active filter techniques are in principle readily compatible with integrated circuits. However, in most cases, the performance characteristics of active filters depend very strongly on the absolute values of circuit components and gain parameters. Therefore, frequency selective integrated circuits often require additional process steps, such as the use of thin film or hybrid technology, to obtain the required absolute tolerance control.

The initial development of active filter techniques predates the advent of integrated circuit technology by a good number of years.[1-4] However, most active filter methods have become economically feasible only after the development of monolithic circuit technology which made low cost gain blocks (such as differential and operational amplifiers) available to the filter designer.

The two basic classes of active filters which are most readily compatible with integrated circuits are those which utilize either linear feedback or phase-lock techniques. Since the basic design principles for these two approaches differ significantly, this chapter is subdivided into two parts, treating each class of design approach separately. The first part of the chapter deals with the conventional active-RC synthesis techniques which use linear feedback. Since a large number of different active-RC synthesis methods exist and are well documented in the literature,[5-9] this part of the chapter is intended more as a state-of-the-art survey of various synthesis techniques, and not an in-depth study of any one particular design example. Whenever possible, the relative merits of each of the approaches will be emphasized with respect to (1) the sensitivity to component tolerances; (2) ease of integrated fabrication; (3) frequency range of operation and selectivity characteristics achieveable in integrated design.

The second part of the chapter covers the application of phase-lock or frequency feedback techniques in the design of frequency selective integrated circuits, with particular emphasis on monolithic integration.

PART I Linear Feedback Techniques

9.1 SENSITIVITY CONSIDERATIONS:

Linear active filter design and synthesis techniques are conceptually well suited to integration. However, from a practical point, the component tolerance and gain sensitivity requirements often impose significant limitations in their application to integrated circuits. The tight component absolute-value tolerance requirement for linear active filters stem from one fundamental fact: The performance characteristics are a very strong function of the system's natural frequencies or "poles," which are determined by the resistor-capacitor (RC) products and the overall loop gain in the feedback circuit. The absolute value of the loop gain can be desensitized at low frequencies by using local fedback around gain stages. However, the absolute value control of a product of two dissimilar circuit elements such as a resistor and a capacitor requires a tight control of the absolute value of each element, and does not benefit from the matching and tracking between "similar" monolithic components. In many cases, this drawback restricts the linear active filters to hybrid rather than monolithic integrated circuits where additional trimming of component values are possible after circuit fabrication. Unfortunately, this also sacrifices some of the inherent batch-processing advantages of integrated circuits.

The required component tolerance or gain control for a given active filter can be directly related to the "sensitivity parameters" of the circuit. Sensitivity is defined as the fractional or percentage change in the performance of the circuit for a fractional change of any one of the independent variables in the network. The sensitivity, S, of a given performance parameter, such as the selectivity, Q, of the active filter with respect to a fractional change in a circuit parameter x is defined as:

$$S_x{}^Q = \frac{\partial Q/Q}{\partial x/x} = (x/Q)\,\frac{\partial Q}{\partial x} \tag{9.1}$$

where x can be either a passive component or a gain parameter.*

The two basic parameters which describe the performance of an active filter are its selectivity Q, and center frequency ω_0. Therefore the Q-sensitivity, S^Q, and the center-frequency sensitivity, S^{ω_0}, are normally the most commonly used sensitivity terms associated with linear active filter. In certain cases, root-sensitivity, which describes a change in the closed-loop poles of the circuit for a fractional change in the open-loop poles, can also be used as a design or optimization criterion.[10]

For a given choice of active filter configuration, the sensitivity parameters allow the designer to express the required component and gain tolerances in terms of filter performance requirements. For this reason, the sensitivity parameters provide an effective means of comparing various filter synthesis techniques, from a practical and economical realizability point of view (for example, see Chapter 7 of Ref. 6).

9.2 CLASSIFICATION OF FEEDBACK METHODS

In designing inductorless filters using linear feedback techniques, one tries to synthesize a predetermined filter response described by the generalized second-order transfer function $T(s)$ as:

$$T(s) = \frac{N(s)}{D(s)} = K\,\frac{(s^2 + a_1 s + a_0)}{s^2 + b_1{}^s + b_0} \tag{9.2}$$

where K is a constant multiplier, and s is the complex frequency variable. In practical filters, the two parameters of importance are the selectivity Q and the center frequency ω_0 of the system. These two parameters can be directly related to the coefficients of the denominator, $D(s)$ in the system transfer function as

* See Chapter 5 of Ref. 5 for detailed description of basic theorems on sensitivity.

$$Q = \frac{\sqrt{b_0}}{b_1} \tag{9.3}$$

and

$$\omega_0 = \sqrt{b_0} \tag{9.4}$$

Many design techniques are available for synthesizing the quadratic transfer function of (9.2) for a given set of coefficients.[11,12] Depending on the manner in which the feedback is applied to generate a pair of complex poles, these numerous approaches can be combined into any one of the following four classes.[13]

1. Single-loop feedback methods
2. Multiloop feedback methods
3. Simulated inductance approach
4. Positive-zero methods

Single-loop feedback circuits, in general, require a lesser degree of design complexity than other classes of active-RC design. They are also well suited to analysis using root-locus techniques. Depending on the polarity of the feedback used, they can be classified further as positive or negative feedback types. Still a third class of single-loop feedback circuits are those which use a passive band-stop network, such as a "bridge-tee" for a "twin-tee" circuit, in feedback around an inverting gain block.

Multiloop feedback techniques offer an added degree of design freedom at the expense of circuit complexity. The use of multiple feedback paths allows one to improve the overall sensitivity of the system with respect to gain and component tolerances. A well known example of this design approach is the so-called "state-variable synthesis" method, which will be discussed later in this chapter.

The simulated inductance approach uses linear feedback techniques to simulate electronically the terminal properties of an inductance. Once one has a means of simulating an inductance, then the actual transfer function $T(s)$ can be readily synthesized using classical, passive filter design techniques. A well known example of this method is the capacitively loaded gyrator approach, which will be covered in a later section of this chapter. In terms of its root-locus characteristics, the simulated-inductance technique is very closely related to single-loop negative feedback circuits.

The positive-zero type frequency selective amplifiers utilize a nonminimum phase* transfer function to obtain the desired selectivity. This class of circuits are derived from the positive-immittance-inverter (PII)

* A nonminimum phase transfer function is one which contains right-hand plane zeros in the complex frequency domain.

configuration.[14] The presence of a right-half plane zero in the system transfer function enables one to obtain a bandpass response using only unity-gain voltage amplifiers. The positive zero approach utilizes two separate feedback paths; however in terms of its root-locus characteristics, it is closely related to single-loop positive feedback circuits.

9.3 NEGATIVE FEEDBACK TECHNIQUES

To synthesize the generalized transfer function of Eq. 9.2, using single-loop negative feedback, one starts with an open-loop system function $H(s)$ having two real poles:

$$H(s) = \frac{\alpha}{(s + p_1)(s + p_2)} \tag{9.5}$$

where α is a gain constant, and $(-p_1)$ and $(-p_2)$ are the open-loop pole locations. If a negative feedback is applied around this system, then the denominator $D(s)$ of Eq. 9.2 can be related to the open-loop poles as

$$D(s) = (s + p_1)(s + p_2) + K \tag{9.6}$$

where K is the overall loop gain. The location of the closed-loop poles, which are the zeros of $D(s)$ can be readily obtained from (9.6) by root-locus techniques. Figure 9.1 shows the typical root-loci for $D(s)$ as a func-

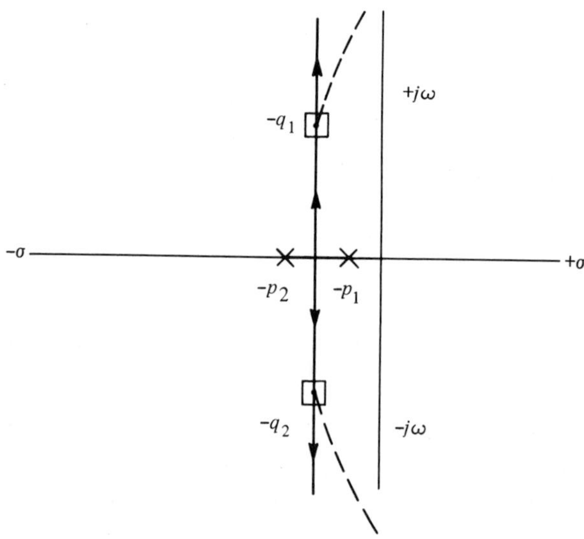

Figure 9.1 Root-locus for a typical negative feedback active filter.

tion of the increasing loop gain K. Note that, for sufficiently high values of gain, a complex pair of closed-loop poles, q_1 and q_2, are formed. From Eqs. (9.3) and (9.6), the Q of the resulting pole pair can be written as

$$Q = \frac{\sqrt{p_1 p_2 + K}}{p_1 + p_2} \tag{9.7}$$

The Q sensitivity of this type of a circuit is relatively low:

$$S_K{}^Q = \frac{K}{2(p_1 p_2 + K)} \leq 1/2 \tag{9.8}$$

Equation (9.8) implies that a 1 percent change in the loop gain appears only as a 0.5 percent change in Q.

The basic drawback with the negative feedback approach is that, to obtain a high Q resonant circuit, a large value of loop gain K is required. Solving Eq. 9.7 for K one can show that the required loop gain goes up proportional to Q^2, i.e.,

$$K = \frac{Q^2(p_1 + p_2)^2}{p_1 p_2} \tag{9.9}$$

Figure 9.2 shows a simplified circuit connection for a negative feedback band-pass amplifier. The unity gain buffer amplifier shown in the Figure is not essential to the circuit but is used to simulate a convenient summing point at the input, for closing the feedback loop. Assuming that the input impedance and the output admittance of the amplifiers are sufficiently high, the open-loop poles, p_1 and p_2, are determined by R_1, R_2, C_1, C_2; and q_1 and q_2 are then set by the inverting amplifier gain K, in accordance with the root-locus of Fig. 9.1.

The high gain requirement of negative feedback techniques sets a severe limitation on their high frequency capability. For example, to get a Q of 10 from the network of Fig. 9.2, with $R_1C_1 = R_2C_2$, one needs an inverting voltage gain of approximately 900 from the output amplifier. Thus, the bandwidth limitations and excess phase shifts associated with

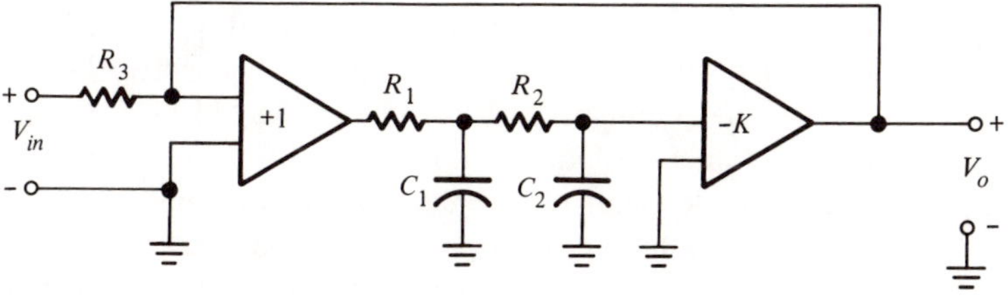

Figure 9.2 A simple active-RC circuit using negative feedback.

the high gain amplifiers normally limit the usefulness of negative feedback approaches to a frequency range below 100 kHz for high Q ($Q > 10$) applications.

At high frequencies or high values of the loop gain, the excess phase lag introduced by the gain stage causes the root-locus of Fig. 9.1 to bend toward the right half plane, as shown by the dotted line, and this can cause instability.

9.4 BAND-STOP NETWORKS IN FEEDBACK

In addition to the basic negative feedback approach described in the previous section, it is also possible to synthesize a band-pass response by using a notch-filter or a band-stop network in feedback around a high gain inverting amplifier. A block diagram of such a circuit configuration is shown in Fig. 9.3(*a*) where the feedback network $N(s)$ exhibits a band-stop response, in the vicinity of the desired filter center frequency ω_0. A basic property of the band-stop networks is to produce a set of complex-conjugate zeros in the complex frequency plane. According to one of the fundamental rules of root-locus, the loci of the closed-loop poles in a feedback network terminate on the zeros of the open-loop transfer function.[15] For a pair of complex zeros in the left-half plane, this results in a root-locus similar to that shown in Fig. 9.3(*b*). In this case the open-loop poles p_1 and p_2, as well as the complex zeros z_1 and z_2, are determined by the voltage transfer function of the band-stop network $N(s)$. The solid line in Fig. 9.3(*b*) shows the loci of the closed loop poles q_1 and q_2 as a function of the increasing loop gain K.

Note that as K is increased indefinitely, the closed-loop poles move toward the open-loop zeros, such that

$$\lim_{K \to \infty} q_i = z_i \tag{9.10}$$

Thus, the complex zeros of the feedback network effectively "anchor" the closed poles, and if K is chosen to be sufficiently large, q_1 and q_2 are set solely by the passive components forming the bandstop network.

Figure 9.4 shows some of the basic notch or band-stop networks which can be used for active filter synthesis applications. The networks of Fig. 9.4(*a*) are the two dual configurations of the basic "bridge-tee" circuit which, for the proper choice of the component values, can produce a set of complex zeros in the left-half plane, along with two poles on the negative real axis. To obtain a high-Q bandpass circuit, the zeros of $N(s)$ must be located very close to the $j\omega$ axis. In the case of bridge-tee networks, complex zeros very near the jw axis are diffcult to obtain since they require

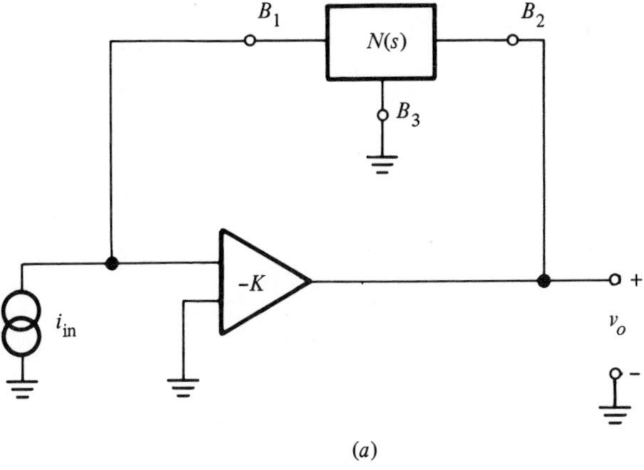

(a)

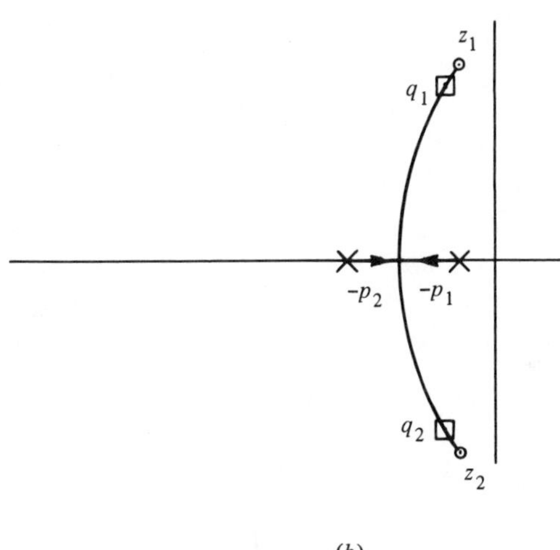

(b)

Figure 9.3 Negative feedback bandpass amplifiers using band-stop network in feedback: (a) system block diagram; (b) root-locus characteristics.

a large spread of component values, i.e., either very large or very small ratios of (C_1/C_2) and (R_1/R_2).

The circuit of Fig. 9.4(*b*) is a distributed equivalent of the lumped bridge-tee networks. It is comprised of a resistor R, with a capacitor C uniformly distributed along it. R_0 is an ordinary lumped resistor. It can be shown[16] that this network exhibits a series of periodic $j\omega$ axis zeros, for

$$R = R_0\, 3\, \frac{2\sqrt{2\pi}}{4} \sinh \frac{3\pi}{4} = 17.786\, R_0 \tag{9.11}$$

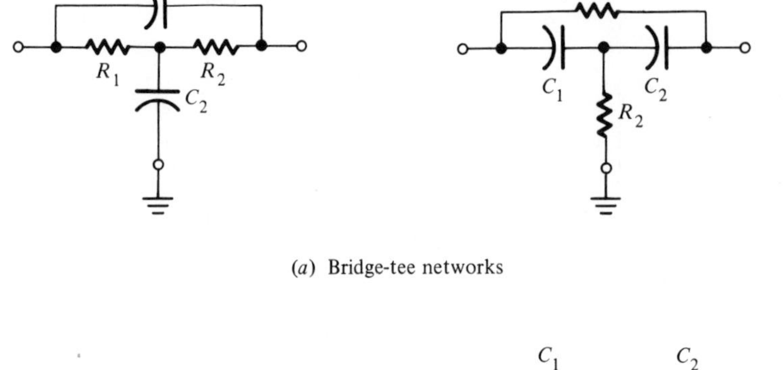

(*a*) Bridge-tee networks

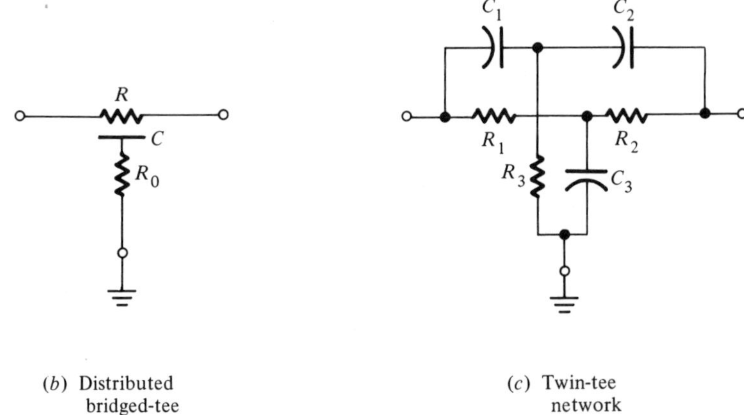

(*b*) Distributed bridged-tee

(*c*) Twin-tee network

Figure 9.4 Band-stop networks suitable for integrated active filter design.

with the first set of zeros located at

$$s = \pm j \frac{9\pi^2}{8} \left(\frac{1}{RC}\right) \tag{9.12}$$

The circuit of Fig. 9.4(*c*), known as the "twin-tee," can be used to produce a pair of complex zeros anywhere in the complex *s*-plane. However, compared with the bridge-tee, it requires a larger number of circuit components. A good description of design procedures and sensitivity considerations for twin-tee networks is given in Ref. 17.

Compared with the basic negative feedback approach of Section 9.3, band-stop networks offer a higher degree of system stability, since the closed-loop poles are no longer likely to move into the right-half plane due to excess phase effects (see dotted line of Fig. 9.1). Therefore, they are better suited to the design of high-Q ($Q > 20$) band-pass filters. However, since the band-stop feedback techniques also require very high amplifier gains, their applications in active filter design are restricted to low frequencies (typically < 100 kHz).

For accurate filter synthesis, band-stop feedback techniques also require a very tight control of the location of the open-loop zeros. The location of z_1 and z_2 in the immediate vicinity of the $j\omega$ axis is a very sensitive function of the absolute values of R's and C's forming the band-stop network. For example, to obtain a Q of 20 with the circuit configuration of Fig. 9.3, using a twin-tee circuit requires the use of three resistors and three capacitors each with an absolute value tolerance of better than ± 0.4 percent. Thus, such a filter approach is in general not compatible with monolithic circuits, but requires extensive use of thin-film or hybrid (thick-film) technology.

9.5 POSITIVE FEEDBACK APPROACH

Positive feedback active filters can be obtained by applying RC feedback around a noninverting (positive gain) amplifier. Many well known network configurations such as the Negative Immitance Converter (NIC) and the Wien-bridge circuits fall into this category. The basic advantage of positive feedback circuits is their low open-loop gain requirements. However, this advantage is often offset by the potential instability and very high gain sensitivity associated with positive feedback.

In general, the positive feedback approach corresponds to a decomposition of the denominator $D(s)$ of the closed-loop response (see Eq. 9.2) into a form,

$$D(s) = (s + p_1)(s + p_2) - Ks \tag{9.13}$$

where $K > 0$ is the active gain parameter.[13] The Q of the closed-loop system can be expressed as

$$Q = \frac{\sqrt{p_1 p_2}}{p_1 + p_2 - K} \tag{9.14}$$

Because of the difference term in the denominator of Eq. (9.14), positive feedback circuits inherently exhibit a very high degree of Q-sensitivity:

$$S_K{}^Q = \frac{K}{p_1 + p_2 - K} \geqq 2\,Q - 1 \tag{9.15}$$

Equation (9.15) implies that in a resonant circuit with a nominal Q of 25, a ± 1 percent change of the system loop gain can cause Q to vary by as much as ± 50 percent.

Figure 9.5 shows the typical root-locus characteristics associated with most positive-feedback type resonant circuits. The closed-loop poles start

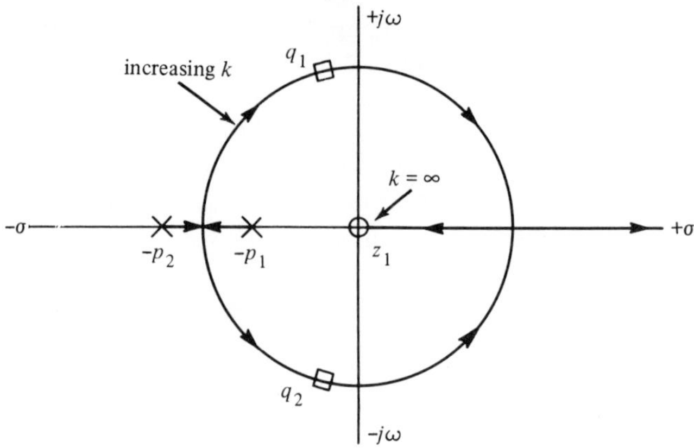

Figure 9.5 Typical root-locus characteristics for positive feedback active filters.

from the open-loop poles p_1 and p_2, become complex, and move in a circle about the origin, as a function of increasing loop gain K. Note that for $K \geqq (p_1 + p_2)$, the closed-loop poles move into the right-half plane, and the system becomes unstable.

Figure 9.6 shows the block diagram of a basic Wien-bridge type positive feedback circuit. The positive gain stage is assumed to be an ideal voltage amplifier with a voltage gain K. The voltage transfer characteristics of the circuit can be expressed as

$$\frac{V_2(s)}{V_1(s)} = \frac{K_0(R_2C_2s + 1)}{(R_1R_2C_1C_2)s^2 + (R_1C_1 + R_2C_2 + (1 - K_0)R_1C_2\ s + 1} \tag{9.16}$$

The amplifier gain K_0 can be related to the gain constant K of (9.13) as

$$K_0 = K(R_2C_1) \tag{9.17}$$

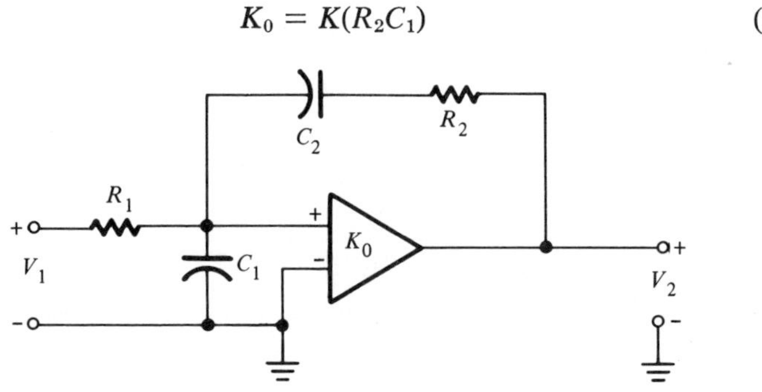

Figure 9.6 Basic circuit configuration for a Wien-bridge type positive feedback band-pass amplifier.

Considering a simplified design example with $R_1 = R_2$ and $C_1 = C_2$ in Fig. 9.6, one can show that the amplifier gain necessary for a given value of Q is

$$K_0 = 3 - \frac{1}{Q} \tag{9.18}$$

The Q-sensitivities of the circuit can be readily derived from Eq. (9.16) as

$$\begin{aligned} S_{K_0}{}^Q &= 3Q - 1 \\ S_{R_1}{}^Q &= S_{R_2}{}^Q = Q + \tfrac{1}{2} \\ S_{C_1}{}^Q &= S_{C_2}{}^Q = 2Q + \tfrac{1}{2} \end{aligned} \tag{9.19}$$

Even though Eq. (9.18) and (9.19) correspond to a particular design example, they illustrate some of the basic properties of positive feedback band-pass circuits. The low value of the required amplifier gain ($K_0 < 3$) makes these types of circuits advantageous for high frequency applications, since the low values of gain can be obtained over a broad range of frequencies. However, the high values associated with the sensitivity parameters of Eq. (9.19) require a very tight control of component values and gain tolerances. For example, with the circuit configuration of Fig. 9.6, to maintain a selectivity specification of $Q = 10 \pm 10$ percent over a 100°C temperature range would require the absolute values R's and C's to be accurate to within ±0.2 percent, and have a temperature coefficient of ≦50 ppm/°C.

9.6 POSITIVE-ZERO AMPLIFIER

The positive-zero frequency selective amplifier is derived from the positive-immitance-inverter (PII) configuration. It utilizes capacitive local feedback around a transconductance amplifier to generate a right-half plane zero in the system transfer characteristics. The presence of a positive zero allows one to synthesize the desired frequency selective response using only unity-gain amplifiers.

Figure 9.7 shows a simplified block diagram of the positive-zero active filter using a differential transconductance amplifier g_m and two unity gain noninverting voltage amplifiers. The differential transconductance block g_m is a voltage-controlled current amplifier shown in Figure 9.8(*a*). In actual design the g_m block can be readily designed as the differential gain stage shown in Fig. 9.8(*b*), where for $R_E \gg V_T/I_E$

$$g_m = 1/R_E \tag{9.20}$$

The open-loop transfer function $H(s)$ for the positive-zero amplifier of

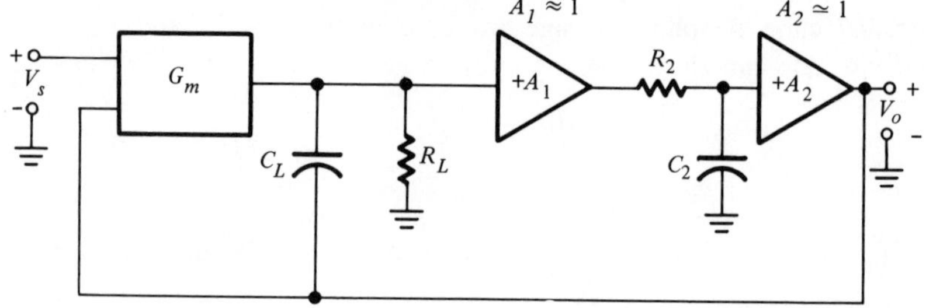

Figure 9.7 Positive-zero amplifier using a differential transconductance block.

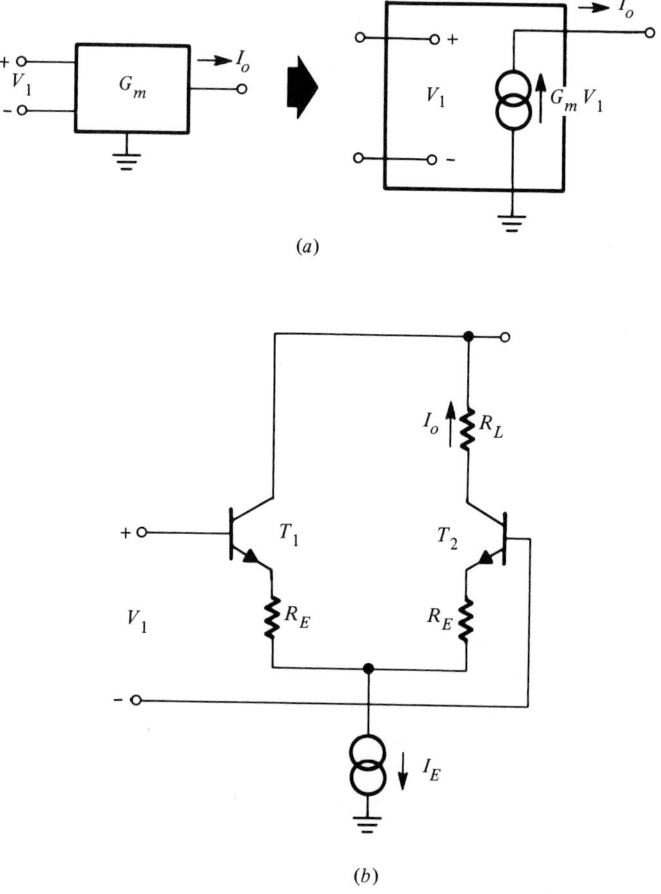

Figure 9.8 The differential transconductance block (*a*), and its circuit implementation.

Fig. 9.7 can be written as

$$H(s) = \frac{-A_1 A_2 g_m R_L\ (1 - sC_L/g_m)}{(1 + sR_2C_2)(1 + sR_LC_L)} \tag{9.21}$$

Note that the positive zero is formed by the feedback capacitance C_L applied around the g_m block. The root-locus of the closed-loop response is shown in Fig. 9.9 as a function of increasing loop gain. Note that these root-locus characteristics are quite similar to those of the positive feedback circuits discussed in Section 9.5, except the root-locus, which forms a circle around the open-loop zero z_i, is now displaced to the right with respect to the origin, since z_1 is now located on the positive real axis instead of the origin. In Fig. 9.9, the solid section of the loci corresponds to the choice of amplifier gains $A_1A_2 \leq 1$. The closed loop poles q_1 and q_2 can be expressed in terms of the circuit parameters as[14]

$$q_1,q_2 = -\tfrac{1}{2}\left[\frac{1 - A_1A_2}{R_2C_2} + \frac{1}{R_LC_L}\right] \pm j\left[\frac{1 + A_1A_2g_mR_L}{R_LC_LR_2C_2}\right]^{\frac{1}{2}} \tag{9.22}$$

As indicated by the loci of Fig. 9.9, or by Eq. (9.22), the positive-zero circuit is unconditionally stable as long as the amplifier gains do not exceed unity. Voltage gains very close to unity can be very tightly controlled and accurately synthesized over a broad frequency range. For example, in many cases a simple emitter-follower stage can be used to simulate the

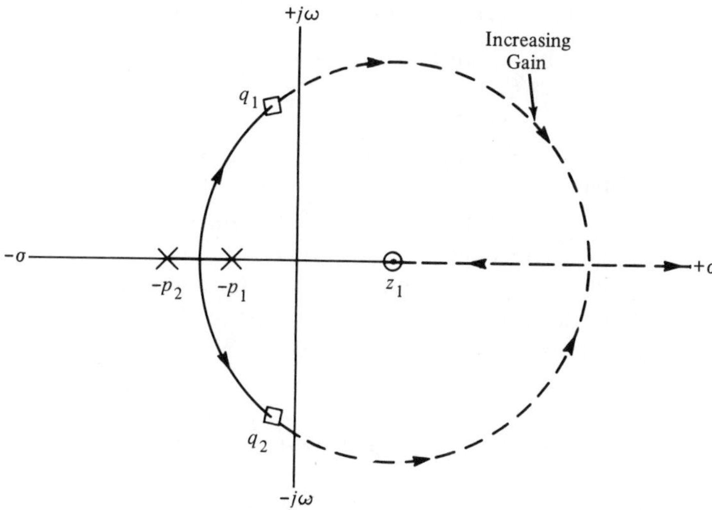

Figure 9.9 Root-locus of positive-zero frequency selective amplifier as a function of amplifier gains (A_1A_2). The solid section of the loci corresponds to $A_1A_2 \leq 1.0$.

unity-gain amplifier stages. Therefore, the positive-zero approach offers a higher frequency capability than most other linear active filters.

The root-locus characteristics of the positive-zero amplifier appear similar to that of the positive feedback Wien-bridge type circuits. However, there is one key difference: In the positive feedback loci of Fig. 9.5, there is a zero at the origin, therefore the loop gain at dc is zero. In the case of the positive-zero circuit, there is a large negative loop gain at dc (see Eq. 9.21) where

$$H(0) = -A_1 A_2 g_m R_L \tag{9.23}$$

which leads to good overall bias stability.

For $A_1 = A_2 = 1$, and $g_m R_L \gg 1$, the selectivity and the center frequency of the positive-zero circuit can be expressed as

$$Q \approx \sqrt{\frac{g_m R_L{}^2 C_L}{R_2 C_2}} \tag{9.24}$$

and

$$\omega_0 = \frac{g_m}{R_2 C_L C_2} \tag{9.25}$$

From Eqs. (9.24) and (9.25), one can readily show that the Q and ω_0 sensitivities with respect to the passive components is quite low:

$$S_{g_m}{}^Q = S_{C_L}{}^Q = (\tfrac{1}{2})\, S_{R_L}{}^Q = -S_{R2}{}^Q = -S_{C2}{}^Q = 1 \tag{9.26}$$

and

$$S_{R_2}{}^{\omega_0} = S_{C_L}{}^{\omega_0} = S_{C_2}{}^{\omega_0} = -S_{g_m}{}^{\omega_0} = -1 \tag{9.27}$$

However, the Q sensitivity with respect to A_1 and A_2 is fairly high:

$$S_A{}^Q \approx 1 + 2\,Q \tag{9.28}$$

which requires a tight control of amplifier gains in the vicinity of unity.

The maximum value of Q obtainable from the basic positive zero circuit (see Eq. 9.24) is somewhat limited by practical design considerations. High values of Q require a large collector load resistance R_L (see Fig. 9.8) for the g_m stage. However R_L cannot be indefinitely increased from the supply voltage and dc bias considerations. The capacitance ratio (C_L/C_2) as well as the transconductance g_m must also be maximized for increased selectivity. However the input capacitance of A_2 and the linearity considerations in the transconductance block set upper bounds for each of these ratios in practical design, and limit the practical values of Q to about 30.

By adding an extra RC phase-shift network, the available Q of the circuit can be greatly increased without significantly altering the basic properties of the circuit. A detailed description of this design technique is given in Ref. 14. The reported results indicate that stable Q values in excess of 50 have been obtained for frequencies up to 2 MHz, using monolithic gain blocks and tantalum thin-film resistors.

9.7 INDUCTANCE SIMULATION—THE GYRATOR

The inductor is the only key circuit component missing from the design tools available to an integrated circuit designer. This drawback could be mostly overcome if one were to simulate the terminal properties of an inductor using only resistors, capacitors, and active devices. If one can economically simulate an inductance, then the well developed and tabulated classical filter synthesis techniques can be readily adapted to integrated circuits. This is the basic idea which has led to the development of the gyrator as a fundamental filter building block. The gyrator can be described as a two-port network with the basic current-voltage relations between the ports described by the following Y matrix:

$$Y = \begin{bmatrix} 0 & g_1 \\ -g_2 & 0 \end{bmatrix} \tag{9.29}$$

where g_1 and g_2 are known as forward and reverse "gyration admittances." In a practical gyrator circuit, finite input and output admittances are also present; thus, the basic admittance matrix of a practical or nonideal gyrator can be expressed as

$$Y = \begin{bmatrix} G_A & g_1 \\ -g_2 & G_B \end{bmatrix} \tag{9.30}$$

where G_A and G_B are the small but finite admittances of the input and the output parts of the gyrator block. The gyrator circuit can be modeled by a set of two voltage-controlled current sources of opposite polarity, as shown in Fig. 9.10.

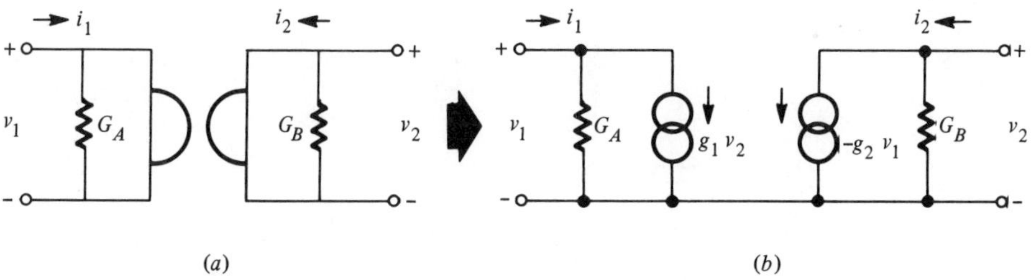

Figure 9.10 A nonideal gyrator and its equivalent circuit.

The gyrator is a subclass of negative feedback circuits, and has the basic sensitivity and high gain requirement characteristics associated with the negative feedback circuits described in Section 9.3. A special case of gyrators is the symmetrical gyrator where $g_1 = g_2$. It can be shown that, even though a symmetrical gyrator can contain a number of active elements internal to the basic circuit block, its terminal properties are passive (i.e., it has a power gain less than unity), and the circuit in unconditionally stable under any kind of passive termination.

The fundamental property of a gyrator is its ability to make a capacitive termination at the output port appear as an effective inductance looking into the input terminal. For example, if the output terminal of Fig. 9.10(*b*) were terminated with a capacitor C_B, the equivalent impedance, seen looking into the input terminal, would correspond to that shown in Fig. 9.11. Thus, a capacitively terminated gyrator behaves as an inductor with a finite loss factor. The effective Q of the inductance is given by[18]

$$Q = \frac{\omega C_B g_1 g_2}{G_A(G_B)^2 + G_B g_1 g_2 + \omega_2 C_B{}^2 G_A} \tag{9.31}$$

The maximum value of Q obtainable at any given frequency is

$$Q_{\max} = \frac{g_1 g_2}{2[g_1 g_2 + G_A G_B]} \sqrt{1 + \frac{g_1 g_2}{G_A G_B}} \tag{9.32}$$

As indicated by Eq. (9.32), to simulate a high Q inductance, it is necessary to maintain the input and output admittances, G_A and G_B, much smaller than the gyration admittances.

Even though the gyrator approach is conceptually well suited to integration, in the actual design of integrated gyrator circuits, one faces a number of practical design problems: Most of the gyrator circuits developed so far require either high-performance complementary transistors with well matched current gains, or a combination of FET and bipolar

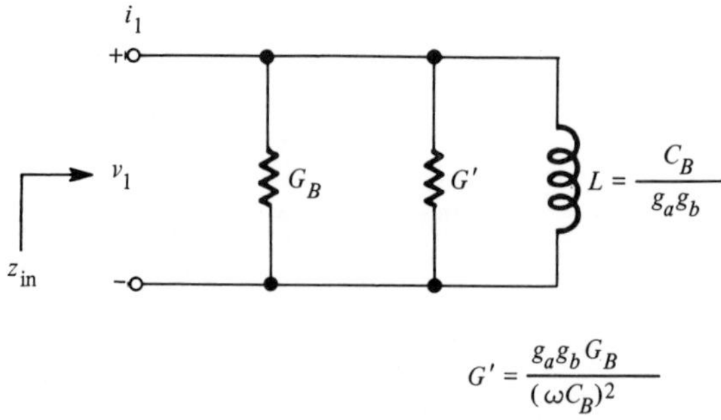

Figure 9.11 Inductance simulation using capacitively terminated gyrator.

devices on the same chip. Various designs have also been suggested using multiple chips, each containing the different types of devices (*npn*'s *pnp*'s or FET's) used in the circuit. Even though these methods are all technically feasible, so far they have been unattractive from yield and cost considerations. The gyrator is very sensitive to the parasitic phase shifts associated with the active sources in the circuit. To minimize these phase shifts and keep the Q of the simulated inductance at an acceptable level, all the bipolar gain stages must be operated at frequencies not much higher than the beta-cutoff frequency. This confines the gyrator applications to low frequencies. Even though some specialized applications of gyrators have been reported in the low mega-hertz range, most gyrator circuits suitable for integration are limited to frequencies below 300 kHz, mostly as a consequence of device limitations.

Figure 9.12 shows the circuit diagram for a practical gyrator circuit which is suitable for monolithic integration.[19] To minimize the input and output admittances, G_A and G_B of Fig. 9.10, the circuit uses *p*-channel JFETs as input devices. These devices are fabricated using an additional *p*-type diffusion to form the gate region of the field-effect transistors (see Fig. 2.26). Note that the two halves of the circuit corresponding to the two voltage-controlled current sources of Fig. 9.10 have similar topology; and the gyration admittances associated with each part of the circuit are

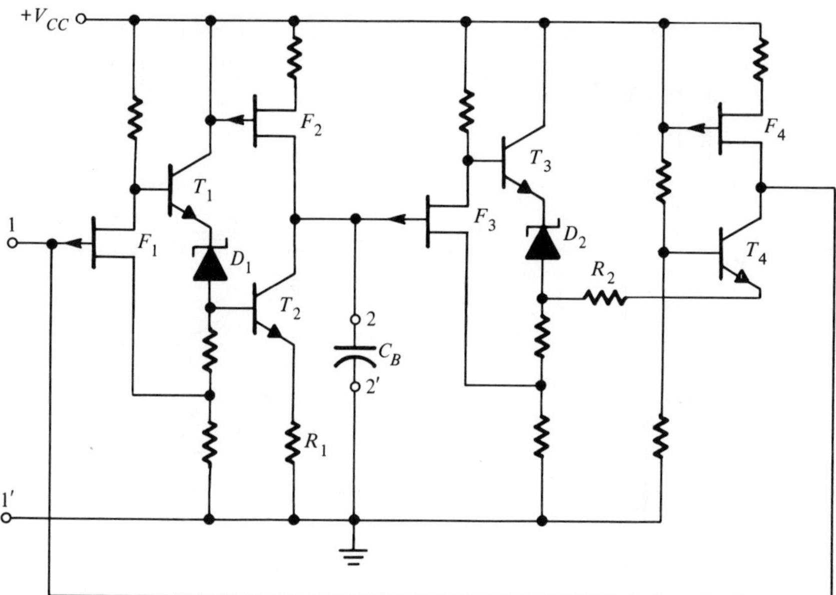

Figure 9.12 Circuit diagram of a gyrator suitable for integration (after Ref. 19).

set by resistors R_1 and R_2 which serve as series feedback resistors on the output transistors T_2 and T_4 as

$$g_1 = \frac{1}{R_1} \quad \text{and} \quad g_2 = \frac{1}{R_2} \tag{9.33}$$

In the actual design, R_1 and R_2 were external to the monolithic chip.

Using the circuit of Fig. 9.12, gyrator Q's of 450 have been obtained at frequencies up to 35 kHz. The circuit can be operated at higher frequencies with somewhat lower Q values. Q's of 240 and 180 have been reported for 250 kHz and 330 kHz operation of the circuit, with a Q variation of less than 25 percent over a 50°C temperature change.[19]

The basic impedance properties obtained from a gyrator can also be simulated using operational amplifiers. Since the monolithic operational amplifier is now available as a low cost building block in many circuit applications, such an approach can have significant advantages over separate gyrator design. Figure 9.13 shows an example of gyrator simulation using two high gain operational amplifiers.[20] Assuming that the gains A_1 and A_2 are very large and the input and output characteristics of the operational amplifiers are very nearly ideal, the input impedance Z_{in} for the circuit can be expressed as

$$Z_{in} = \frac{Z_1 Z_3 Z_5}{Z_2 Z_4} \tag{9.34}$$

Note that this type of gyrator simulation also involves a positive immittance inversion, and is closely related to the positive-zero synthesis technique described in the preceding section.[14] In this simulated gyrator circuit of Fig. 9.13, if one sets

$$Z_1 = Z_2 = Z_3 = Z_5 = R \quad \text{and} \quad Z_4 = (1/j\omega_c) \tag{9.35}$$

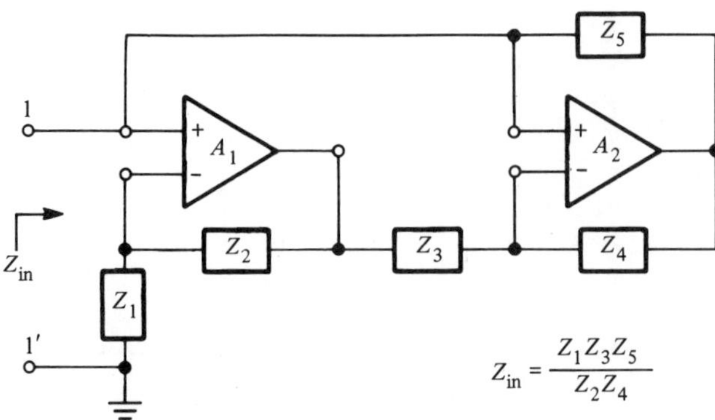

Figure 9.13 Gyrator simulation using two operational amplifiers.

then the input impedance becomes inductive:

$$Z_{in} = j\omega_c CR^2 \tag{9.36}$$

All the gyrator or simulated inductor techniques described so far are restricted to simulating "grounded" inductors, i.e., one terminal of the equivalent inductor is connected to an ac ground point. In certain filter applications, a "floating" rather than a grounded inductor is needed. A number of design techniques are available to simulate floating inductors using a cascade of gyrators.[11,18] These methods require at least two gyrators to simulate an ungrounded inductor and are generally impractical for integrated circuit applications due to cost, complexity, and power dissipation considerations.

9.8 STATE-VARIABLE SYNTHESIS METHOD

The state-variable synthesis method is derived from analog simulation techniques used in analog computers. The basic block in an analog computer is the operational amplifier, used either as a "summing amplifier" or as an "integrator" (see Fig. 5.5). Since the operational amplifier is now available as a low cost monolithic building block, these basic synthesis techniques, originally limited to analog computer applications, can be extended into active filter design.

In state-variable synthesis, one divides an nth order system into n separate state variables where each successive variable is related to the preceding one by the mathematical operation of integration. Then, by proper summing of each of the variables generated in this manner, one obtains the desired overall transfer function. Figure 9.14 shows a multifunction active filter system which can be used to synthesize the generalized second-order transfer function of Eq. 9.2.[21] In this case operational amplifiers K_1 and K_4 are used as input and output summing amplifiers; K_2 and K_3 are used as integrators. In general, to synthesize on n^{th} order transfer function, $(n + 2)$ operational amplifiers would be required. Thus, in discrete design, such a synthesis approach can be expensive; however, if all operational amplifiers can be fabricated monolithically, the cost and complexity of the design can be greatly reduced. The operational amplifiers used in the system are in general noncritical. They are required to have open-loop voltage gains of the order of 60 to 70 dB, and can be designed to have low power consumption.

The generalized second-order filter block of Fig. 9.14 provides three independent outputs. The transfer functions associated with each of these outputs have the same denominator polynomial, $D(s)$:

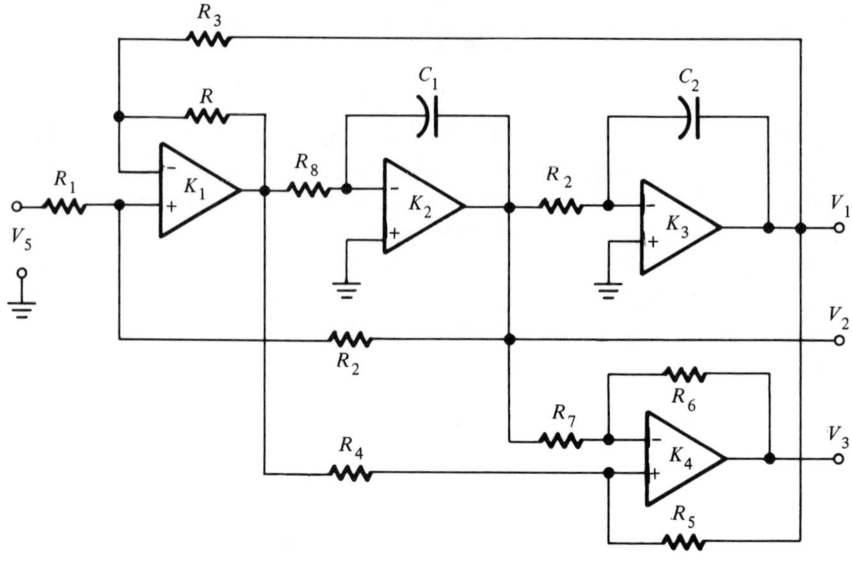

Figure 9.14 Second-order filter building block using state-variable synthesis.

$$D(s) = s^2 + \left[\frac{R_1(R + R_3)}{(R_1 + R_2)R_3}\left(\frac{1}{\tau_1}\right)\right] s + \frac{R}{R_3\tau_1\tau_2} \tag{9.37}$$

where $\tau_1 = R_8C_1$ and $\tau_2 = R_9C_2$. However, the numerator $N(s)$ associated with each output is different:

$$N_1(s) = s^2 + \left[\frac{(R_4 + R_5)R_6}{R_5(R_6 + R_7)}\right]\frac{s}{\tau_1} + \frac{1}{\tau_1\tau_2} \tag{9.38}$$

$$N_2(s) = s/\tau_1 \tag{9.39}$$

and

$$N_3(s) = 1 \tag{9.40}$$

Thus, these three outputs are referred to as "high-pass," "band-pass," and "low-pass" terminals, respectively, since they emphasize different regions of the frequency spectrum at the output. If one chooses

$$R_1 = R_3 = R_5 = R_6 = R_0 \tag{9.41}$$

and

$$\tau_1 = \tau_2 = \tau_0 \tag{9.42}$$

the center frequency and the selectivity of the generalized second-order filter can be written as

$$\omega_0 = \frac{1}{\tau} \sqrt{R/R_0} \tag{9.43}$$

and

$$Q = \sqrt{R/R_0} \left(\frac{R_0 + R_2}{R + R_0} \right) \tag{9.44}$$

The particular advantage of the state-variable technique is its low sensitivity with respect to parameter changes. For example, the sensitivity of ω_0 or Q with respect to the component values of R's and C's is equal to or less than unity. If the amplifier gains are assumed to be finite, one can show that the gain sensitivity of Q is of the order of

$$S_{K_1}{}^{Q} \approx Q/K_1 \qquad \text{and} \qquad S_{K_2}{}^{Q} \approx Q/K_2 \tag{9.45}$$

Thus, Q-sensitivity can be kept quite low with reasonable amplifier gains among the active filter methods examined so far. The state-variable synthesis method shows the least amount of sensitivity to component or gain tolerances. Therefore, it is best suited for synthesizing stable high-Q filters. However, it has two inherent drawbacks:

1. It requires large amplifier gains for stable high-Q circuits; thus it is mainly limited to applications in the audio frequency range.
2. It is somewhat costly from circuit complexity and power dissipation considerations since it requires $(n + 2)$ operational amplifiers for synthesizing n^{th} order transfer function.

9.9 AN OVERVIEW OF LINEAR ACTIVE FILTER TECHNIQUES

A diversity of filter synthesis techniques are available to the linear active filter designer. Conceptually all of these techniques are well suited to integration, since all the components necessary for design can be fabricated using monolithic or hybrid technology. The previous sections of this chapter have reviewed some of these basic synthesis techniques in terms of their use in analog integrated circuit design. Particular attention has been paid to the relative circuit complexity, loop-gain requirements, and sensitivity parameters for each of the classes of circuits considered.

The choice of any one particular active-RC design technique over the others is often a difficult one, and depends very strongly on the specific application. Therefore, in almost all practical cases, active filter design tends to be a "committed" rather than a "general-purpose" design, i.e., it is custom designed for a specific function. A number of hybrid integrated designs have been suggested as general purpose building blocks₂[2,23] however, none has yet gained widespread acceptance to become an "industry-standard." This is in contrast to other classes of analog integrated circuits such as operational amplifiers, regulators, or multipliers which have gained wide customer acceptance.

There are three basic limitations to integrated active-RC filters techniques: (1) frequency range: stability considerations limit the use of most of these filters to below 100 kHz; (2) sensitivity: the selectivity and the center frequency are often very sensitive functions of active gains or absolute values of feedback components; (3) cost: typically four precision components (two R's and two C's) are required for each complex pole pair. These components can only be fabricated in integrated form using thin-film and hybrid technology, thus adding to the cost and complexity of design.

In certain applications, it is possible to overcome some of these disadvantages of active-RC filters by using alternate design approaches which differ significantly from the basic linear feedback techniques described so far. One specific example of such an approach is the use of a phase-locked loop (PLL) system to obtain desired selectivity function. This is described in the second part of this chapter.

PART II Phase-Lock Techniques

9.10 THE PHASE-LOCKED LOOP (PLL)

In the design of frequency selective integrated circuits using phase-lock techniques, one utilizes the frequency selective signal processing properties of a phase-locked loop (PLL) system. The basic concept of a PLL has been known since the early 1930's. However, due to its cost and complexity in nonintegrated system design, the application of the PLL has been limited to precision measurements requiring a high degree of noise immunity and very narrow bandwidths.

Figure 9.15 shows the block diagram of a basic phase-locked loop system. The PLL is a feedback system comprised of a phase comparator, a low-pass filter and an error amplifier in the forward signal path, and a voltage controlled oscillator (VCO) in the feedback path. A detailed analysis of the PLL as a quasi-linear feedback system is available in the literature [24,25] and will not be repeated here. However, from a qualitative point of view, the basic principle of operation of a PLL can be briefly explained as follows: With no signal input to the system, the error voltage, $v_d(t)$ of Fig. 9.15, is equal to zero. Then the VCO operates at a set fre-

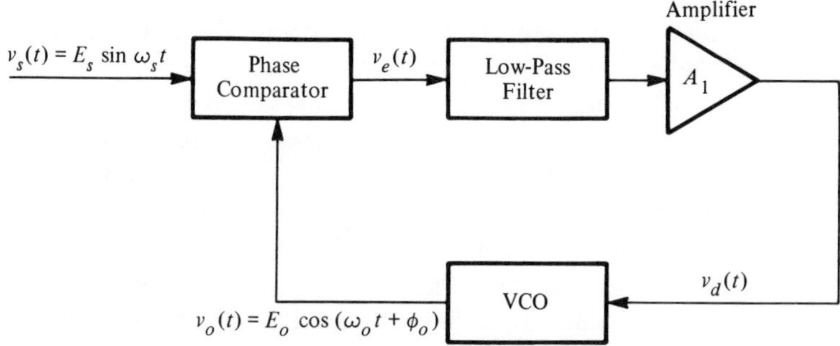

Figure 9.15 Block diagram of a phase-locked loop (PLL) system.

quenty, $\omega_0 = 2\pi f_o$, known as its "free-running frequency." If an input signal, $v_s(t)$, is applied to the system, the phase comparator compares the phase and the frequency of the input with the VCO frequency and generates an error voltage, $v_e(t)$, related to the frequency and the phase difference between the two signals. This error voltage is then filtered, amplified, and applied to the control terminals of the VCO. In this manner, the control voltage, $v_d(t)$, forces the VCO frequency to vary in a direction which reduces the frequency difference between f_0 and the input signal. If the input frequency f_s is sufficiently close to f_0, the feedback nature of the PLL causes the VCO to synchronize or "lock" with the incoming signal. Once in lock, the VCO frequency is identical to the input signal, except for a finite phase difference. This net phase difference ϕ_o is necessary to generate the corrective error voltage v_d to shift the VCO frequency from its free-running value to the input signal frequency f_s, and thus keep the PLL in lock. This self-correcting ability of the system also allows the PLL to "track" the frequency changes of the input signal once it is locked. The range of frequencies over which the PLL can *maintain* lock with an input signal is defined as the "lock range" of the system. This is always larger than the band of frequencies over which the PLL can *acquire* lock with an incoming signal. This latter range of frequencies is known as the "capture range" of the system.

The capture process is highly complex, and does not lend itself to simple mathematical analysis. However, a heuristic and highly qualitative description of the capture mechanism may be given as follows: Since frequency is the time derivative of phase, the frequency and the phase errors in the loop can be related as:

$$2\pi \Delta f(t) = \gamma \frac{d\phi_o(t)}{dt} \tag{9.46}$$

where $\Delta f(t)$ is the instantaneous frequency separation between the signal and the VCO frequencies, and γ is a constant of proportionality.

If the feedback loop of the PLL were opened, say between the low-pass filter and the VCO control input, then for a given setting of f_0 and f_s the phase comparator output would be a sinusoidal beat-note at a fixed frequency Δf. If f_s and f_0 were sufficiently close in frequency, then this beat-note would be passed by the low-pass filter, with negligible attenuation. Now suppose that the feedback loop is closed by connecting the low-pass filter output to the VCO control terminal. Then the VCO frequency would be modulated by the beat-note, and when this happens, Δf itself becomes a function of time. If during this modulation process, the VCO frequency moves closer to f_s (i.e., decreasing Δf), then $(d\phi_o/dt)$ decreases and the output of the phase comparator becomes a slowly varying function of time. Similarly if the VCO is modulated *away* from f_s, $(d\phi_o/dt)$ increases and the error voltage becomes a rapidly varying function of time. Therefore, under this condition, the beat-note waveform no longer looks sinusoidal; instead it looks like a series of aperiodic "cusps," shown schematically in Fig. 9.16. Because of its assymetry, the beat-note waveform contains a finite dc component which pushes the "average" value of the VCO toward f_s, thus decreasing Δf. In this manner the beat-note frequency rapidly decreases toward zero, the VCO frequency drifts toward f_s, and the lock is established. When the system is in lock, Δf is equal to zero and only a steady state dc error voltage remains.

The total time taken by the PLL to establish lock is called the "pull-in" time. Pull-in time depends on the initial frequency and phase difference between the two signals, as well as on the overall loop gain and the low-pass filter bandwidth. Under certain conditions, the pull-in time may be shorter than the period of the beat-note; and the loop can lock without an oscillatory error transient.

In the operation of the loop, the low-pass filter serves a dual function: first, by attenuating the high frequency error components at the output of the phase comparator, it enhances the interference rejection characteristics;

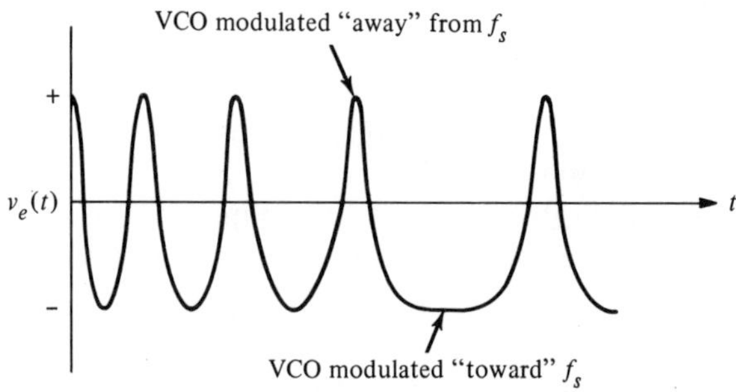

Figure 9.16 Asynchronous error beat-note during capture process.

secondly, it provides a short-term memory for the PLL and ensures a rapid recapture of the signal if the system is thrown out of lock due to a noise transient. Since the low-pass filter attenuates the high-frequency error voltage within the loop, it directly controls the capture and the transient-response characteristics of the PLL. The reduction of the filter bandwidth has the following effects on system performance:

1. The capture process becomes slower, and the pull-in time increases.
2. The capture range decreases.
3. Interference rejection properties of the PLL improve since the error voltage caused by an interfering frequency is attenuated further by the low-pass filter.
4. The transient response of the loop, i.e., the response of the PLL to sudden changes of the input frequency within the capture range, becomes underdamped.

The last point also brings about a practical limitation of the low-pass loop filter bandwidth and roll-off characteristics from stability considerations. These points will be explained further in the following section.

9.11 BASIC DESIGN PARAMETERS FOR A PLL SYSTEM

The phase-locked loop exhibits frequency selective response characteristics centered about the VCO free-running frequency. The two parameters which describe the selectivity characteristics are the "lock" and the "capture" ranges defined in the preceding section. These characteristics can be readily related to the parameters of the individual blocks shown in Fig. 9.15. The time dependence of the phase error $\phi_o(t)$ can be expressed by the following differential equation:[26]

$$d\phi_o/dt = K_o K_d A_1[f(t) * \sin \phi_o] \tag{9.47}$$

where the asterisk denotes convolution and,

K_o = VCO voltage to frequency conversion gain in radians per volt/second.

K_d = conversion gain of phase comparator (in volts/radian)

$f(t)$ = impulse response of the low-pass filter

The sine ϕ_o dependence in Eq. (9.47) comes about because most practical phase-comparator circuits have an output voltage which is proportional to either the sine or the cosine of the phase error (see Eq. 7.45).

Even for a simple one-pole low-pass loop filter, Eq. (9.47) leads to a nonlinear second-order differential equation, and cannot be solved in a closed form. The lock range, $\Delta\omega_L$, corresponds to the maximum value of

$(d\phi_o/dt)$ for which a steady state solution of (9.47) is possible. Assuming unity dc gain for the low-pass filter, one can show that[25]

$$\Delta\omega_L = K_o K_d A_1 = \text{total dc loop gain} \tag{9.48}$$

Since the capture range $\Delta\omega_c$ denotes a transient condition, it is not as readily derived as the lock range. However, an approximate parametric expression for the capture range can be written as[27]

$$\Delta\omega_c \approx K_o K_d A_1 |F(j\Delta\omega_c)| \tag{9.49}$$

where $|F(j\Delta\omega_c)|$ is the low-pass filter amplitude response at $\omega = \Delta\omega_c$. It should be noted that, since at all times $|F(j\Delta\omega_c)| \leq 1.0$, the capture range is always smaller than the lock range. If a simple lag filter is used (see Fig. 9.18(*a*)) the capture range equation can be approximated as

$$\Delta\omega_c \approx \sqrt{\frac{\Delta\omega_L}{\tau_1}} \tag{9.50}$$

where τ_1 is the low-pass filter time constant.

From Eqs. (9.48) and (9.49), it is seen that the gain A_1 of the dc amplifier within the loop provides an independent means of controlling the loop gain, and the lock range. Similarly, for a given value of the loop gain, the capture range is set by the choice of low-pass filter response $F(j\omega)$.

When the PLL is in lock, nonlinear capture transients are no longer present. Therefore, under lock condition, it can be approximated as a linear control system, as shown in Fig. 9.17, and can be analyzed using Laplace transform techniques. In this case, it is convenient to use the net phase error within the loop as the system variable, with ϕ_s and ϕ_o denoting the relative phase between the signal input and the VCO output. Note that since the VCO converts a voltage to a frequency, and since phase is the integral of frequency, the VCO functions as an integrator in the feedback loop. In the linear model of Fig. 9.17, it is assumed that the net phase shift $(\phi_o - \phi_s)$ is sufficiently small such that

$$\sin(\phi_o - \phi_s) \approx \phi_o - \phi_s \tag{9.51}$$

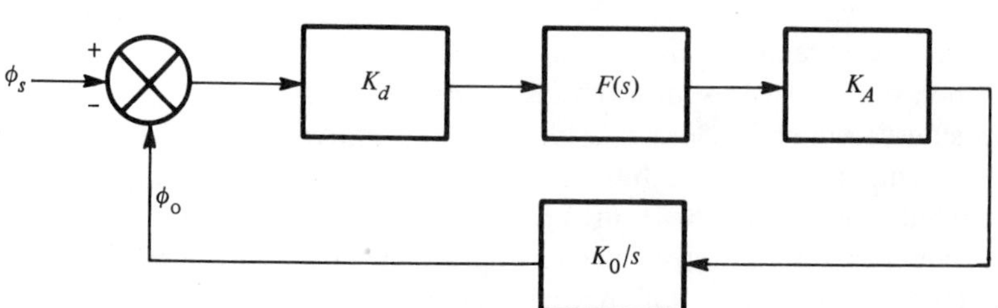

Figure 9.17 Linearized model of the PLL as a negative feedback system.

From the Figure, the open loop transfer function for the PLL can be written as

$$T(s) = \frac{K_T F(s)}{s} \tag{9.52}$$

where K_T is the total loop gain ($K_T = K_o K_d A_1$) and $F(s)$ is the low-pass filter transfer function. Using linear feedback analysis techniques, the closed-loop transfer characteristics $H(s)$ can be related to the open-loop performance as

$$H(s) = \frac{\phi_o(s)}{\phi_s(s)} = \frac{T(s)}{1 + T(s)} \tag{9.53}$$

and the roots of the characteristic system polynomial can be readily determined by root-locus techniques. Figure 9.18 shows the PLL root-loci for the linearized model of Fig. 9.17, as a function of increasing loop gain K_T for single-pole lag and lag-lead type filters. In each case, the open-loop pole at the origin is due to the integrating action of the VCO.

With reference to the root-locus characteristics of Fig. 9.18(*a*), one can make the following observations:

1. As the loop gain K_T increases, for a given choice of τ_1, the imaginary part of the closed-loop poles increases; thus, the natural frequency of the loop increased and it becomes more and more under damped.
2. If the filter time constant τ_1 is increased, the real part of the closed-loop poles becomes smaller, and the loop damping is reduced.

As in any practical feedback system, excess phase shifts or non-dominant poles associated with the blocks within the PLL can cause the loci to bend toward the right-half plane, as shown by the dotted lines in 9.18(*a*). This is likely to happen if either the filter time constant or the loop gain is too large, and can cause the loop to break into sustained oscillations. The stability problem can be eliminated by using a lag-lead type filter as shown in 9.18(*b*). By proper choices of R_2 this type of filter confines the root-locus to the left-half plane and ensures stability. However, this type of filter also reduces the interference rejection characteristics of the system since the high frequency error components within the loop are now attenuated to a lesser degree.

When the PLL is in lock, a frequency shift at the input is transferred to a voltage level shift at the VCO control terminal. Figure 9.19 shows the typical frequency-to-voltage transfer characteristics of the PLL. The input is assumed to be a sine-wave whose frequency is swept slowly over a broad frequency range; and the vertical scale is the corresponding loop error voltage. In the top figure the input frequency is being gradually increased. The loop does not respond to the signal until it reaches a frequency f_1,

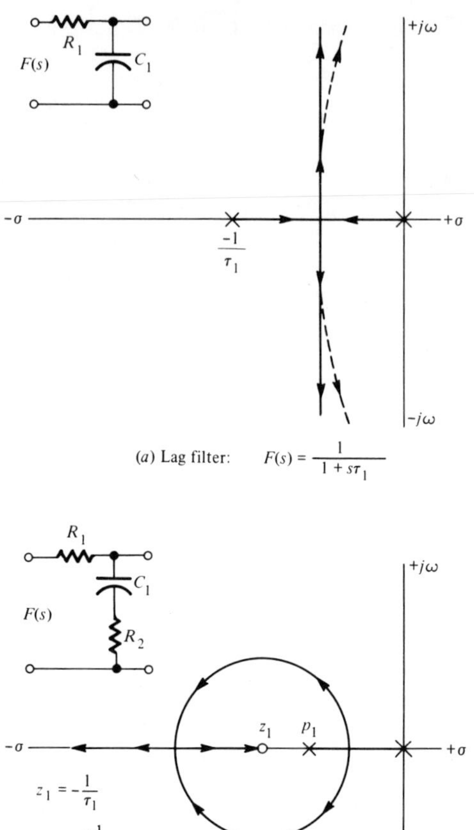

Figure 9.18 PLL root-locus for various low-pass filters.

corresponding to the lower edge of the capture range. Then, the loop suddenly locks on the input, causing a negative jump of the loop error voltage; as the input frequency is increased further V_d varies with frequency with a slope equal to the reciprocal of VCO gain ($1/K_0$) and goes through zero as $f_s = f_o$. The loop tracks the input until the input frequency reaches f_2, corresponding to the upper edge of the lock range. Then the PLL loses lock, and the error voltage drops to zero. and the VCO frequency returns to f_o. If the input frequency is now swept slowly back, the cycle repeats

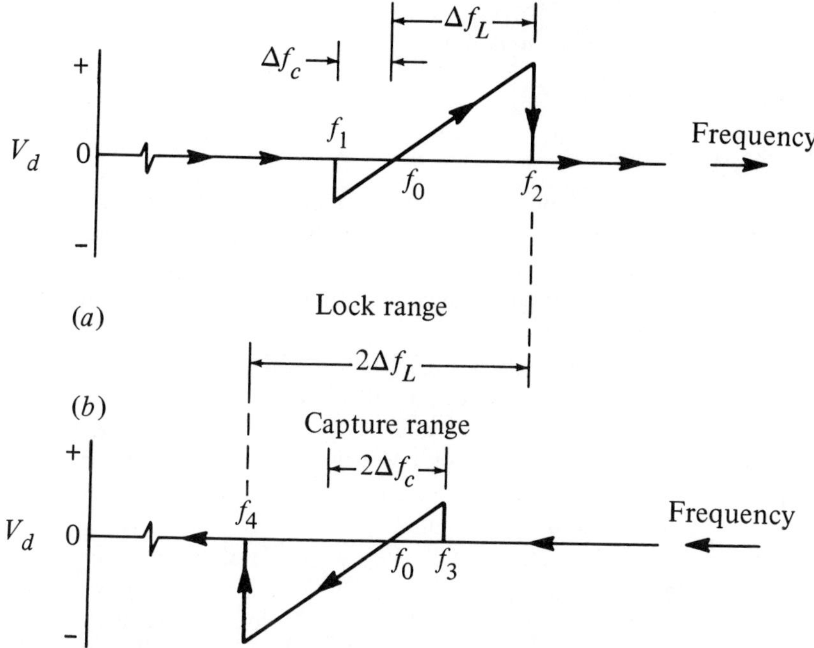

Figure 9.19 Typical PLL frequency-to-voltage transfer characteristics: (*a*) slowly increasing input frequency; (*b*) decreasing frequency.

itself as shown in Fig. 9.19(*b*). The loop recaptures the signal of f_3 and traces it down to f_4. The frequency spread between (f_1, f_3) and (f_2, f_4) corresponds to the total capture and lock ranges of the system, i.e.,

$$f_3 - f_1 = 2\Delta f_C \qquad \text{and} \qquad f_2 - f_4 = 2\Delta f_L \tag{9.54}$$

Note that, as indicated by the transfer characteristics of Fig. 9.19, the PLL system has an inherent selectivity about the center frequency set by the VCO free-running frequency, f_o; and it would only respond to the input signal frequencies which are separated from f_o by less than Δf_C or Δf_L, depending on whether the loop starts with or without an initial lock condition. It is also worth noting that the linearity of the frequency-to-voltage conversion characteristics for the PLL are solely determined by the VCO conversion gain. Therefore, in most applications, the VCO is required to have a highly linear voltage-to-frequency transfer characteristic.

9.12 APPLICATIONS OF A PLL

The phase-locked loop is a versatile system block which is suitable for a variety of frequency selective demodulation, signal conditioning, or frequency synthesis applications. Some of these basic applications are briefly described below:

1. FM Demodulation

If the PLL is locked on a frequency modulated (FM) signal, the VCO tracks the instantaneous frequency of the input. Then the filtered error voltage $v_d(t)$, which constrains the VCO to maintain lock with the input signal, corresponds to the demodulated output. In this case, the linearity of the demodulated output is determined by the VCO voltage-to-frequency conversion characteristics (see Fig. 9.19). The PLL can be used for detecting either wide-band (high deviation) or narrow-band FM signals with a higher degree of linearity than can be obtained by other FM detection means. It is worth noting that for FM deviation purposes, the PLL functions as a self-contained receiver system since it combines the functions of frequency selection and demodulation.

In the case of frequency-shift keyed (FSK) data transmission, the digital information is transmitted by switching the input frequency between any one of the two discrete input frequencies, corresponding to a digital "one" and a digital "zero," respectively. When the PLL is locked on an FSK input signal, the error voltage $v_d(t)$, which is in the form of discrete voltage steps, corresponds to the demodulated binary output.

It should be noted that since the PLL is in lock during the FM demodulation process, the frequency response as well as the rise time of the demodulated output can be readily predicted from the root-locus plots of Fig. 9.18

2. Frequency Synchronization

Using the phase-locked loop system, the frequency of a relatively poor oscillator such as the VCO can be phase-locked with a low level but highly stable reference signal. Then, the VCO output reproduces the input signal frequency at the same per-unit accuracy as the input reference, but at a much higher power level. In some applications, the synchronizing signal can be in the form of a low duty cycle burst at a specific frequency. Then the PLL can be used to regenerate a coherent CW reference frequency, by locking on this short synchronizing pulse. A typical example of such an application is the phase-locked chroma-reference generators in color TV receivers.

In digital systems, the PLL can be used for a variety of synchronization functions. For example, two system clocks can be phase-locked to each other such that one can function as a backup for the other; or it can be used in synchronizing disc or tape drive mechanisms in information storage and retrieval systems. In pulse-code modulation (PCM) telemetry receivers or in repeater systems, the PLL is used for bit synchronization.

3. Signal Conditioning

By proper choice of the VCO free-running frequency, the PLL can be made to lock on any one of a number of signals present at the input. Then the VCO output reproduces the frequency of the desired signal while greatly attenuating the undesired frequencies or side bands present at the input. If the loop bandwidth is sufficiently narrow, the signal-to-noise ratio at the VCO output can be much higher than that at the input. Thus the PLL can be used as a noise filter for regenerating weak signals buried in noise.

4. Frequency Multiplication and Division

By inserting a frequency divider into the feedback loop, between the VCO output and the phase comparator input, the PLL system can function as a frequency-selective frequency multiplier. A block diagram of this configuration is shown in Fig. (9.20) where N is the frequency divider modulus.

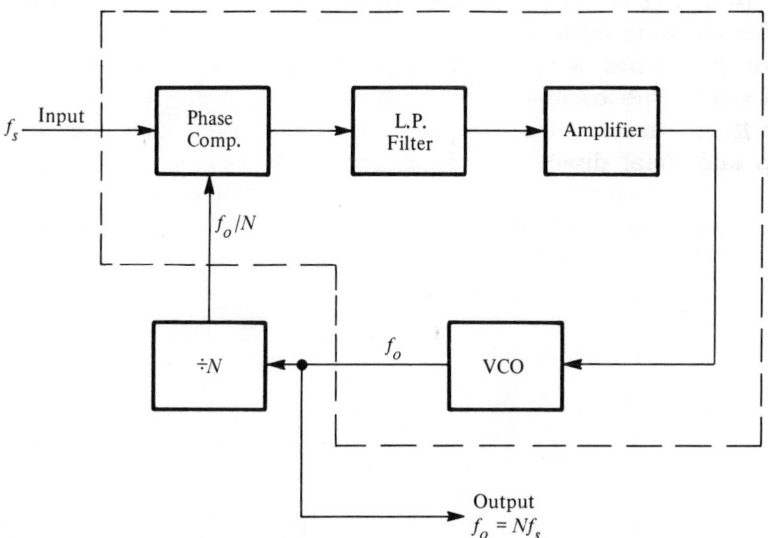

Figure 9.20 Frequency multiplication by using a frequency divider in PLL feedback loop.

When the system is in lock, the two inputs to the phase comparator are at the same frequency, and $f_o = Nf_s$.

Under certain conditions, frequency multiplication can also be achieved without the use of a frequency divider network, by operating the PLL in its "harmonic-locking" mode. The principle of harmonic-lock can be briefly explained as follows: If the VCO output is nonsinusoidal, it will contain a number of harmonics in addition to its fundamental frequency. In other words, the output of a nonsinusoidal VCO is effectively a composite signal containing frequency components at integral multiples of VCO fundamental frequency. The same is also true for a nonsinusoidal input signal, such as a pulse train. If the VCO free-running frequency is set to be close to the nth harmonic of a harmonic-rich input signal, the VCO fundamental can be made to synchronize with the nth harmonic of the input. Then, under this condition, the PLL operates in a harmonic-lock mode, and the VCO frequency is exactly n times the input frequency, or $f_o = nf_s$.

Similarly, if the VCO produces a harmonic-rich output waveform, the mth harmonic of the VCO output can be synchronized with the input fundamental. Then, under this condition, the VCO fundamental is a subharmonic of the input frequency, i.e., $f_o = f_s/m$. When the PLL is operated in the harmonic-locking mode, the spacing between the adjacent harmonics in the frequency spectrum decreases rapidly as the harmonic order n or m is increased. This in turn, increases frequency stability requirements for the VCO free-running frequency, to enable the system to differentiate between adjacent harmonics. In integrated phase-locked loop systems which use multivibrator type oscillators,[28] thermal drifts of VCO frequency usually restrict the harmonic-lock operation of the system to values of n or $m \leqq 10$. An additional disadvantage of harmonic-locking at large values of n or m is that the phase-detector gain K_d decreases inversely with the harmonic order, thus decreasing both the lock and the capture ranges of the system at higher harmonics.

5. Frequency Translation

The PLL system can be used to translate the frequency of a highly stable but fixed frequency reference oscillator by a small amount in frequency. This can be achieved by adding a mixer and a low-pass filter stage to the basic PLL, as shown in Fig. (9.21). In this case, the reference input f_R and the VCO input f_o are applied to the inputs of the mixer stage. The mixer output is made up of the "sum" and the "difference" components of f_o and f_R. The sum component is filtered by the first low-pass filter. The translation or offset frequency f_1 is applied to the phase comparator, along with

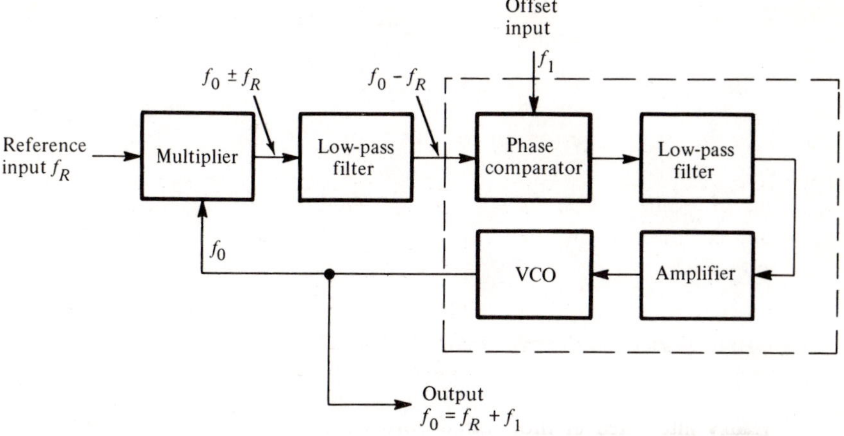

Figure 9.21 Frequency translation or "frequency-offset" loop.

the $f_R - f_o$ component of the mixer output. When the system is in lock, the two inputs of the phase comparator are at identical frequency, i.e.,

$$f_0 - f_R = f_1 \qquad \text{or} \qquad f_0 = f_R + f_1 \tag{9.55}$$

6. AM Detection

The PLL can be used as a coherent detector for demodulating amplitude modulated (AM) signals. In this mode of operation, the PLL locks on the carrier of the AM signal and produces a reference signal at the output of the VCO, which has the same frequency as the AM carrier, but no amplitude modulation. Then, by multiplying this coherent reference signal with the modulated input signal and low-pass filtering the output of the multiplier, one can obtain the demodulated information. A block diagram of such a system is shown in Fig. (9.22). Since the PLL only responds to

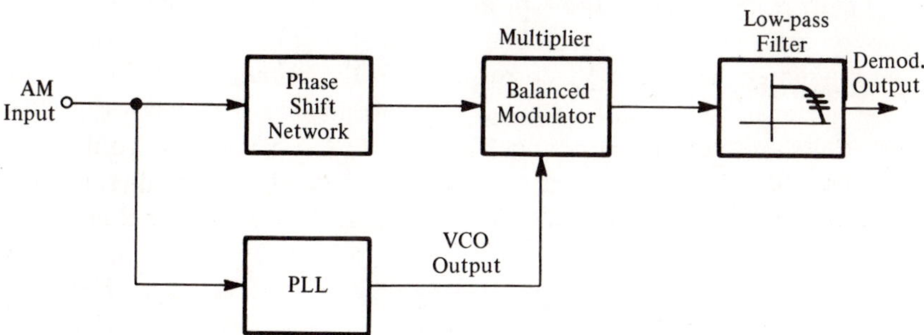

Figure 9.22 Coherent AM detection using a phase-locked loop.

carrier frequencies very close to VCO frequency f_o the phase-locked AM detector system also exhibits a high degree of selectivity, cntered about f_o.[29] The phase-shift network shown in the block diagram of Fig. (9.22) is a noncritical RC network which is used to offset the 90° phase shift introduced by the PLL. The reason for this phase shift will be described in the following section.

The phase-locked AM detection method shown in Fig. (9.22) is a coherent detection technique; therefore, it offers a higher degree of noise immunity than conventional peak-detector type AM dmodulators.

9.13 BUILDING BLOCKS FOR A MONOLITHIC PLL

Each of the basic blocks forming the PLL system of Fig. 9.15 can be readily integrated in monolithic form, with present-day IC technology. Therefore the entire PLL system itself is well suited to integration. The key blocks which determine the characteristics of the system are the phase comparator and the VCO sections. In this section some basic circuit configurations will be described for the VCO and the phase comparator sections. These particular circuit topologies are chosen because they are readily compatible with monolithic bipolar technology; and their operations rely mainly on the matching and the thermal tracking of monolithic components, rather than on the tight control of absolute value tolerances.

1. The Phase Comparator

The simplest phase comparator circuit suitable for monolithic integration is the switch-type phase detector which is shown schematically in Fig. 9.23(*a*) This type of a detector operates as a synchronous switch which is opened and closed by the reference input, and effectively "chops" the signal input at the same repetition rate as the reference drive. Normally, the reference drive is supplied by the VCO output.

Figure 9.23(*b*) shows the typical output waveforms for the switch-type phase comparator, for a sinusoidal input and a square-wave drive signal. The filtered error voltage V_d corresponds to the average value of the output waveform, and is shown as the shaded area in the waveforms. The error voltage is zero when the net phase shift ϕ_o between the two inputs is 90°. This 90° phase shift is a common property of all switch-type phase comparator circuits, and results in a comparator gain expression of the form,

$$K_d = K_A \cos \phi_o \tag{9.56}$$

where K_A is a constant of proportionality at a given input signal level. In a

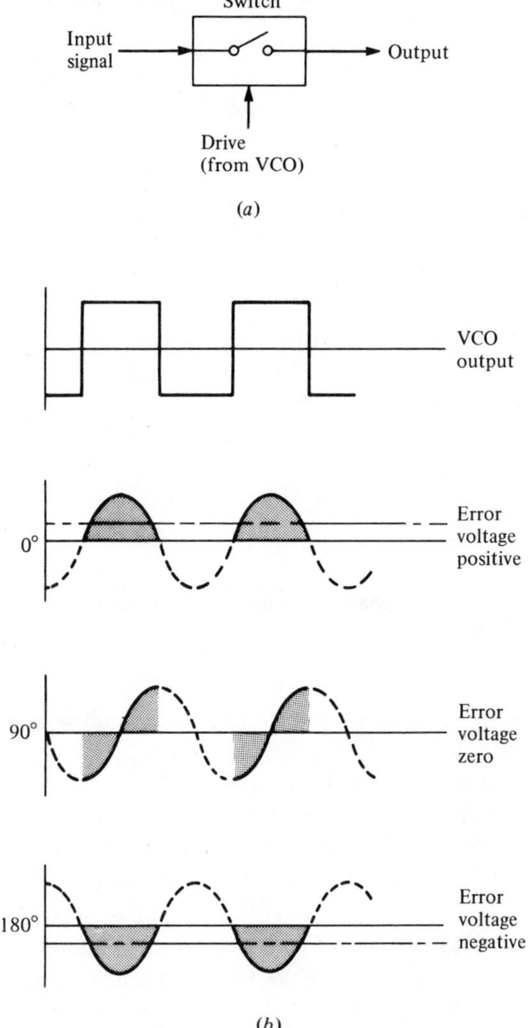

Figure 9.23 Operation of a switch-type phase detector: (*a*) block diagram; (*b*) typical waveforms.

PLL system using this type of a phase comparator when the PLL is in perfect lock condition (i.e., $v_d(t) = O$), the VCO output is at quadrature phase with the input.

The switch-type phase comparator circuits can be readily derived from the balanced modulator circuits discussed in Chapter 7 (see Sections 7.5 and 7.6). Figures 7.3 and 7.10 show two balanced modulator configura-

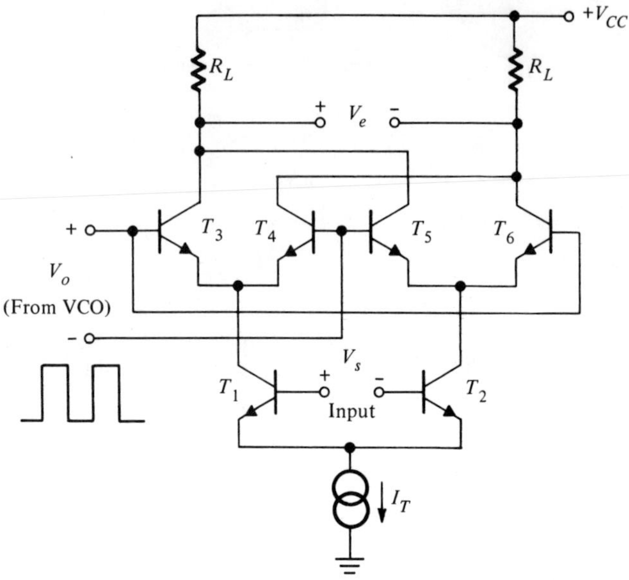

Figure 9.24 A phase comparator circuit based on the balanced modulator configuration.

tions suitable for monolithic integration. The circuit of Fig. 7.3 is normally preferred over that of Fig. 7.10 since it provides higher conversion gain and can be built with all bipolar transistors.[30]

Figure 9.24 shows the balanced modulator circuit of Fig. 7.3 in phase comparator application. In this circuit, the input signal $v_s(t)$ is applied to the bases of T_1 and T_2, and controls the partitioning of bias current between these two devices. The high-level VCO output is applied to cross-coupled transistor pairs (T_3, T_4) and (T_5, T_6); and causes these devices to function as two sets of single-pole double-throw switches actuated by the VCO waveform. If $v_s(t)$ and $v_o(t)$ are at the same frequency, the dc output voltage V_e is related to the phase difference ϕ_o between the two signals as

$$V_e = \frac{g_m R_L E_s}{\pi} \cos \phi_o \tag{9.57}$$

where g_m is the transconductance of T_1 or T_2, and E_s is the amplitude of $v_s(t)$.

The low-pass filter can be readily incorporated into the phase-comparator circuit of Fig. 9.24 by connecting a series combination of a capacitor C_1 and resistor R_2 across the output terminals of the phase detector. This results in a low-pass filter function, $F(s)$, given as

$$F(s) = \frac{1 + sR_2C_1}{1 + sC_1(2\ R_L + R_2)} \tag{9.58}$$

In an actual design, this low-pass filter would normally be left external to the monolithic circuit, to facilitate a higher degree of flexibility.

2. The Voltage Controlled Oscillator

In the design of a PLL the voltage-controlled oscillator is usually the most critical block since the frequency stability and the FM demodulation characteristics of the system are normally determined by the VCO performance. For maximum versatility, the VCO is required to have the following desirable properties:

1. Linear voltage to frequency conversion.
2. Good frequency stability (low thermal and long term drift).
3. High frequency capability.
4. High voltage-to-frequency conversion gain.
5. Wide tracking range.
6. Ease of tuning (frequency of oscillation determined by a minimum number of circuit components).

In addition to these requirements, to be suitable for monolithic integration, the VCO circuit should contain no inductors.

Figures 9.25 and 9.26 show two basic oscillator configurations which fulfill most of the requirements listed above. The circuit of Fig. 9.25 is an integrator and Schmitt-trigger combination where the timing capacitor C_o is alternately charged and discharged by a voltage controlled current source I_1.[31] The Schmitt trigger, comprised of T_4 through T_7, senses the voltage level V_A across C_o and turns the switch transistor T_3 off or on to initiate the charge and discharge cycles, respectively. The frequency of oscillation f_o can be expressed as

$$f_o = \frac{V_c g_m}{2\ C_o(V_2 - V_1)} \tag{9.59}$$

where g_m is the transconductance of the voltage-controlled current source, and V_2 and V_1 are the upper and lower trip levels for the Schmitt trigger. This type of an oscillator can provide either a triangular-wave (at node A) or a square-wave (at node B) output. Note that transistors T_1 and T_2 form a diode-connected current sink (see Fig. 4.2) and thus force the charging and the discharging currents to be equal. This results in a 50 percent duty cycle of the VCO output waveform. In the actual design of the VCO circuit of Fig. 9.25, the voltage-controlled current source can be designed using any one of the circuit configurations discussed in Chapter 4 (see Figs. 4.17

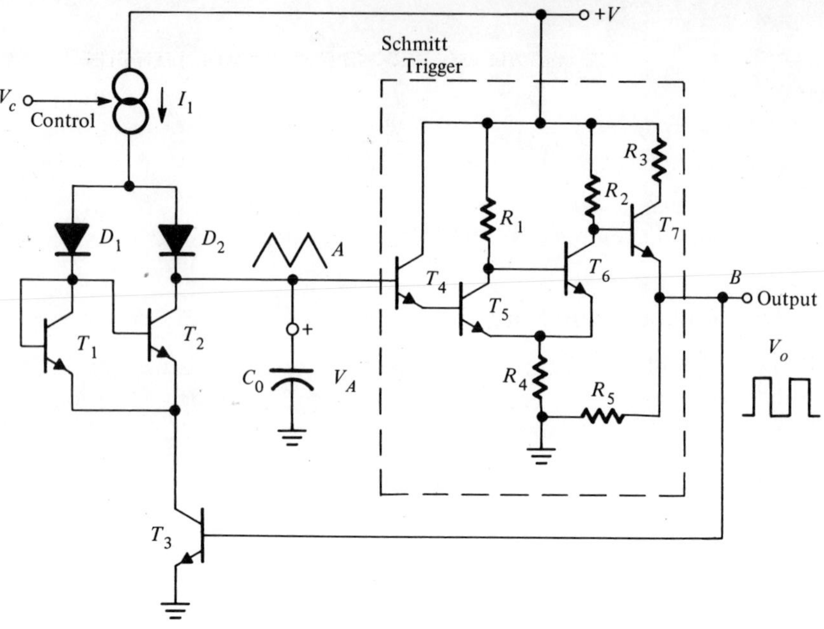

Figure 9.25 A voltage-controlled oscillator using an integrator and Schmitt-trigger combination.

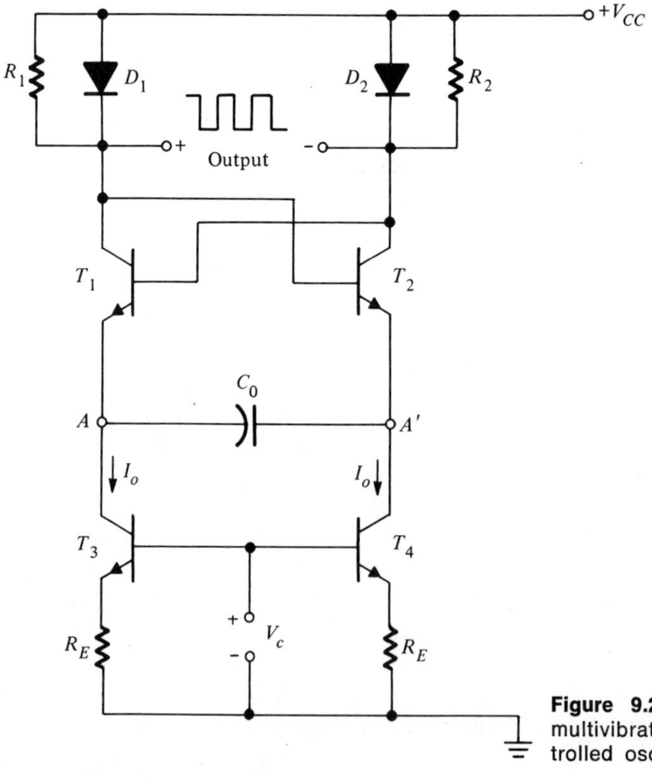

Figure 9.26 An emitter-coupled multivibrator type voltage-controlled oscillator.

and 4.18) with an external resistor used to set the bias current level I_1, for zero control voltage. Since the current source uses a *pnp* transistor, the high frequency capability of the circuit is somewhat limited ($f_o \leqq 1$ MHz). However, since the current I_1 and switching levels V_1 and V_2 can be made to have very low temperature coefficients using the matching and thermal-tracking of monolithic components, this type of an oscillator offers a high degree of temperature stability (typically $\partial f_o/\partial T < \pm 200$ ppm/°C) over a broad temperature range.

The VCO circuit of Fig. 9.26 is basically an emitter-coupled multivibrator, with cross-coupled transistors T_1 and T_2 forming the positive feedback gain stage.[32] At any one time, either T_1 or T_2 is on and the timing capacitor C_o is alternately charged and discharged by a constant current I_o provided by the voltage-controlled current sources T^3 and T_4. The frequency of oscillation can be expressed as

$$f_o = \frac{I_o}{4\,C_o V_{BE}} \tag{9.60}$$

where V_{BE} is the transistor base-emitter voltage. Diodes D_1 and D_2 clamp the output swing to one V_{BE} drop below the positive supply voltage, and make the output amplitude independent of the control voltage. The circuit provides a balanced square-wave output across diodes D_1 and D_2. The conversion gain K_0 can be written as

$$K_o = \frac{1}{4\,C_o R_E V_{BE}}\ \text{Hz/V} \tag{9.61}$$

Since the VCO of Fig. 9.26 is a nonsaturating circuit using all *npn* transistors, it offers higher frequency capability than that of Fig. 9.25, and can operate up to 60 MHz. As shown by Eq. 9.60, the frequency of the emitter-coupled VCO is inversely proportional to V_{BE}, which has a strong negative temperature coefficient ($\partial V_{BE}/\partial T \approx -2$mV/°C or ≈ -3000 ppm/°C). In the actual VCO design, this temperature drift can be compensated by introducing an equal temperature dependence on the current I_o so as to keep the ratio (I_o/V_{BE}) independent of temperature. Using this type of compensation, the temperature coefficient of f_o can be kept as low as ±400 ppm/°C over a temperature range of −55 to +125°C.[28]

It should be noted that both of the VCO circuits described in this section offer linear control characteristics (i.e., f_o proportional to V_c) and can be tuned to any desired frequency by means of a single timing capacitor C_o.

9.14 A MONOLITHIC PLL SYSTEM

The basic blocks described in the previous section can be readily fabricated and interconnected in monolithic form, to function as a self-

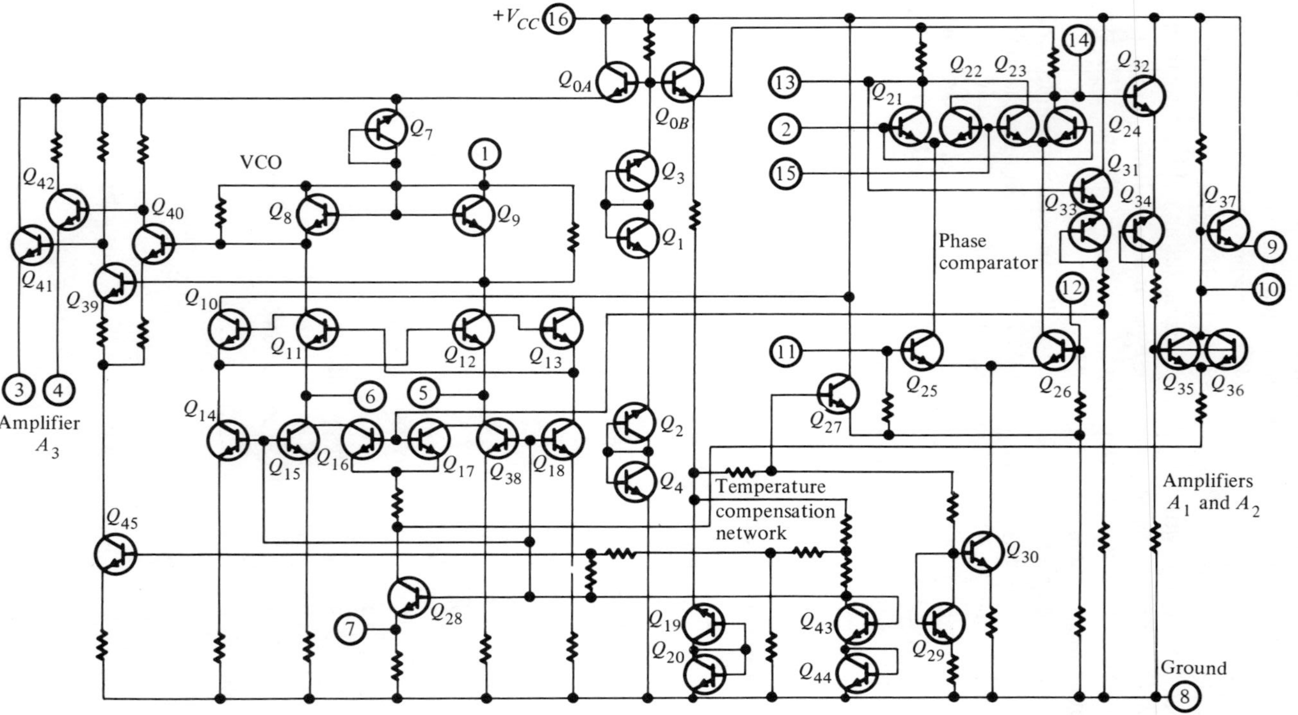

Figure 9.27 Circuit schematic of a monolithic PLL system.[28]

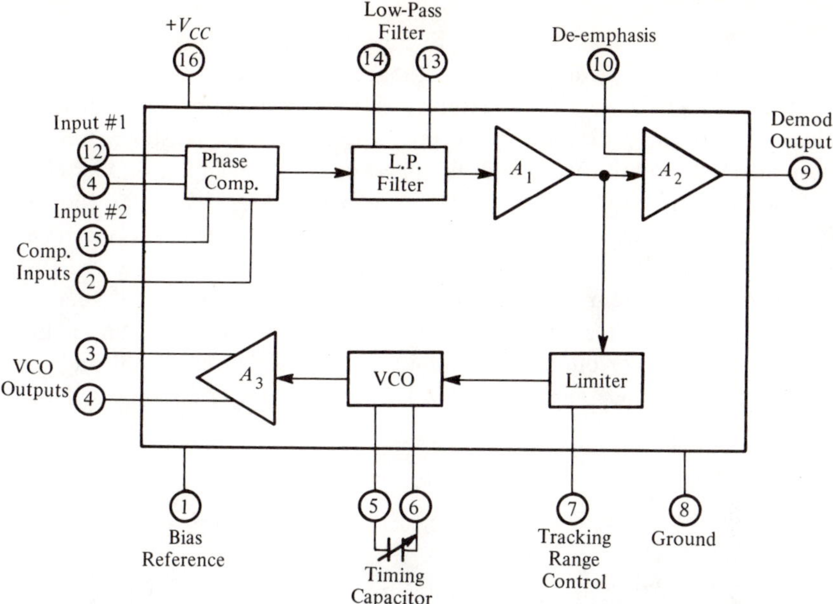

Figure 9.28 Functional block diagram of the monolithic PLL circuit shown in Figure 9.27.

contained phase-locked loop system. The actual circuit diagram of such a monolithic system is shown in Fig. 9.27. The corresponding block diagram of the circuit is given in Fig. 9.28, in terms of its monolithic circuit package. The basic blocks forming the overall system are outlined in the circuit diagram of Fig. 9.27. Note that the externally accessible terminals of the circuit are designated with the numbers corresponding to the terminals on the system block diagram. As described earlier, the phase comparator section is formed by a balanced modulator circuit, comprised of transistors Q_{21} through Q_{26}. The VCO is designed as an emitter-coupled multivibrator, similar to that described in Fig. 9.26. However, in this case a differential control-voltage input is used to minimize the dc offset or drift problems. The frequency of the VCO is determined by means of an external capacitor C_o connected across terminals 5 and 6. This frequency can be varied from a fraction of a cycle to in excess of 30 MHz by proper choice of C_o. A temperature compensating bias network is also included on the chip to minimize the frequency drift of the VCO. This compensation network varies the current levels within the circuit to compensate for the transistor V_{BE} changes with temperature (see Eq. 9.60) and thus keeps the temperature drift of f_o to less than 500 ppm/°C.

The particular example shown in Figs. 9.27 and 9.28 was designed as a general purpose building block, suitable for any one or a combination of circuit functions listed in Section 9.12. For maximum versatility, the phase-locked feedback loop is not internally connected. Instead, the loop is opened between the VCO and the phase comparator inputs so that either a divide-by-n circuit or a mixer circuit can be inserted into the loop for frequency-multiplication or frequency-translation applications (see Figs. 9.20 and 9.21). In addition to the basic blocks necessary to form the PLL, the monolithic system also contains two additional amplifiers A_2 and A_3. A_2 is used as a buffer amplifier to increase the demodulated output level for FM or FSK detection; A_3 amplifies the VCO output signal and provides digital interface capability with conventional logic circuits. A limiter

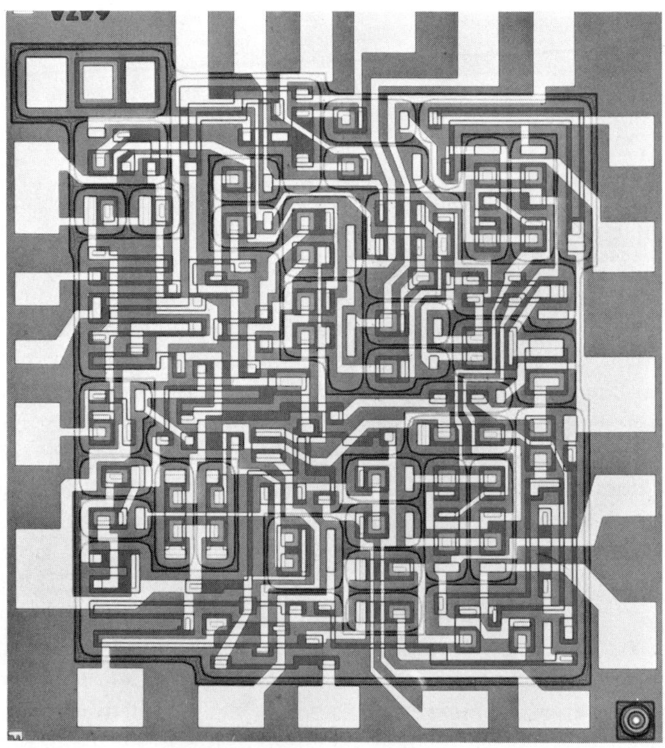

Figure 9.29 Topological layout of the PLL circuit chip size: 67 × 72 mils. *(Photo: Signetics.)*

block is incorporated into the loop (see Fig. 9.28) to control the maximum amplitude of the loop error voltage applied to the VCO. This limiter block is directly incorporated into the VCO by controlling the current partitioning between the common-collector transistor pairs (Q_{15}, Q_{16}) and (Q_{17}, Q_{38}). The threshold of this limiter block can be adjusted by means of an external bias, thus allowing one to control the total tracking range of the VCO. In the present design, the tracking range of the VCO can be externally adjusted from ± 1 to ± 25 percent of the VCO frequency.

The low-pass filter can be formed by connecting a series resistor and capacitor combination across output terminals 13 and 14 of the phase comparator. This results in a lag-lead type filter with the filter transfer function $F(s)$ given by Eq. 9.58.

Figure 9.29 shows the topological layout of the PLL circuit of Fig. 9.27. The die size of the monolithic chip is 67×72 mils. To enhance the high frequency capability, the monolithic chip is fabricated using dielectric isolation techniques rather than conventional diffused-junction isolation.

Performance Characteristics

The monolithic PLL circuit of Fig. 9.27 is designed to operate with a single power supply, within the 15 to 24 V range, and has a nominal power dissipation of 160 mW at $V_{CC} = 18$ V. The system can operate over the full military temperature range (−55 to +125°C) and can handle input signals over an 80 dB dynamic range (from 250 μV to 2 V).

Figure 9.30 shows an oscillogram of the PLL frequency-to-voltage conversion characteristics of the monolithic PLL at 10 MHz center fre-

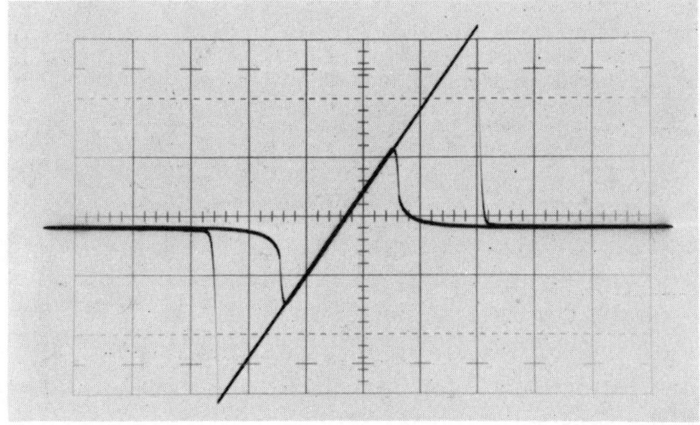

Figure 9.30 Oscillogram of PLL frequency-to-voltage transfer characteristics at $f_0 = 10$ MHz. (Scale: vertical = 0.5 V/division, horizontal = 500 kHz/division.)

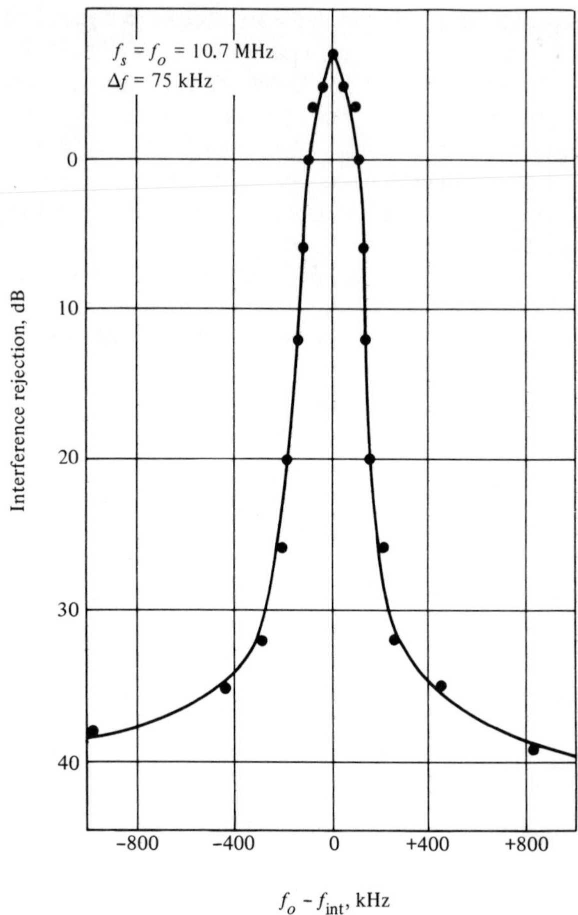

Figure 9.31 Interference rejection characteristics of the monolithic PLL in FM demodulation.

quency. The inner and outer traces in the Figure correspond to the capture and the lock ranges of the system (i.e., a superposition of the transfer characteristics sketched in Fig. 9.19). Figure 9.31 shows the typical interference rejection characteristics of the monolithic PLL as an FM demodulator at 10.7 MHz center freequency with a 15 kHz low-pass loop filter. The shift selectivity of the interference rejection characteristics exhibit a form factor* of 2.2 for 3 to 30 dB roll-off, which compares favorably with the skirt selectivity of a three-stage FM IF-strip. Figure 9.32

* "Form factor" for a given attenuation range is defined as the ratio of the system bandwidth for the higher attenuation to that for the lower attenuation.

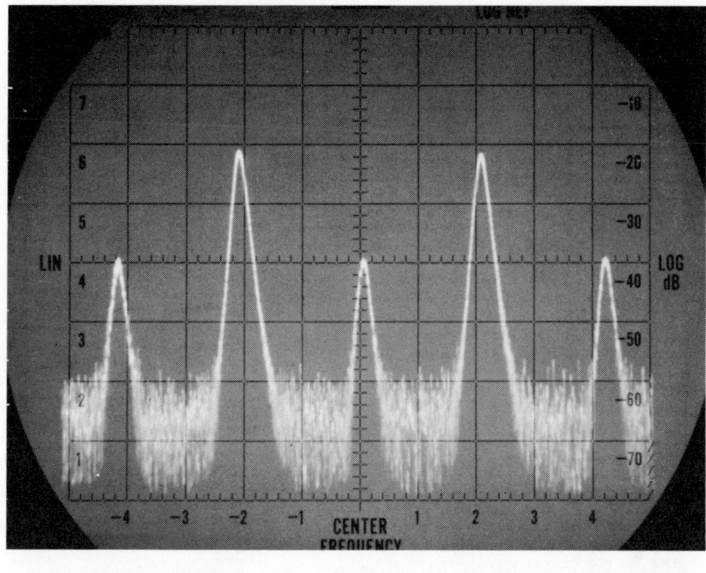

(*a*)

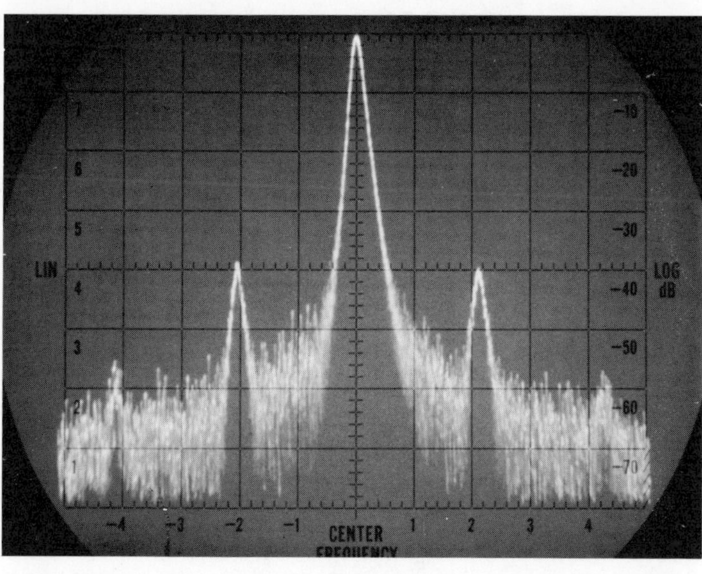

(*b*)

Figure 9.32 Performance of the integrated PLL as a signal conditioner: (*a*) input spectrum with the desired signal (center) at 1.0 MHz, and two undesired signals at 0.96 and 1.04 MHz; (*b*) VCO output spectrum. (Scale: center frequency = 1.0 MHz; horizontal = 20 kHz/division, vertical = 10 dB/division.)

shows an oscillogram of the PLL input and output spectrum in signal conditioning applications. The input signal spectrum to the PLL system, shown in Fig. 9.32(*a*), is made up of a desired signal at 1 MHz with two undesired sidebands at ± 40 kHz from the desired frequency. In this case, the undesired signal levels are 20 db *higher* than the desired signal. Figure 9.32(*b*) shows the VCO output spectrum with the PLL locked on the desired signal. In this case a 5 kHz low-pass loop filter is used. Note that at the VCO output, the undesired side bands are 40 dB *below* the desired signal level.

9.15 A COMPARISON OF PLL AND ACTIVE-RC TECHNIQUES

At this point, having examined the basic properties and the performance characteristics of linear feedback (active-RC) techniques and the PLL filters, it may be worthwhile to pause briefly and compare the pros and cons of the two approaches with respect to monolithic integration. From the preceding discussion, the following advantages of the phase-locked loop filters have become apparent:

1. *High frequency capability:* The monolithic PLL can operate at frequencies in excess of 50 MHz; most active-RC techniques using integrated components are limited to a frequency range below 100 kHz.
2. *Independent control of selectivity and center frequency:* The center frequency is set by the VCO free-running frequency, and the selectivity is determined by the low-pass loop filter. This eliminates the "alignment" problem associated with cascading conventional filter stages.
3. *Fewer external components:* Compared to monolithic active filters, the integrated PLL generally requires fewer external energy storage or precision components. For example, to obtain a skirt selectivity comparable to that given in Fig. 9.16 using active-RC filters would require at least eight capacitors and as many precision resistors whose absolute values must be controlled to better than 0.5 percent.
4. *Ease of tuning:* The PLL can be tuned to any desired frequency by the proper choice of the VCO free-running frequency. This frequency is normally set by a single external component, and can be continuously adjusted from a fraction of a cycle to in excess of 50 MHz.

However, there is one significant point which must be considered: active-RC filters, at least in theory, form a direct replacement for conventional LC-tuned filters. The PLL, on the other hand, is a somewhat specialized filter; therefore, it is *not* a direct replacement for conventional filters in all applications. Compared with LC or active-RC filters, the PLL has the following drawbacks:

1. *Lack of amplitude information:* The PLL responds only to the *frequency* of the input signal and not the amplitude, as long as the amplitude of the input is high enough to maintain lock. Thus, it filters out only the frequency information and not the amplitude.
2. *Responds to harmonics:* The PLL systems suited to integration tend to respond to the harmonics and the subharmonics of the input, through the "harmonic-lock" effect described earlier. This is useful for frequency multiplication and division, but degrades the interference rejection of the system for parasitic signals which are harmonically related to the desired input.
3. *Difficult to AGC:* Since the PLL responds only to the frequency and not to the amplitude of a given input, it is difficult to derive a control signal from the PLL for automatic gain control (AGC) applications. The AM detection capability of the PLL (see Fig. 9.8) can be used for "squelch" or "muting" type AGC, but is not suitable for conventional gain control applications.
4. *Synthesis techniques not yet developed:* Because of its nonlinear capture characteristics, no detailed synthesis procedures exist for a PLL filter design with a predetermined interference rejection characteristic.

9.16 OTHER INDUCTORLESS FILTER TECHNIQUES

Inductorless filter synthesis is a very broad topic which stretches far beyond the realm of integrated circuits. The aim of this chapter has been to examine analog circuit design and signal processing techniques which are suitable for integration with the present-day integrated circuit technology. For the sake of brevity the techniques covered have been limited to only two classes of inductorless filters: Active-RC and the phase-locked loop circuits. This is because these classes offer the greatest promise for integrated circuits, and particularly the monolithic IC's, in the immediate future. However, various other kinds of inductorless filters also offer some potential promise and compatibility with integrated circuit technology, and should be mentioned here for completeness.

One particular class of inductorless filters which show potential promise in the hybrid circuit area are the crystal[33] and the ceramic[34] resonators, and surface wave filters,[35] the last one being particularly useful in the UHF and microwave frequencies.

Another class of frequency selective filters which also show potential feasibility for integrated circuits are the n-path filters, comprised of a number of parallel signal flow channels which are switched or commutated in and out of the signal path at discrete time intervals.[36] n-path filters are basically frequency-translation systems where the transfer function of a low-pass filter can be translated to a bandpass response with the center frequency determined by the sampling frequency, and the selectivity determined by the number of parallel paths present in the system. n-path filters can be reduced to continuous, rather than switched or sampled, systems by replacing the sampling switches with analog multipliers or modulators.[37]

Last, but not the least, among the classes of inductorless filters not covered in this chapter are digital filters.[38] These filters process the signals in time rather than frequency domain, and obtain the desired selectivity by performing in proper sequence the mathematical operations of addition (or subtraction), multiplication, and time delay. Digital filters basically operate as numerical calculators, and can be multiplexed to operate on either a number of multiple input channels or to process a single input a multiple number of times. As a complete system, digital filters are not directly compatible with analog integrated circuits. However, they have become economically feasible because of the advent of low cost digital integrated circuits and particularly the large-scale integration (LSI) technology.

REFERENCES

Part I

1. H. H. Scott, "A New Type of Selective Circuit and Some Applications," *Proc. IRE,* **26** (Feb., 1938): 226–235.
2. J. G. Linvill, "RC Active Filters," *Proc. IRE,* **42** (March, 1954): 555–564.
3. R. P. Sallen and E. L. Key, "A Practical Method for Designing RC Active Filters," *IRE Trans. Ckt. Theory,* **CT–2** (March, 1955): 78–85.
4. W. E. Newell, "Tuned Integrated Circuits—A State-of-the-Art Survey," *Proc. IEEE,* **52** (Dec. 1964): 1603–1608.
5. S. K. Mitra, *Analysis and Synthesis of Linear Active Networks,* Wiley, New York, 1969.
6. R. W. Newcomb, *Active Integrated Circuit Synthesis,* Prentice-Hall, Englewood Cliffs, N.J., 1968.
7. P. M. Chirlian, *Integrated and Active Network Analysis and Synthesis,* Prentice-Hall, Englewood Cliffs, N.J., 1967.
8. K. L. Su, *Active Network Synthesis,* McGraw-Hill, New York, 1965.
9. *Handbook of Operational Amplifier Active-RC Networks,* Burr-Brown Research Corp., Tucson, Ariz., 1966.
10. A. A. Gaash, R. S. Pepper and D. O. Pederson, "Design of Integrable

Desensitized Frequency Selective Amplifiers," *IEEE J. Solid State Ckts.,* **SC–1**(1) (Sept., 1966): 29–35.

11. G. S. Moschytz, "Inductorless Filters: A Survey—Part II: Linear Active and Digital Filters," *IEEE Spectrum,* Sept., 1970, 63–75.
12. I. M. Horowitz and G. R. Branner, "A Unified Survey of Active RC Synthesis Techniques," *Proc of Natl Electronics Conf.,* **23** (Dec., 1967).
13. S. K. Mitra, "Filter Design Using Integrated Operational Amplifiers," WESCON Technical Paper 4.1, Aug., 1969.
14. G. A. Rigby and D. G. Lompard, "Integrated Frequency Selective Amplifiers for Radio Frequencies," *IEEE J. Solid State Ckts.,* **SC–3** (Dec., 1968): 417–422.
15. J. J. D'Azzo and C. H. Houpis, *Feedback Control System Analysis and Synthesis* McGraw-Hill, New York, 1960, pp. 130–171.
16. W. M. Kaufman, "Theory of a Monolithic Null Device and Some Novel Circuits," *Proc. IRE,* **48** (Sept., 1960): 1540–1545.
17. G. S. Moschytz, "A General Approach to Twin-T Design and Its Application to Hybrid Integrated Linear Active Networks," *Bell System Tech.* J. **49** (July-Aug., 1970): 1105–1149.
18. S. K. Mitra, "Synthesizing Active Filters," *IEEE Spectrum,* Jan., 1969, pp. 47–63.
19. R. G. Howe and C. A. Kleingartner, "Silicon Monolithic Gyrator Using FETs," WESCON Technical Paper 4.3, Aug., 1969.
20. R. H. S. Riordan, "Simulated Inductors Using Differential Amplifiers," *Electronics Lett.,* **3** (Feb., 1967): 50–51.
21. W. J. Kerwin, L. P. Huelsman and R. W. Newcomb, "State-Variable Synthesis for Insensitive Integrated Circuit Transfer Functions," *IEEE J. Solid State Ckts.,* **SC–2** (Sept., 1967): 87–92.
22. G. S. Moschytz and W. Thelen, "Design of Hybrid-Integrated Filter Building Blocks," *IEEE J. Solid State Ckts.,* **SC–5** (June, 1970): 99–107.
23. G. Hurtig, "Positive Results from Negative Feedback," *Electronics,* March, 1969, pp. 96–102.

Part II

24. A. J. Viterbi, *Principles of Coherent Communication,* McGraw-Hill, New York, 1966.
25. F. M. Gardner, *Phase-Lock Techniques,* Wiley, New York, 1966.
26. A. B. Grebene and H. R. Camenzind, "Frequency Selective Integrated Circuits Using Phase Lock Techniques," *IEEE J. Solid-State Circuits,* **SC–4** (Aug., 1969): 216–225.

27. G. S. Moschytz, "Miniaturized RC Filters Using Phase-Locked Loop," *Bell System Tech. J.,* **44** (May, 1965): 823–870.

28. A. B. Grebene, "The Monolithic Phase-Locked Loop—A Versatile Building Block," *IEEE Spectrum,* March, 1971, pp. 38–49.

29. L. P. Chu, "Phase-Locked AM Radio Receiver, *IEEE Trans. Broadcast TV Receivers,* Dec., 1969, pp. 300–308.

30. A. Billotti, "Applications of a Monolithic Analog Multiplier," *IEEE J. Solid State Ckts.,* **SC–3** (Dec., 1968): 373–380.

31. J. A. Mattis and H. R. Camenzind, "A New Phase-Locked Loop with High Stability and Accuracy," Signetics Corp. App. Note, Sept. 1970.

32. A. B. Grebene and G. A. Rigby "Phase-Locked Integrated Circuits," *NEREM Record,* **11** (Nov., 1969): 86–87.

33. R. A. Sykes, W. L. Smith and W. J. Spencer, "Monolithic Crystal Filters," *IEEE Int. Conf. Record,* March, 1967, pp. 78–93.

34. F. Sauerland and W. Blum, "Ceramic IF Filters for Consumer Products," *IEEE Spectrum,* Nov., 1968, pp. 112–126.

35. H.M. Gerard, "Interdigital Surface-Wave Transducers," *Proc. NEC,* **26** (Dec., 1970): 675–680.

36. L. E. Franks and I. W. Sandberg, "An Alternative Approach to the Realization of Network Transfer Functions: The N-Path Filter," *Bell System Tech. J.,* **39** (Sept., 1960): 1321–1350.

37. G. A. Rigby, "An Integrated Selective Amplifier Using Frequency Translation," *IEEE J. Solid State Ckts.,* **SC–1** (Sept., 1966): 39–44.

38. J. F. Kaiser, *Digital Filters, Systems Analysis by Digital Computer* (J. F. Kaiser and F. F. Kuo, eds.), Wiley, New York, 1966, pp. 218–285.

10 Digital/Analog Interface Circuits

Digital and analog signal processing correspond to the two fundamental, yet distinctly different modes, of information handling. In the analog case, the signals are handled in a nonquantized, continuously variable manner; in digital processing they are quantized in binary "bits," i.e., in ONEs and ZEROs. In many cases it is necessary to interface these two fundamental means of signal processing; and to convert the data from digital to analog or vice versa. This is accomplished by the use of digital-to-analog (D/A) or analog-to-digital (A/D) converter circuits.

In their natural state, all variables (such as current, voltage, pressure, distance, time, etc.) appear in analog form. However, for signal transmission and computation purposes, they are often handled in a digital manner. Therefore, the A/D and D/A converters can be considered to be a class of "coding" and "decoding" devices, respectively. Figure 10.1 shows a "black-box" description of the two classes of converters. The input to a D/A converter is a digital "word" of a prescribed number of "bits"; and the output is an analog voltage level uniquely corresponding to the input word. Conversely, the analog input applied to the A/D converter results in a digital word of a prescribed number of bits.

The interface circuits between analog and digital signals often require very close matching and tracking of circuit components and

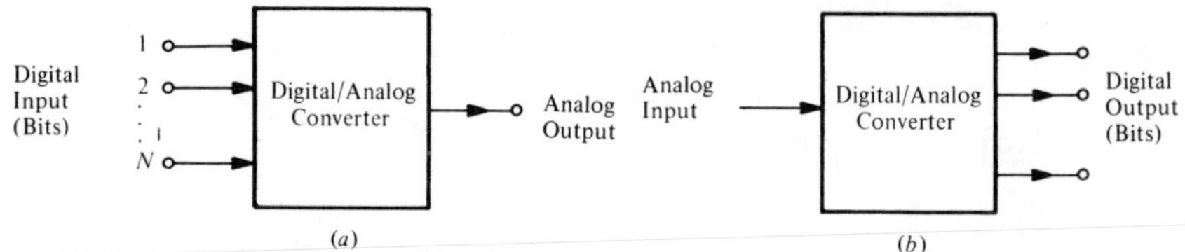

Figure 10.1 Functional block diagrams of digital/analog interface circuits: (*a*) D/A converter; (*b*) A/D converter.

parameters. This is particularly true in the case of D/A converters where the accuracy of the analog output has to be maintained over a broad temperature range.

A large number of D/A and A/D conversion techniques are available and are well covered in the literature.[1-5] The purpose of this chapter is to examine some of these techniques from the point of view of integrated circuits; and more specifically monolithic integration. Since the D/A and A/D conversion techniques are quite different, the chapter is divided into two parts, treating each class of converter circuits separately. The D/A converters fall more readily into the category of "analog" circuits; whereas A/D converters fit more closely as "digital" circuits. Furthermore, D/A converters are, in general, more readily suited to monolithic integration. Therefore, a correspondingly larger part of the chapter is devoted to digital-to-analog conversion.

PART I Digital-to-Analog Converters

10.1 PRINCIPLES OF D/A CONVERSION

The D/A converter can be considered as a decoding device which accepts a digitally coded signal D and an analog reference P as inputs and provides an analog output A related to the input as:

$$A = PD \tag{10.1}$$

D is a digital word of given number of "bits" and can be represented as

$$D = \frac{b_1}{2^1} + \frac{b_2}{2^2} + \frac{b_3}{2^3} + \cdots + \frac{b_N}{2^N} \tag{10.2}$$

where N is the total number of bits and b_1, b_2, etc. are the bit coefficients which are quantized to be either ONE or ZERO. Thus, in terms of an arbitrary reference quantity P and the analog output A, the generalized transfer function of a D/A converter can be written as

$$A = P[b_1 2^{-1} + b_2 2^{-2} + \cdots + b_N 2^{-N}] \tag{10.3}$$

The actual implementation of a D/A system contains four separate parts: A reference quantity (normally a voltage) corresponding to parameter P of Eq. (10.3); a set of binary switches to simulate the binary coefficients $b_1, b_2, \cdots b_N$; a resistive weighting network; and an output summing means. An actual D/A configuration incorporating all four of these essential components is shown in Fig. 10.2. In this case, the relative weights of the bit currents $I_1, I_2, \cdots I_N$ are set by a binary-weighted resistor network. An operational amplifier with high input impedance and a large inverting gain A_1 is used as a means of summing the individual bit currents and generating the corresponding analog voltage. The total analog current I_o appearing at the summing node, i.e., the inverting input of the operational amplifier, is related to the input reference as

$$I_o = \frac{2\,V_{\text{REF}}}{R}\,[b_1 2^{-1} + b_2 2^{-2} + \cdots + b_N 2^{-N}] \tag{10.4}$$

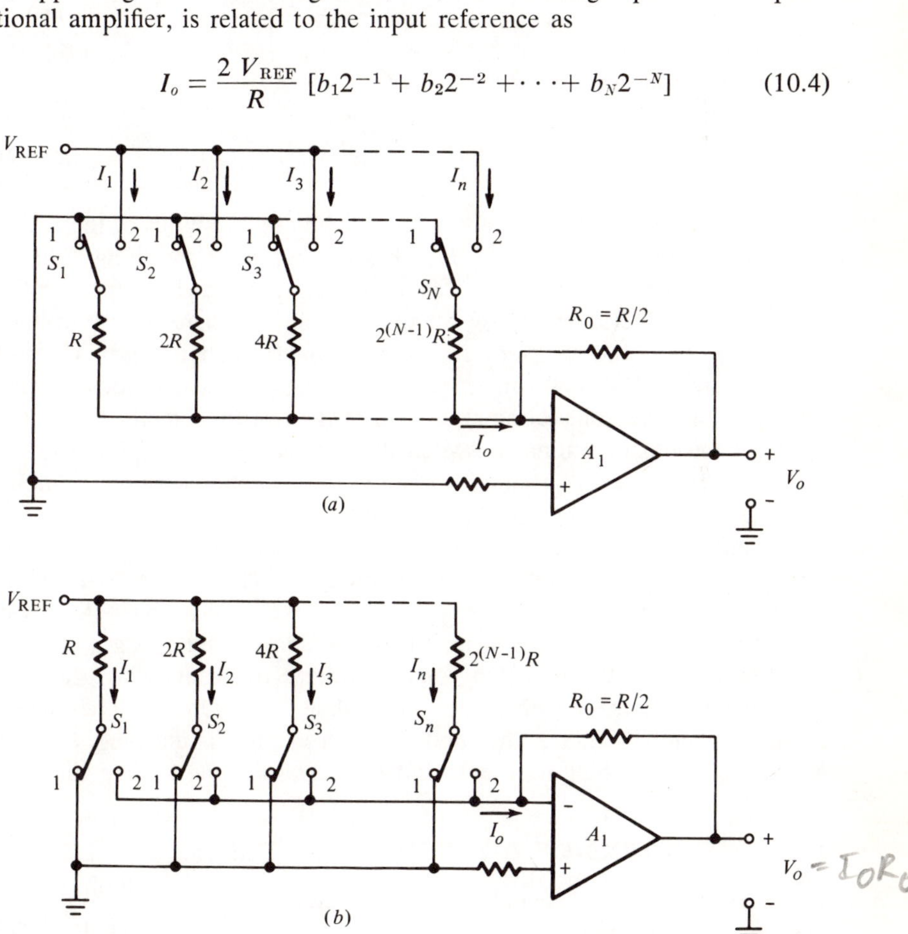

Figure 10.2 D/A converter configurations using (*a*) voltage switching; (*b*) current switching.

where the binary coefficients b_1, b_2, $\cdots$ b_N are either ZERO or ONE depending on the corresponding switch S_j being in position 1 or 2 in the Figure.

The output voltage V_o is directly proportional to I_o:

$$V_o = -I_oR_o = -V_{REF}[b_1 2^{-1} + \cdots + b_N 2^{-N}] \qquad (10.5)$$

where the op amp feedback resistor R_o, which serves as a scale factor, is set equal to $(R/2)$ for convenience.

As shown by Eq. (10.5), for a given number of bits N, the output exhibits (2^N) discrete voltage levels, ranging from 0 to a maximum value of

$$(V_o)_{\max} = V_{REF}\left[\frac{2^N - 1}{2^N}\right] \qquad (10.6)$$

with the minimum change, given as

$$(\Delta V_o)_{\min} = \frac{V_{REF}}{2^N} \qquad (10.7)$$

The binary bit coefficients are determined by the positions of the corresponding switches in the Figure. One has the option of switching either a voltage or a current in the circuit, as a function of the digital input. In the circuit of Fig. 10.2(*a*) voltage switching is employed where the net voltage across any one of the weighting resistors is switched either to ground or to V_{REF}. Figure 10.2(*b*) shows an alternate switching arrangement for the same circuit. In this case, one terminal of each of the resistor remains connected to V_{REF}; the other terminal is switched between the actual ground (position 1) and the "virtual ground" formed at the operational amplifier input. This method of switching is called "current switching."

In most applications, and particularly in integrated circuits, current-switching is normally preferred over voltage switching because it offers significant speed advantages. The network nodes have parasitic stray capacitances associated with them. Therefore, the sudden changes of node voltages during voltage switching creates voltage transients which have to die down before the circuit settles to its final state. On the other hand, during current switching, the node voltages remain unchanged. This minimizes the switching transients and the corresponding settling time.

10.2 DEFINITIONS OF D/A CONVERTER TERMS

The performance criteria and the design parameters for D/A converters are significantly different than other classes of circuits discussed so far. Therefore it is useful at this point to review some of the basic parameters and the terminology used in describing the D/A converter performance. Some of these terms are defined below:

Resolution: Indicates the number of possible analog levels available. It is normally expressed as the total number of input bits the converter will handle. For N-bit resolution, the converter must be capable of producing (2^N) discrete analog output levels.

Accuracy: A measure of the deviation of the analog output level from its predicted value. It can be expressed as a percentage of full scale output V_{FS}, or as a number of bits, or as a fraction (normally ½) of the least significant bit (LSB). N-bit accuracy means a maximum possible error V_E of

$$V_E \leq V_{FS}(2^{-N}) \tag{10.8}$$

Similarly, for an N-bit converter, ½ LSB accuracy means

$$V_E \leq V_{FS}\left(\frac{1}{2}\right)\left(\frac{1}{2^N}\right) = V_{FS}\left[\frac{1}{2^{N+1}}\right] \tag{10.9}$$

Note that resolution and accuracy are not necessarily equal. For example, one can have a 12-bit resolution with 10-bit accuracy, or vice versa.

Settling Time: Total time measured from a digital input change to the time analog output reaches its new value within a specified error band (typically ±½ LSB). Settling time determines their overall response speed of the converter.

Most Significant Bit (MSB): The digital input bit carrying the highest numerical weight; or the analog level shift corresponding to this bit. In a binary-weighted converter MSB creates an output level shift equal to ($V_{FS}/2$).

Least Significant Bit (LSB): The digital input bit carrying the lowest numerical weight; or the analog level shift associated with this bit. This is the smallest possible analog step (see Eq. (10.7) and is equal to ($V_{FS}/2^N$).

Two additional terms which are sometimes used to describe D/A converter are "linearity" and "monotonicity." These are illustrated in Fig. 10.3. When a binary count sequence is applied to the input bits, the output of the converter should be a staircase voltage with one LSB step. The uniformity (uniform width and height) of these steps determines the linearity of the output. The amount of "nonlinearity" allowable at the output is determined by the accuracy specification. The worst kind of nonlinearity is the nonmonotonicity of the output where the output voltage level does not change monotonically with the binary input code. Nonmonotonic behavior may come about due to accumulation of bit errors. The linearity and the monotonicity characteristics tend to degrade as the converter speed is increased, due to the inbalance in bit time constants.

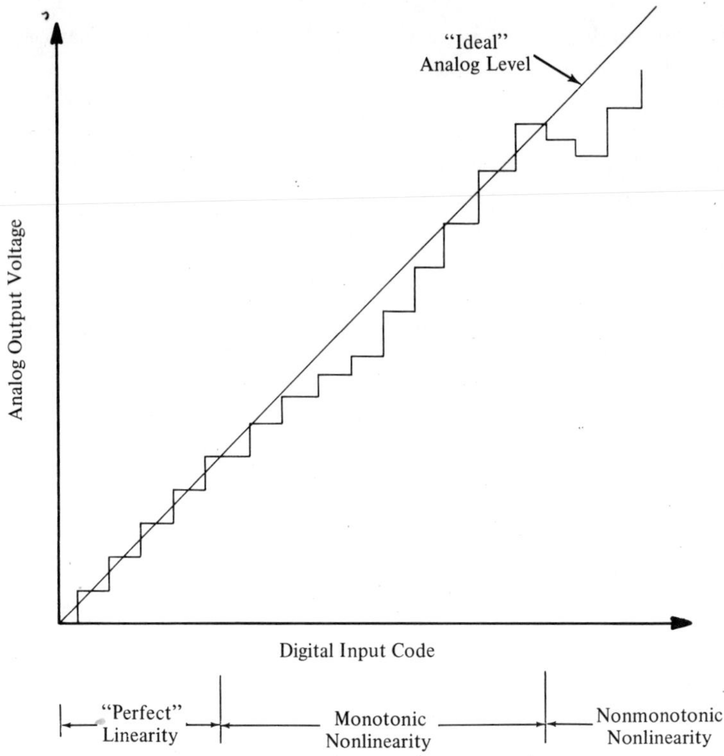

Figure 10.3 Graphical illustration of different non-linearities present at the analog output.

10.3 RESISTOR LADDER NETWORKS

In the D/A converter configurations shown in Fig. 10.2 the current-weighting function is achieved by using N parallel, independent branches in the weighting network which have respective impedance levels of R, 2 R, 4 R, 8 R, etc. In this type of a resistor network, the spread of resistor values increases very rapidly as the number of bits increases, such that the resistance of the MSB branch is related to that of the LSB branch as

$$\frac{R_{\text{MSB}}}{R_{\text{LSB}}} = \frac{1}{2^{N-1}} \tag{10.10}$$

Thus, for example, for 8-bit resolution, one needs a set of precision resistors covering a range of resistor values from R to 128 R. In monolithic or thin-film circuits such a wide range of resistor values are difficult to obtain with sufficient precision without resorting to expensive selection and trimming processes (the latter in the case of deposited resistors).

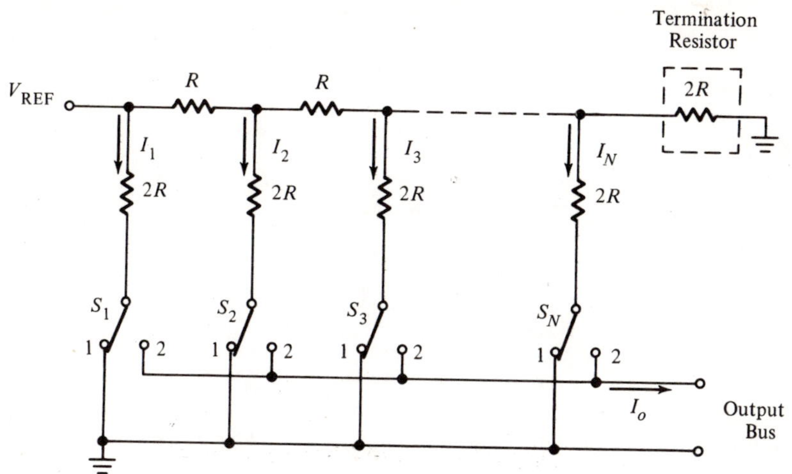

Figure 10.4 *R*-2 *R* Ladder Network.

An alternate resistor configuration which eliminates the large component spread of the binary-weighted resistor networks of Fig. 10.2 is the *R*-2 *R* ladder network shown in Fig. 10.4. In this type of a network the binary division of the currents I_1, I_2, etc. is achieved by successive partitioning of current between shunt (2 *R*) and the series (*R*) branches. Thus, the branch currents still satisfy the binary relationship

$$I_1 = 2\, I_2 = 4\, I_3 = \cdots = 2^{N-1}\, I_N \tag{10.11}$$

while maintaining the resistor values within an easily attainable (2/1) ratio. The *R*-2 *R* ladder requires twice as many resistors as the binary-weighted resistor network and must be properly terminated by a termination resistor (see Fig. 10.4). A particular advantage of the *R*-2 *R* ladder is that it can be used in conjunction with equal value current sources to perform the D/A conversion, as shown in Fig. 10.5. In monolithic circuits where excellent matching can be obtained between devices of the *same geometry* and operating at the *same current level,* equal-current D/A conversion offers significant advantages,[6] particularly for circuits having a complexity of 8 bits or more.

For monolithic integrated D/A converters, the *R*-2 *R* ladder is generally preferred over the binary-weighted ladder for circuits of 6-bit or higher complexity because it offers a lesser spread of component values.[7-9] However, the weighted-resistor ladders are better suited for design of 4-bit converter sections which can then be cascaded, as building blocks to form higher-order converters.[10-11] Figure 10.6 shows a method of combining two

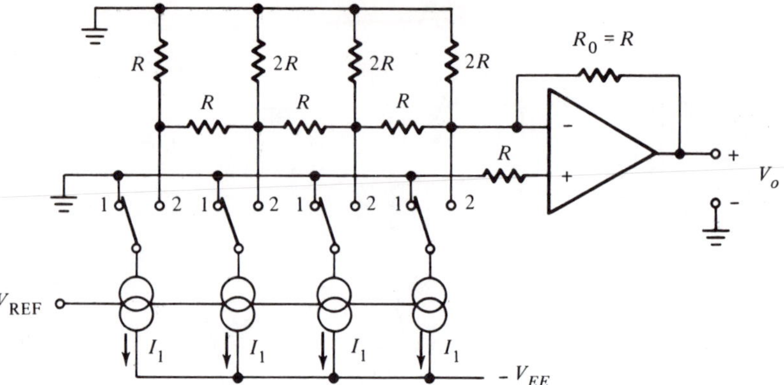

Figure 10.5 A four-bit D/A converter using *R*-2 *R* ladder and equal value current sources.

4-bit converter sections through a (16/1) current divider network (resistors R_1 and 15 R_1 of Fig. 10.6). In this configuration, with all switches closed, current I_A and I_B in each section of the resistor networks will be equal, i.e.,

$$I_A = I_B = I_1[1 + (1/2) + (1/4) + (1/8)] \tag{10.12}$$

where I_1 is the MSB current. The presence of (16/1) divider between the two networks forces the total output current to be

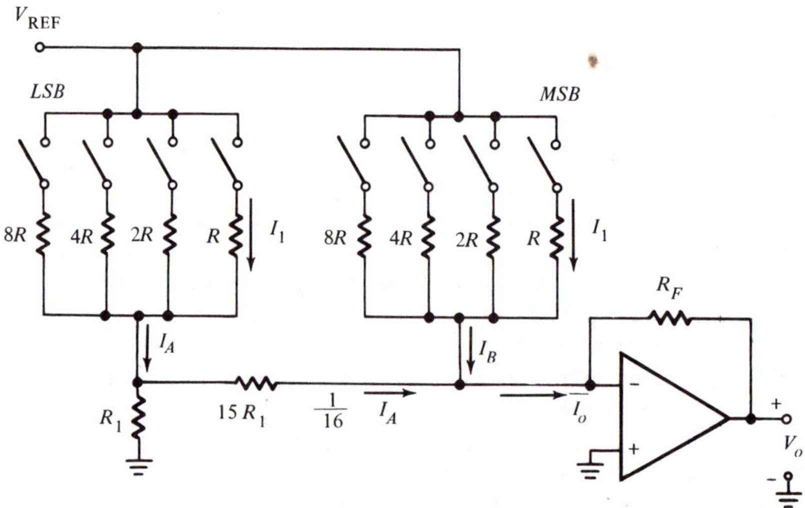

Figure 10.6 Cascading two 4-bit binary-weighted ladders to get 8-bit resolution.

$$I_o = I_B + (1/16)I_A = I_1 \sum_{N=1}^{8} \frac{1}{2^{N-1}} \tag{10.13}$$

which is equivalent to the output of a 8-bit binary-weighted resistor network.

The function of the resistor network is to ensure a binary division of current between each successive branch of the ladder. The absolute values of the currents are of secondary importance, but their ratios must be maintained within very close tolerances. Consequently, the absolute values of the ladder resistors are not important; however, their *matching* and *thermal tracking* properties must be maintained within very critical tolerances. These tolerance requirements will be discussed further in Section 10.5.

To ensure close matching and tracking of the resistor values, in the design of monolithic D/A converters additional care must be exercised. To facilitate close thermal coupling the resistors are grouped together in the layout, and wide resistor geometries (resistor width > 20 μ) are used to minimize the photomasking errors. A low temperature coefficient of resistance is also necessary to facilitate close thermal tracking. Some of the basic resistor types available for monolithic D/A converter applications are listed in Table 10.1. It should be noted that the matching tolerances specified in the table are for 30 μ wide identical resistors in close proximity and at least 5 squares long. Therefore, these values are somewhat better than those for ordinary (10 μ wide) resistors listed in Table 3.3.

TABLE 10.1 TYPICAL CHARACTERISTICS OF MONOLITHIC RESISTORS

Resistor Type	Sheet Resistivity (kΩ/sq)	Matching, % (sigma limit)	Temp. Coef. (ppm/°C)
Ni-Cr	0.04-0.4	0.3	50
Tantalum	0.2-5	0.3	50
Ion Implanted	0.2-10	0.3	250
Diffused (*p-type*)	0.1-0.2	0.5	1800

10.4 CURRENT SWITCHES

Most of the parasitics in monolithic circuits are capacitive. Therefore, current-switching is almost always preferred over voltage-switching for monolithic D/A converter applications. A number of current-switching schemes have been developed which are readily suitable for monolithic integration. Some of these circuit configurations will be reviewed in this section.

To be suitable for D/A converter applications, a current switch circuit must have the following desirable properties:

High Speed: Rapid switching with minimum settling time or transients. In order to reduce the effects of parasitic capacitances, the voltage swings at the switching node must be kept to a minimum.
Good Buffering: It should provide good isolation between the digital switching signals and the analog portion of the circuitry.
Low Reverse Leakage: Leakage current through the switch in the "OFF" state should be negligibly small.

The current switching action is accomplished by forward or reverse-biasing a diode or a transistor junction. Since the reverse leakages associated with silicon *p-n* junctions under normal operating conditions can be made negligibly small compared to bit currents, the low leakage requirement normally does not present a design problem, except at elevated temperatures.

Figure 10.7 shows two practical current-switch configurations commonly used in monolithic D/A design. In the circuit of 10.7(*a*),[12] the bit current I_n is carried by transistor T_3, which is permanently biased from a stable voltage reference. The current through T_3 is determined by V_{REF}, V_{BE} drop of T_3, and the voltage drop due to the resistor network in the emitter circuit. Therefore, T_3 acts as a constant-current sink, carrying a current I_n inrespective of the polarity of the digital input. When the digital input is low, corresponding to a digital zero, transistors T_1, T_2, and the diode D_1 are off; therefore, the entire bit current I_n is drawn from the output bus, through the forward-biased diode D_2. When a positive pulse (digital ONE) is applied to the base of T_1, T_1, T_2, and D_1 are brought into conduction, the voltage at node A is raised, and diode D_2 is turned off. Thus, the bit current I_n drawn by T_3 no longer flows through the output bus; instead it is switched to transistors T_1 and T_2.

In the switch circuit of Fig. 10.7(*b*) a *pnp* transistor is used as the switching element.[6] In this device can be either a lateral or a vertical *pnp* transistor using the device structures described in Chapter 2 (see Fig. 2.18). In this case the current-switching is accomplished by biasing the bit transistor T_1 in the OFF state and thus completely shutting off bit current I_n. With a digital ZERO applied to the binary input terminal, T_2 is biased off, and bit current I_n is conducted through T_1. When a digital ONE is applied to the input, T_2 is brought into conduction, which in turn raises the dc

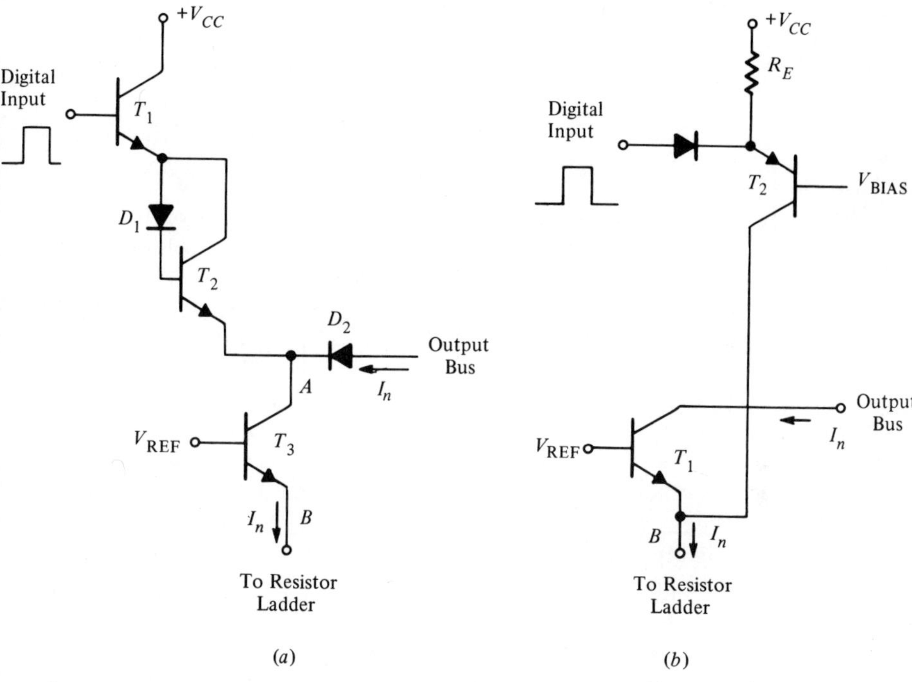

Figure 10.7 Practical current switch circuits.

level at the emitter of T_1 and turns it off. In this type of a switch, the current coming out of node B when T_1 is off and T_2 is on is *not* equal to the bit current. Therefore, this type of a current switch is only useful with binary-weighted resistor ladders (see Fig. 10.6) or with equal current D/A converters (see Fig. 10.5).

Figure 10.8(*a*) shows an alternate current-switch configuration suitable for integration. In this case an emitter-base avalanche diode Q_1 is used in the switching branch for dc level shifting. With a binary ZERO input, Q_1 and T_1 are both off and T_1 conducts the bit current I_n supplied by the output bus. When a digital ONE is applied to the base of T_1, both T_1 and Q_1 are brought into conduction. This raises the voltage level at node A, and turns T_2 off, reducing I_n to zero. Except for the use of an *npn* rather than a *pnp* transistor, the basic switch circuit of Fig. 10.8(*a*) is similar to that of Fig. 10.7(*b*).

In the switch circuits of Figs. 10.7 and 10.8(*a*), the switching node experiences the full voltage swing or level shift associated with the binary input. Since the steering diode D_2 of Fig. 10.7(*a*), or the base-emitter junctions of the transistors T_1 and T_2 of Figs. 10.7(*b*) and 10.8, have a finite

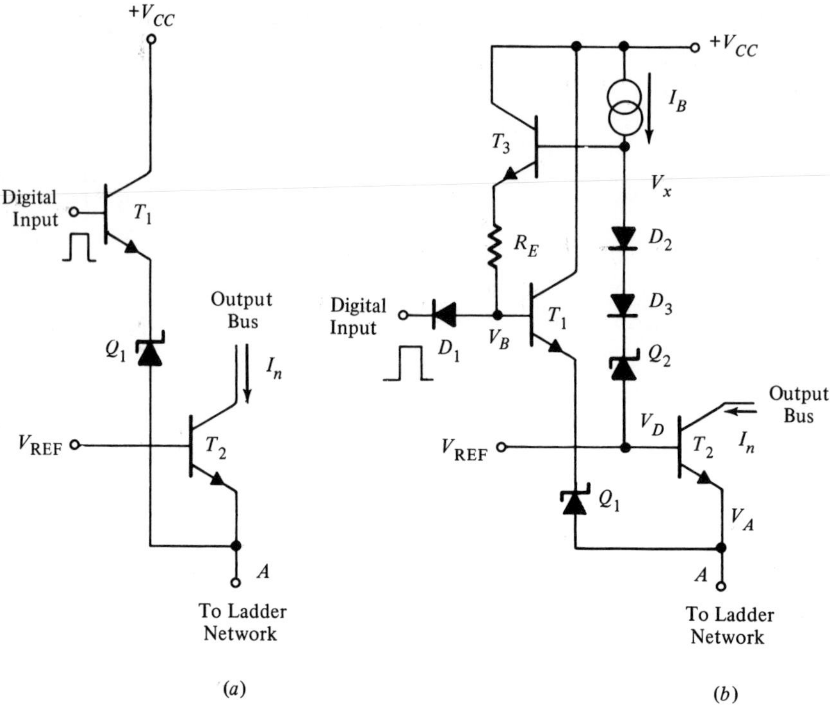

Figure 10.8 A high speed current switch configuration:[13] (*a*) simplified diagram; (*b*) actual circuit.

junction capacitance associated with them, these large voltage swings at the switching node can lead to excessive switching delays. To utilize the full speed advantage of current-, rather than voltage-, switching it is necessary that these voltage swings at the switching node be minimized. This can be done by using a voltage clamping circuit. Figure 10.8(*b*) shows a modified version of the basic switch circuit, utilizing such a voltage clamp.[13] The operation of the circuit can be explained as follows:

With a binary ZERO input, D_1 is on, T_1 and Q_1 are off, and T_2 conducts. Thus, the current switch is closed. With a binary ONE input, D_1 is off, T_1 and Q_1 conduct, and T_2 is cutoff; thus the switch is opened, and $I_n = 0$. When T_1 is conducting, the base of T_1 is at a potential V_B, independent of the binary input signal amplitude, where

$$V_B \approx V_X - V_{BE} \tag{10.14}$$

V_X is the dc bias level at the base of T_3. The voltage V_A at the emitter of T_2 is

$$V_A = V_X - 2V_{BE} - V_Z \tag{10.15}$$

where V_Z is the voltage drop across the avalanche diode Q_1. If the voltage drop across D_1 and D_2 is the same as the transistor V_{BE}, and Q_1 is identical to Q_2, then the voltage V_D at the base of T_2 is equal to

$$V_D = V_X - 2V_{BE} - V_Z \tag{10.16}$$

which is identical to the emitter voltage V_A. Thus, when T_2 is off, the maximum reverse bias across its base-emitter junction is clamped to zero volts, irrespective of the digital input level or the power supply voltage. Since all the V_Z and V_{BE} drops drift in the same manner with temperature changes, the clamp level across T_2 is also insensitive to temperature. Results reported in the literature[13] indicate that up to a factor of five improvement can be obtained in the settling time by using the clamping circuit of Fig. 10.8(*b*) to limit the voltage swings.

10.5 BASE CURRENT AND V_{BE} COMPENSATION

Figure 10.9 shows the basic circuit configurations for weighted-current type D/A converters, using weighted resistors or R-2 R type ladder networks. For simplicity, only a four-bit section is shown in the Figure. In this type of a converter, each bit transistor T_1, T_2, T_3, etc. carry one-half the amount of current carried by the preceding stage. Thus, for example, T_4 carries only one-eighth the current carried by T_1. Therefore, each bit transistor would exhibit a different V_{BE} drop unless care is taken to scale the device geometries so as to maintain the *same* current density across each base-emitter junction. Similarly, in the converter circuits of Fig. 10.9, the bit currents are summed or measured at the collectors of the bit transistors; whereas, their level is set by the weighting resistors in the emitter circuit. Threfore, the actual and the measured bit currents differ by the base current of the bit transistor. Considering the MSB current in Fig. 10.9, the measured current I_1 differs from the actual bit current I_1' as

$$I_1 = I_1' - I_{B1} \tag{10.17}$$

where I_{B1} is the base current of T_1. This error source can be compensated by subtracting a current approximately equal to I_{B1} out of I_1'. This is called "base current compensation," and can be implemented as shown in Fig. 10.10, for the case of a weighted resistor ladder.

From Fig. 10.10, the bit current I_1 for transistor T_1 can be expressed as

$$I_1 = \frac{V_{\text{REF}} - V_{BE}}{R} + (I_{B1} - I_{B1}') \tag{10.18}$$

If, by proper choice of R' the current level in T_1' is set equal to I_1, then

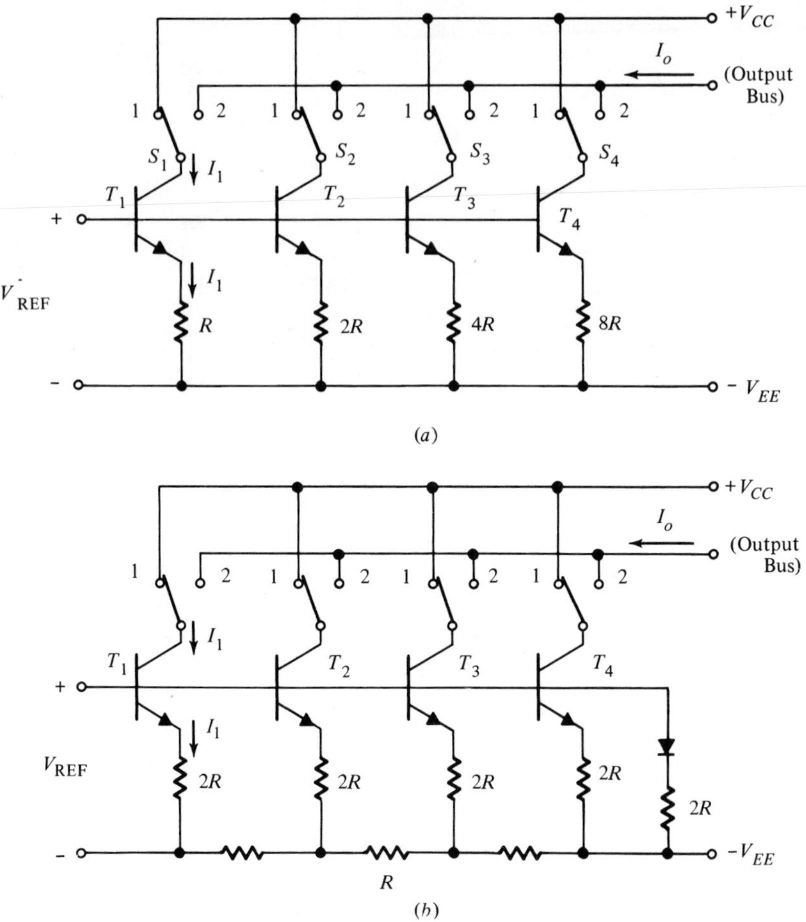

Figure 10.9 Four-bit D/A converter circuits using weighted-current sources.

$I_{B1} \approx I_{B1}'$, and the second term in Eq. 10.18 becomes negligibly small. A similar compensation can be provided for the second bit transistor T_2, as shown in the Figure, thus

$$I_2 = \frac{V_{REF} - V_{BE_2}}{2\,R} \tag{10.19}$$

If the emitter area of T_1 is chosen to be twice the emitter area of T_2, the two V_{BE} drops are equal and

$$I_1 = 2\,I_2 \tag{10.20}$$

In Fig. 10.10, the V_{BE} and the base current compensation are shown for

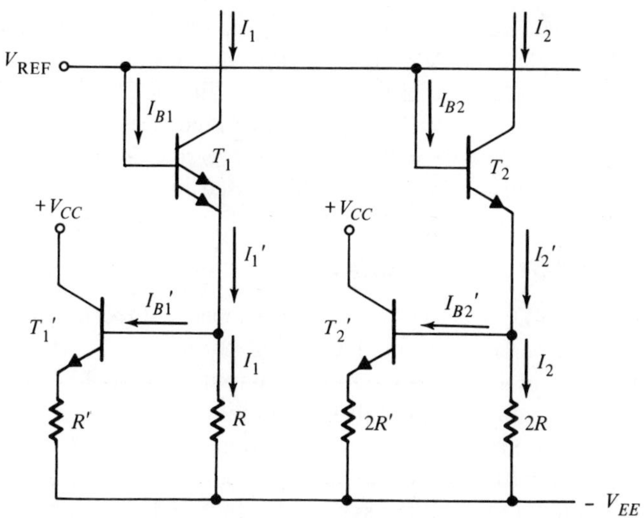

Figure 10.10 Compensation for base current and V_{BE} variations.

only the first two bit stages. However, this compensation scheme can be readily extended to the rest of the converter stages.

The practical limit on the V_{BE} compensation by area scaling is set by the device size considerations. To maintain equal current density, the emitter areas of the MSB and the LSB transistors must be related as

$$\frac{(\text{Area})_{\text{MSB}}}{(\text{Area})_{\text{LSB}}} = 2^{N-1} \tag{10.21}$$

The lower bound on the emitter area is set by photomasking tolerances. As the bit count N is increased, the emitter area of the MSB transistor gets prohibitively large. For example, for an 8-bit converter, an area mismatch of 128:1 is required.

As shown in Eqs. (10.18) and (10.19), V_{BE} appears in each of the current level expressions. This, in turn, introduces a temperature dependence on each of the bit current levels. In the actual design, this dependence is compensated by introducing an equal drift into the reference voltage by placing a diode in series with the reference, to form a new reference source V'_{REF} where

$$V'_{\text{REF}} = V_{\text{REF}} + V_{BE} \tag{10.22}$$

To ensure thermal tracking of temperature coefficients, the diode connected in series with V_{REF} must operate at the same current density as the bit transistors.

The design of a stable voltage reference is one of the key steps in the design of a complete D/A converter. For this purpose any one of the numerous voltage reference circuits discussed in Chapter 4 or Chapter 6 can be used, depending on the voltage level and accuracy requirements. Since the subject of voltage sources and voltage references have been covered earlier, they will not be explicitly discussed in this chapter.

10.6 ACCURACY CONSIDERATIONS

The accuracy of a D/A converter is stated in terms of the maximum allowable error in the analog output level. The output error is the difference between the measured and the predicted output level, and is normally expressed as a percentage of the full scale output voltage V_{FS}. As discussed in Section 10.2, accuracy is *not* the same thing as resolution. However, a converter circuit with high accuracy but poor resolution, or vice versa, offers very limited appeal to the system designer. Therefore, from practical design considerations, both the accuracy and the resolution specifications are normally chosen to be comparable. Typical accuracy requirement from a D/A converter circuit is of the order of ±½ LSB. This implies that the total allowable output error voltage ΔV_E for an N-bit converter with a full-scale output swing of V_{FS} is

$$\Delta V_E = \pm 1/2 \text{ LSB} = \pm \frac{V_{FS}}{2^{N+1}} \tag{10.23}$$

Thus, the total percentage error, ϵ_t, allowable at the output is

$$\epsilon_t = \pm \frac{100}{2^{N+1}} \% \tag{10.24}$$

for ½ LSB accuracy. Thus, the accuracy requirement increases very rapidly as the bit count is increased. For example, a 4-bit converter with ±½ LSB accuracy can allow a total error of ±3.12 percent at the output; but for 6- and 8-bit converters, the allowable error decreases to ±0.78 and ±0.195 percent, respectively.

The total error ϵ_t can be separated into two parts:

$$\epsilon_t = \epsilon_{to} + \epsilon_t(T) \tag{10.25}$$

where ϵ_{to} is an initial "offset" error, independent of temperature; and $\epsilon_t(T)$ is the temperature dependent portion of the total error arising from temperature drifts and imperfect thermal tracking of matched components. In D/A converters with 6-bit or higher resolution, an offset adjustment means is normally provided to null out ϵ_{to}.

Every D/A converter normally contains four basic sections: voltage reference, binary switches, resistor network, and current-summing amplifier. Each of these sections in turn contribute a component to the total output error:

$$\epsilon_t = \epsilon_R + \epsilon_S + \epsilon_L + \epsilon_A \tag{10.26}$$

where ϵ_R = reference error
ϵ_S = switch error
ϵ_L = resistor ladder network error
ϵ_A = summing amplifier error

The switch error is significant in circuits using voltage switches, due to the finite ON resistance of switching devices. In current-switching D/A circuits, the leakage currents associated with the OFF state of the switch can be a potential problem if the circuit is operated at low current levels (i.e., LSB current of $<10\ \mu$A). In most designs where fairly high current levels are maintained, the leakage currents are negligible except at very high temperatures. Thus, for monolithic designs using current-switching, ϵ_S is usually neglected. The remaining three components of error are roughly equal in their significance. In an actual circuit design, the upper bound on ϵ_t is set by the accuracy requirement (see Eq. 10.24). Thus, one is forced to work backward from ϵ_t of Eq. 10.24 and estimate the maximum bounds on each of the error contributions of Eq. 10.26. A pessimistic "worst case" analysis method is to assume that all errors are additive and thus allocate equal shares of maximum allowable error to each of the contributors (neglecting ϵ_S). Thus, in percentage basis

$$\epsilon_R \approx \epsilon_L \approx \epsilon_A \leq \frac{\epsilon_t}{3} = \frac{100}{(3)2^{N+1}}\ \% \tag{10.27}$$

A more optimistic approach is to assume that all error sources are totally uncorrelated, and that they add in an rms manner, i.e.,

$$\epsilon_t^2 = \epsilon_R^2 + \epsilon_L^2 + \epsilon_A^2 \tag{10.28}$$

which leads to an overall error allocation of

$$\epsilon_R \approx \epsilon_L \approx \epsilon_A \leq \frac{\epsilon_t}{\sqrt{3}} \tag{10.29}$$

In most design cases, the rms error allocation is somewhat optimistic and the real situation lies somewhere between Eqs. (10.27) and (10.29).

It should be noted that the equal error allocation of Eqs. (10.27) and (10.29) is reasonable but somewhat arbitrary since the exact nature of error sources depends on the specific design choices. In certain cases, the reference or the summing amplifier, or both, can be left external to the integrated circuit leaving ϵ_S and ϵ_L as the major sources of error for the monolithic portion of the design. Each of the error components also exhibits a finite temperature dependence; therefore, their contribution to ϵ_{to} and $\epsilon_t(T)$ should be considered separately.

At this point it is instructive to examine typical accuracy and temperature-tracking requirements for a practical design example. For illustration purposes, consider the case of a monolithic 8-bit D/A converter with ±½ LSB accuracy over a temperature range of −55 to +125°C. The accuracy requirement for each section will be examined separately assuming rms-type equal error allocation (Eq. 10.29).

Reference Error: Assuming that the initial reference offset error, ϵ_{Ro}, can be nulled out by a zero-scale adjustment, only the temperature dependent portion of reference error, $\epsilon_R(T)$, need to be considered. For 8-bit accuracy, this implies that

$$\epsilon_R \approx \epsilon_R(T) \leqq \pm \frac{0.195}{\sqrt{3}} \% \tag{10.30}$$

To satisfy this requirement, a voltage reference with a temperature dependence of ±10 ppm/°C is required.

Ladder Error: The resistor ladder error ϵ_L is due to the matching and temperature tracking errors associated with the ladder network. The matching error due to the ladder resistors cannot be ruled out since the latter are constantly switched in or out of the circuit as a function of the digital input word. Therefore, in calculating the total error allocation for the ladder network, both the *matching* and the *tracking* errors must be considered. The most significant error contribution comes from the resistors associated with the MSB and the next significant bit, and each successive bit in turn contributes a diminishing fraction of error. The MSB is responsible only for ½ V_{FS}; therefore, the total matching and tracking error between MSB and the next bit can be as large as 2 ϵ_L, i.e., ≈ ±0.2 percent.[14] Assuming that the matching and the tracking errors are approximately equal, then the matching error in the first two bit resistors must be less than 0.1 percent, and their temperature-tracking error less than ±10 ppm/°C for 8 bit, ±½ LSB accuracy. This kind of matching and tracking cannot be obtained with diffused resistors. It normally requires the use of deposited thin-film resistors.[6]

Amplifier Error: The summing amplifier has three sources of error contributions, due to the input bias current, I_{Bo}, input offset voltage, and the temperature drift of the input offset. A general design procedure is to maintain I_{Bo} at a level ≦5 percent of LSB current. In an 8-bit converter with MSB current of 1 mA, the LSB current is approximately 8 μA. Thus, the amplifier bias current should be <400 nA over the operating temperature range.

Assuming that the summing amplifier is operating at unity gain configuration with 5 V full-scale output, the total allowable output error due to the amplifier offset is ±0.1 percent of V_{FS}, or ±5 mV. A typical monolithic operational amplifier exhibits an input offset voltage of ±3 mV, and a thermal drift of ±15 μV/°C, which can produce a worst-case error of ±4.5 mV over the temperature range. Therefore, the summing amplifier performance requirements are readily compatible with monolithic technology.

10.7 MONOLITHIC DESIGN EXAMPLES

A large number of integrated D/A converters are currently available from over 200 separate manufacturers.[7] These devices range from low resolution units (4 to 6 bits) to high accuracy precision circuits (10 to 12 bits). Most of these designs are presently built in a modular form, using hybrid integrated circuits. However, in the low and medium accuracy (4 to 8 bits) D/A converter field monolithic designs are rapidly gaining wide acceptance. In this section some practical monolithic design examples will be examined to illustrate the capabilities of monolithic D/A converters. The two specific design examples to be considered are the Harris Semiconductor HI-1080 8-bit D/A Converter, and the Presicion Monolithics DAC-01 6-bit D/A Converter. These particular examples are chosen because they give a good representation of circuit accuracy and complexity that can be economically fabricated in large volumes, with the current, state-of-the-art, monolithic IC technology.

Figure 10.11 shows a simplified circiut diagram for a monolithic 6-bit D/A converter[8] in terms of the circuit package. In this particular design, all the basic sections of the converter including the voltage reference and the summing amplifier are contained in the monolithic chip. The basic converter scheme uses the weighted-current switching configuration of Fig. 10.9(*b*). Current switches S_1 through S_6 are designed using a circuit configuration similar to that of Fig. 10.7(*a*). The *R*-2 *R* ladder is formed using conventional base-diffused resistors. The internal reference source V_{REF} sets the base bias for bit transistors T_1 through T_6.

In most D/A converter applications the output voltage is required to vary between zero and the full-scale output V_{FS}, depending on the digital input. This type of operation where the output level does not change polarity is called "unipolar" operation. In certain applications it is necessary to have the output be "bipolar," i.e., have a symmetrical swing, from $(-V_{FS}/2)$ to $(+V_{FS}/2)$. The circuit of Fig. 10.11 can be made to operate in either unipolar or bipolar mode, depending on the interconnection of the circuit terminals. This added flexibility is provided by the offset current source I_x and the added bit transistor T_7 in Fig. 10.11.

When the circuit is used in unipolar mode, terminal 12 is grounded,

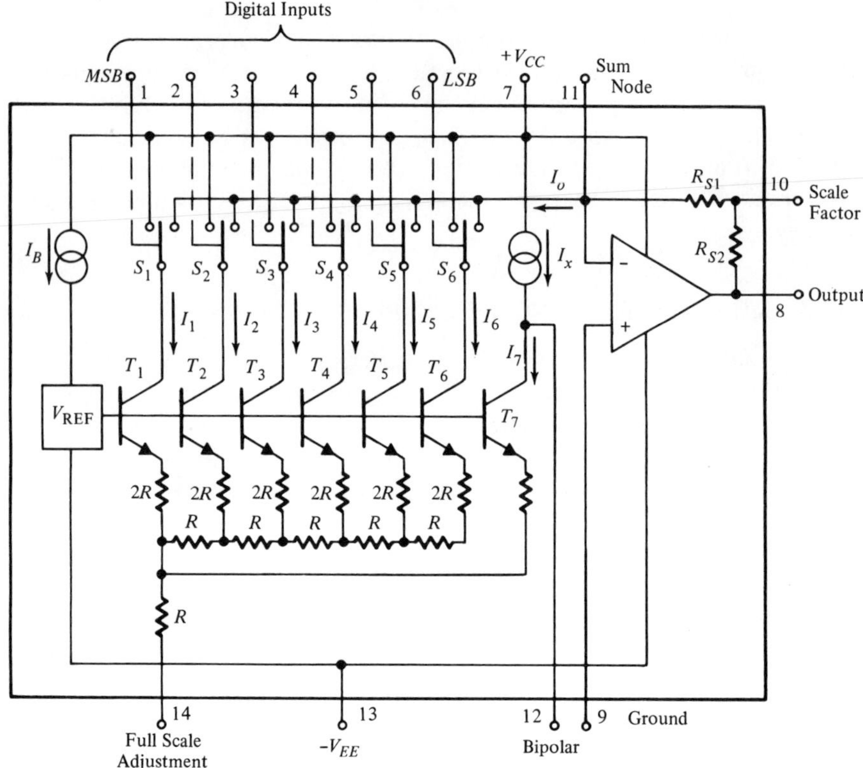

Figure 10.11 Simplified circuit diagram for a 6-bit monolithic D/A converter.[8]

shunting the offset current source I_x to ground. Thus, the output voltage varies from zero to $\pm V_{FS}$, as a function of the bit current I_o. Output voltage is equal to V_{FS} when I_o is maximum and equal to I_{FS}. The bipolar operation is obtained using an offset technique, by connecting terminal 12 to the summing node (terminal 11), instead of ground. This forces an offset current equal to $(I_x - I_7)$ into the summing node, such that

$$V_o = [I_o - (I_x - I_7)]R_F \tag{10.31}$$

where R_F is the total feedback resistance across the summing amplifier. If one can set $(I_x - I_7)$equal to $I_{FS}/2$, then a net scale offset of $V_{FS}/2$ can be obtained at the output.

$$\frac{I_{FS}}{2} = I_{MSB} - \frac{I_{LSB}}{2} \tag{10.32}$$

One can obtain a net output offset of $I_{FS}/2$ by setting $I_x = I_{MSB}$ and $I_7 = I_{LSB}/2$. This is the design technique used in the circuit of Fig. 10.11 to provide bipolar operation, i.e., a current equal to the most significant bit current I_{MSB} is generated in the offset current source I_x, and a small cor-

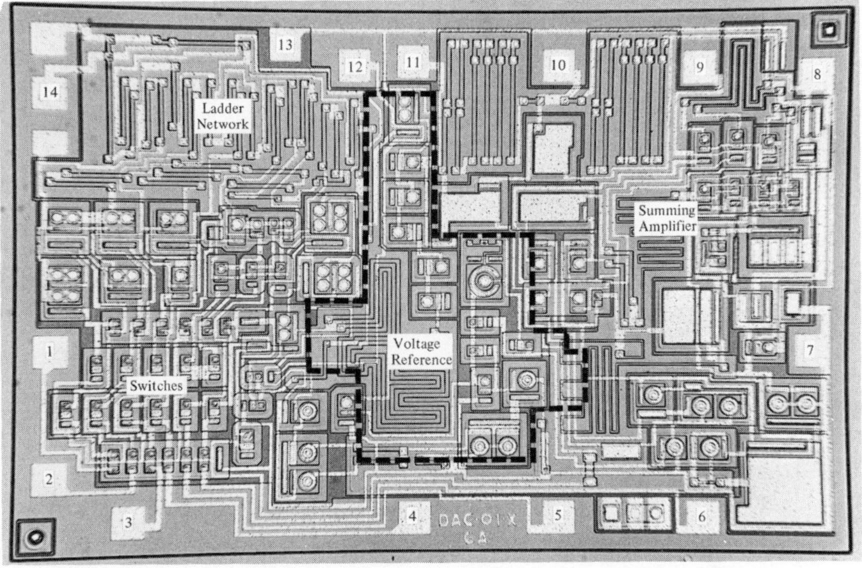

Figure 10.12 Circuit layout for the monolithic D/A converter of Figure 10.11 (chip size: 70 mils × 106 mils). (*Photo: Precision Monolithics, Inc.*)

rection current I_7, equal to ½ I_{LSB}, is subtracted from it to provide the correct offset current given by Eq. 10.32.

Figure 10.12 shows the monolithic chip photograph for the complete 6-bit D/A converter system of Fig. 10.11. The circuit chip, measuring 70 × 106 mils, uses only diffused components. The basic terminals in the circuit layout are also identified. The voltage reference circuit takes up the center portion of the layout, as outlined within the dotted area. The ladder network and the current switches take up the left half of the circuit layout. The bit transistors associated with the first four bits (T_1 through T_4 of Fig. 10.11) have their emitter areas scaled so as to maintain the same V_{BE} drop. To improve matching, large geometry transistors with circular emitters are used for $T_1 - T_4$. Base current compensation is also utilized (see Fig. 10.10) to maintain close tolerances among the first four bit currents. The right-hand side of the chip is taken up by the summing amplifier. The summing amplifier is internally compensated and has the following typical characteristics:

Input bias current	100 μA
Input offset voltage	7.5 mV
Open loop again	15,000
Slew rate	30 V/μ sec

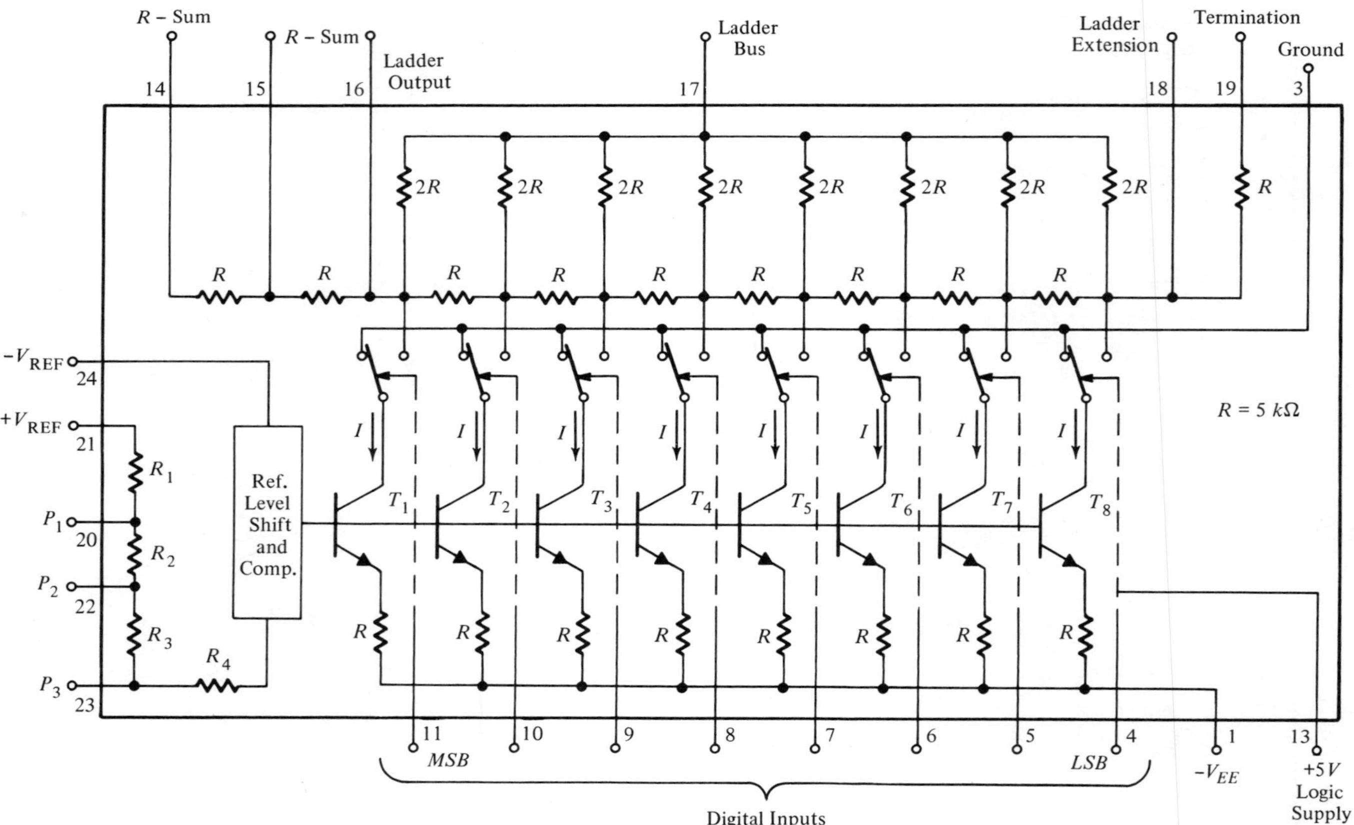

Figure 10.13 A monolithic 8-bit digital-to-analog converter (HI-1080).

The performance characteristics of the complete monolithic D/A converter are briefly summarized below:

Resolution	6 bits
Accuracy	±½ LSB
Full scale drift	80 ppm/°C
V_{FS} (adjustable)	0 to 10, ±5, ±10 V
Settling time	1.5 μsec
Power consumption	200 mV
Supply voltage	±15 V
Operating temperature range	−55 to +125°C

Figure 10.13 shows the simplified schematic for a commercially available 8-bit monolithic D/A converter.[6] In this case, the summing amplifier and the voltage reference are left external to the monolithic chip. The numbers associated with the circuit terminals in the Figure correspond to the external leads for a 24-pin package. This circuit is designed using the equal-value current source method, similar to that shown in Fig. 10.5. Thus, current source transistors T_1 through T_8 all carry the same current I. The current switches are designed using a circuit configuration similar to that of Fig. 10.7(*b*). The circuit is fabricated using dielectric isolation techniques, and the *pnp* transistors necessary for the current switches use a vertical, rather than a lateral, device structure. This type of a *pnp* can be readily obtained using dielectric isolation techniques (see Fig. 2.18) and provides significant speed advantages over the lateral devices.

The circuit can be operated with either positive or negative polarity external reference, and the unusual reference terminal is grounded. The taps P_1, P_2, and P_3 in the reference circuit are used for the fine adjustment of the full-scale output voltage. An internal level shift and compensation circuit converts the input reference to a desired dc level to bias the bit transistors, and also compensates for transistor V_{BE} variations. The ladder output also has provisions for internal resistors of value R and 2 R which can be connected in feedback around an external summing amplifier. The use of equal-value current sources with the R-2 R ladder has the advantage of keeping all resistor and current levels within a factor of two of each other. However, since all bit transistors now conduct a current equal to the MSB current, the overall power dissipation is increased significantly.

Figure 10.14 shows the chip photograph for the 8-bit converter. Thin-film (Ni-Cr) resistors are used to keep the resistor matching and tracking errors within the ±½ LSB tolerance. The R-2 R ladder network is shown in the upper half of the circuit layout and is comprised of a number of identical resistors of value R, which are then interconnected to form the ladder. In this particular design, R is chosen to be 5 kΩ. The width of the resistors are chosen to be sufficiently large ($\approx 30\ \mu$) to minimize photomasking errors.

The bottom half of the circuit layout is taken up by the current switches and the equal-value current sources. The current-source resistors are chosen to have the same value and geometry as the ladder resistors, and are located in close proximity to the current sources. These resistors are shown as thin-film strips in between the current-source transistors in the lower half of the layout. The reference level shift and compensation circuit is located along the left edge of the chip layout. The respective terminals of the circuit are also identified in the photograph.

The typical performance characteristics of the 8-bit D/A converter (HI-1080) are briefly summarized below for -55 to $+125°C$ operation:

Resolution	8 bits
Accuracy	$\pm\frac{1}{2}$ LSB
Full-scale drift	20 ppm/°C
Settling time	0.75 μsec
Power consumption	450 mW
Supply voltage	+5 V, -15 V

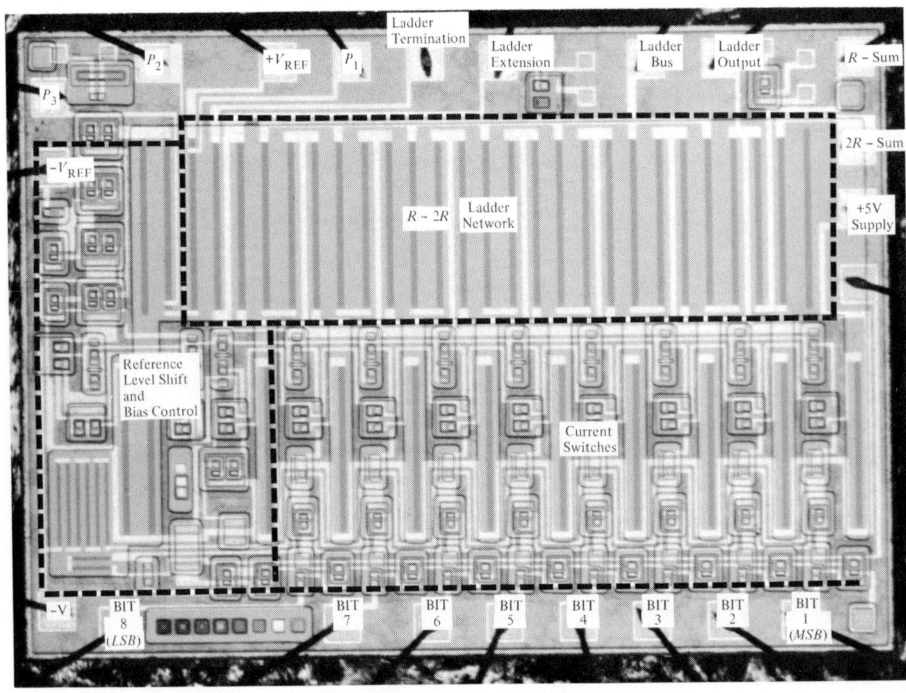

Figure 10.14 Topological layout for the 8-bit D/A converter. (*Photo: Harris Semiconductor.*)

PART II Analog-to-Digital Converters

10.8 PRINCIPLES OF A/D CONVERSION

The function of an analog-to-digital or A/D converter is to convert an analog voltage into a digital word. The D/A converter circuits described in the preceding sections are basically "decoding" devices where a digital word is decoded into an analog output. The A/D converters in turn perform the converse of this operation; i.e., they "encode" a given analog voltage into a digital output of predetermined bit length.

In an A/D converter, the input analog voltage V_A is approximated as a binary fraction of a reference voltage V_{REF}. Thus, the output of the converter, corresponds to a digital word D, given as

$$D = \frac{V_A}{V_{REF}} = b_1 2^{-1} + b_2 2^{-2} + \cdots + b_N 2^{-N} \tag{10.33}$$

where N is the length of the digital word in bits, and b_1 are the binary bit coefficients having a value of either ONE or ZERO. The bit coefficients which form the digital output can be obtained from the output of the A/D converter either simultaneously in the form of N parallel outputs (see Fig. 10.1(*b*), or can be sequentially displayed at the same output terminal. These output formats are called "parallel" and "serial" outputs. In serial output format, the bit coefficient b_1, corresponding to MSB, normally is calculated and displayed first, then followed by bits of successively decreasing importance.

In encoding an analog voltage V_A into a binary coded output given by Eq. (10.33), one effectively "quantizes" V_A into any one of a finite number of discrete levels separated by one LSB of the digital word. This results in a finite resolution or "quantization" error in the A/D conversion process, which can have a maximum value of ±½ LSB. In terms of an arbitrary analog voltage $V_A < V_{REF}$, this leads to a quantization error, ΔV_A, where

$$0 \leqq |\Delta V_A| \leqq \frac{V_{REF}}{2^{N+1}} \tag{10.34}$$

Note that quantization error is inherent to the digital encoding process, and is therefore present in any A/D converter.

During the conversion process, the analog input V_A is sampled, and its digital counterpart is generated a finite time interval later due to the finite conversion rate of the A/D converter. This time taken to complete the conversion of an analog input to a digital word is referred to as the conversion or "aperture" time for the system. How fast the A/D conversion must be performed is determined by the frequency content of the analog input and by the required accuracy of conversion, or by a combination of these two factors.

If the input analog signal varies as a function of time, the presence of a finite aperture time can lead to an additional error in the encoded input. For example, if the input is a linearly varying function of time, the aperture error ΔV_x can be related to the analog input as

$$\Delta V_x = \left(\frac{dV_A}{dt}\right) t_x \tag{10.35}$$

where t_x denotes the aperture time. Thus, as the frequency content of the input increases, the aperture error due to finite conversion rate increases quite rapidly.

The analog-to-digital conversion function can be performed in a wide variety of ways;[15] however, not all of these methods are suitable for integrated circuits. In implementing the A/D conversion with integrated circuits, the designer faces a fundamental problem: All the A/D conversion techniques require both analog and digital circuitry. Since high speed digital and analog circuits cannot, in general, be fabricated in monolithic form, due to different processing requirements, A/D converter design is more readily suited to hybrid, rather than monolthic, integration, where the analog and the digital parts of the system can be partitioned into separate chips.

The A/D conversion techniques suitable for integration can be classified into the following categories:

1. Parallel feedback converters
2. Serial feedback converters
3. Indirect converters
4. Simultaneous converters

In the following sections, each of these classes of converters will be reviewed briefly.

10.9 PARALLEL FEEDBACK A/D CONVERTERS

Parallel feedback type A/D converters operate by using a D/A converter block in a feedback loop. Most of the presently available A/D converters fall into this category. There are three major types of parallel feedback A/D converter circuits: tracking (or servo) converter; staircase converter; successive-approximations type converter.

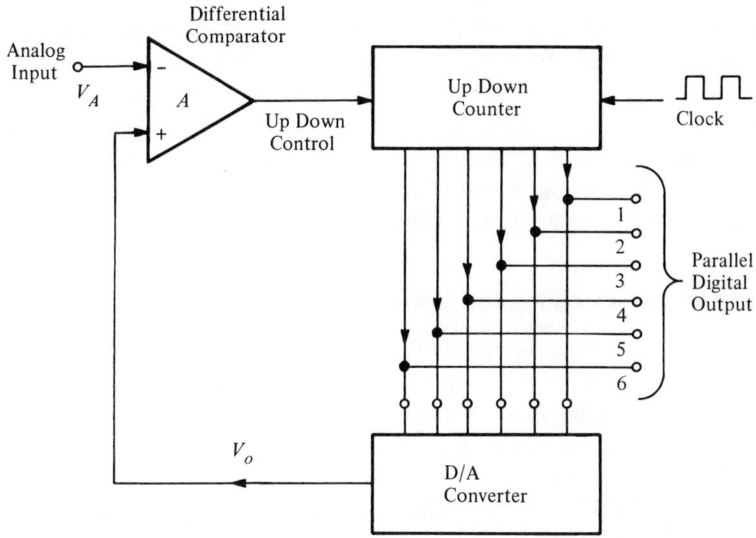

Figure 10.15 Tracking-type analog-to-digital converter using a D/A converter in feedback.

Tracking Converter

Figure 10.15 shows a simplified system block diagram for a tracking or servo type A/D converter. It employs a binary up-down counter to generate a binary output which in turn drives a D/A converter in feedback around a differential comparator. Starting from zero, the system counts the clock pulses, and the binary count is then converted to an analog voltage V_o at the output of the D/A converter and compared to the actual analog input V_A. When V_o reaches V_A, the comparator switches and stops the count. The counter output necessary to set $V_o = V_A$ is the digital word corresponding to the analog input voltage. Note that, in this case, the reference voltage V_{REF} is included as an integral part of the D/A converter. For full-scale digital output, this type of converter is quite slow since $(2^N - 1)$ clock cycles need to be counted. However, it has a very rapid small signal response. For example, if the input varies only a small amount, the output can follow it within a few clock cycles. The name, a "tracking" or "servo" type converter, stems from this property.

Staircase Converter

This is a simplified version of the tracking converter. Instead of an up-down counter, a simple binary type is used. In each conversion step the counter is reset to zero and starts counting the clock pulses. This creates a "staircase" output from the D/A converter. The count is stopped when the D/A converter output equals the analog input and the last count present in the counter corresponds to the digital word for the analog input. Unlike the tracking converter, the staircase type A/D converter resets itself to zero after every conversion step. Thus, it does not have the rapid small signal response properties of a tracking converter.

Successive-Approximation Converter

This type of converter approximates the analog input by successively trying a digital ONE in each successive bit of a feedback D/A converter, starting with MSB. A block diagram of such a system is shown in Fig. 10.16. In the first step of conversion, a ONE is tried for the MSB. If $V_A > V_o$, and the comparator output remains unchanged, ONE is retained; otherwise it is replaced by a ZERO and a ONE is tried in the next bit. In this manner the approximation process continues until all bits are calculated. The

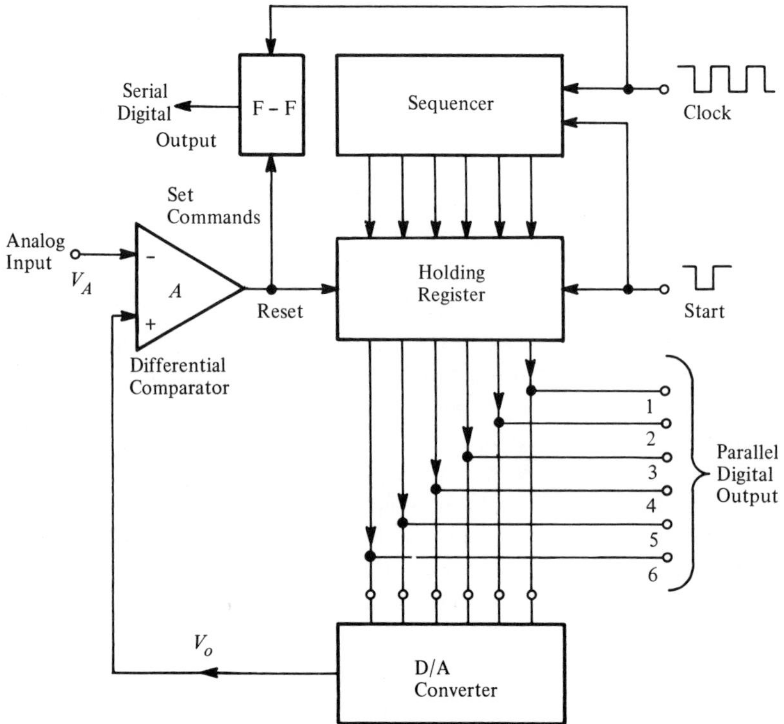

Figure 10.16 Successive-approximation type A/D converter.

decision making logic, concerning whether to leave a trial ONE or replace it with a ZERO, is performed by the comparator and the sequencer blocks, as shown in Fig. 10.16. In this type of a converter each bit of the digital word is calculated one at a time starting with the MSB. The set and reset commands of the comparator in each approximation step indicate the value of the corresponding bit. Thus, this output corresponds to the digital word in a serial form. It can be shown that, for N-bit resolution, at most N successive approximations are needed to reach the final value.[4] Since each approximation takes only one clock cycle, the aperture time of this type of a converter is quite short, i.e., always less than or equal to N clock cycles. This is a very significant speed improvement over the other two types of parallel feedback converters; however, it is obtained at the expense of increased logic complexity. At the beginning of each conversion cycle, the holding register is set to zero and the complete conversion sequence is repeated. Therefore this type of a converter does not have the small signal response capability of a tracking or servo converter.

The successive-approximation type converter is the most versatile of the three A/D conversion schemes discussed so far, since it offers the highest conversion rate and also provides parallel output capablity.

10.10 SERIAL FEEDBACK A/D CONVERTERS

Serial feedback A/D converters do not require the use of a D/A converter in the feedback path. Instead, the binary value of each successive bit is computed by feeding back the calculated value of the previous bit. There are two classes of serial feedback converters: circulating converters and charge equilizing converters. Both of these are well covered in the literature.[4,16] In this section only the circulating converters will be reviewed briefly.

The circulating-type serial feedback A/D converter operates by sampling the analog input V_A at the beginning of the conversion cycle and retaining the value of V_A in a sample-and-hold circuit only long enough to calculate the value of the MSB bit coefficient, b_1. This is done by subtracting one-half of the internal reference voltage, $V_R/2$, from V_A. Thus, $b_1 = 1$ if $V_A \geqq V_R/2$, and $b_1 = 0$ if $V_A < V_R/2$.

After calculating b_1, a new internal voltage V_2 is generated from the value of b_1 and the input analog voltage V_A, still present at the input sample-and-hold circuit as

$$V_2 = 2(V_A - b_1V_R/2 + \overline{b_1}V_R/2) \tag{10.36}$$

where $\overline{b_1}$ is the complement of b_1. This internal voltage V_2 now replaces the actual analog voltage V_A in the sample-and-hold circuit and the conversion process proceeds to b_2. The second bit coefficient b_2 is calculated by comparing V_2 to $V_R/2$; $b_2 = 1$ if $V_2 \geqq V_R/2$ and $b_2 = 0$ if $V_2 < V_R/2$. After b_2 is calculated, a third internal voltage is generated as

$$V_3 = 2(V_2 - b_2V_R/2 + \overline{b}_2V_R/2) \tag{10.37}$$

which replaces V_2 in the sample-and-hold circuit and the conversion proceeds to the next bit, until the desired number of bit coefficients are calculated. After the Nth bit is computed, the system is reset to its initial state and the next conversion cycle is initiated by sampling input voltage V_A. Note that in this type of conversion each intermediate bit coefficient b_i is determined by comparing an internally generated intermediate voltage level V_i with the fixed reference $V_R/2$. The intermediate voltage level V_i is in turn calculated from the recurrence relationship:

$$V_i = 2(V_{i-1} - b_{i-1}V_R/2 + \overline{b}_{i-1}V_R/2) \tag{10.38}$$

The digital output of the converter is obtained serially, with MSB first. In this type of a converter only N conversion steps are necessary to obtain N-bit resolution. However, each conversion step is relatively slow, since a number of logic functions have to be performed and a new intermediate voltage V_i has to be generated prior to calculating the appropriate bit coefficient.

10.11 INDIRECT A/D CONVERTERS

The indirect converters operate by initially transferring the analog voltage to an intermediate signal form (such as frequency, pulse width, etc.) and then converting them into a binary word. Indirect converters in general require much simpler circuitry than other types of converters. However, their conversion rate is in general slower than other classes of converters.

One of the most commonly used indirect A/D converter configurations is the "ramp-comparator" or the "pulse-width modulator" converter system of Fig. 10.17. The operation of the circuit can be explained as follows: At the start of conversion, cycle switch S_1 is opened and C_1 is charged with a constant current I_1 generating a ramp voltage V_B where

$$V_B = (I_1/C_1)t \tag{10.38}$$

At the instant S_1 is opened, the counter is activated and counts the incoming clock pulses. This count continues until the ramp voltage V_B reaches the analog input level V_A. Then, the comparator stops the counter, and the last count from the binary counter corresponds to the digital output. After the conversion is completed, the programmer resets the counter

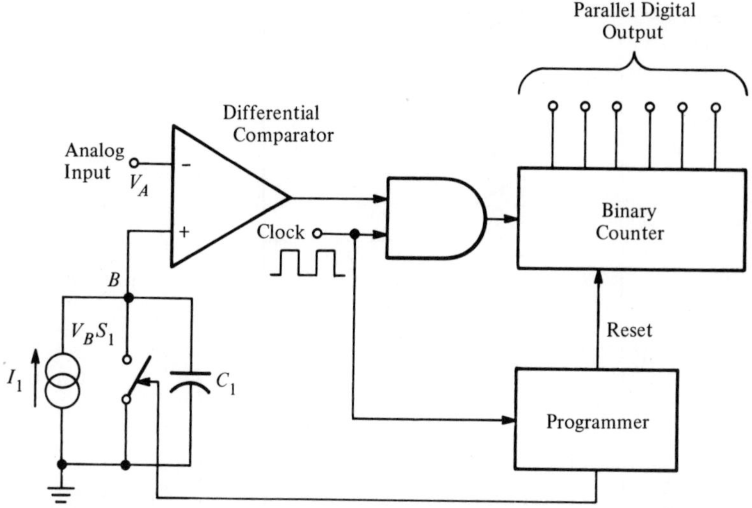

Figure 10.17 Indirect A/D conversion using ramp comparison.

to zero, and closes S_1 and the system is ready for the next conversion cycle.

In this type of a converter $(2^N - 1)$ clock pulses must be counted to obtain a full-scale output for N-bit resolution. This results in a very slow conversion rate. Indirect conversion techniques in general trade off response speed for reduced complexity.

10.12 SIMULTANEOUS CONVERTERS

Simultaneous or parallel A/D conversion is by far the fastest and conceptually simplest conversion. This type of a converter uses a separate analog comparator with a fixed reference for every quantization level in the digital word, from zero to full scale. Then, the outputs of these comparators are appropriately interconnected by the encoding logic circuitry to produce a parallel digital output.

Figure 10.18 shows a block diagram of a simultaneous A/D converter system. For N-bit resolution $(2^N - 1)$ separate comparators and separate reference levels are needed. Thus, the system complexity increases very rapidly as the number of parallel bits are increased. In this type of converter all input bits are processed simultaneously; therefore, the entire encoding operation can be performed within one clock cycle.

For converter sections beyond 4 bits (i.e., 16 comparators and 16 separate voltage references), the simultaneous converters become too imprac-

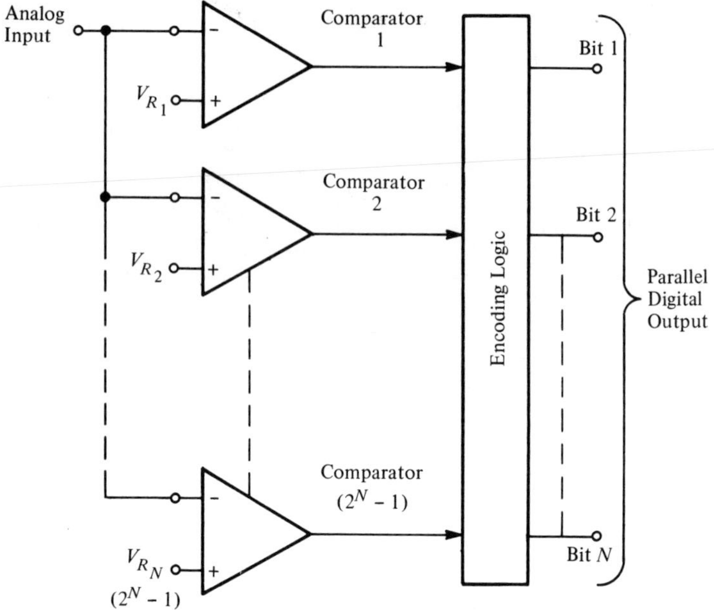

Figure 10.18 Block diagram of a simultaneous (parallel) A/D converter.

tical due to overall circuit complexity. However, it is often advantageous to use 2- or 3-bit simultaneous converter sections as building blocks in forming other classes of A/D converter systems to speed up the overall conversion rate.

REFERENCES

1. D. F. Hoeschele, *Analog-to-Digital and Digital-to-Analog Conversion Techniques,* Wiley, New York, 1968.
2. H. Schmid, *Electronic Analog/Digital Conversions,* Van Nostrand Reinhold, New York, 1970.
3. H. Schmid, "D/A Conversion," *Electronic Design,* **22** (Oct., 1968): 49–88.
4. H. Schmid, "A/D Conversion—Part I," *Electronic Design,* **25** (Dec., 1968): 49–72.
5. H. Schmid, "A/D Conversion—Part II," *Electronic Design,* **26** (Dec., 1968): 57–76.
6. D. Jones, "Digital to Analog Converter Application Note No. 511," Nov., 1970.
7. "Current Status of D/A Converters," *EEE Magazine,* Sept., 1970, pp. 43–46.

8. D. J. Dooley, "A Complete Monolithic 6-Bit D/A Converter," Int. Solid State Ckts. Conf., *Digest Tech. Papers,* **14** (Feb., 1971): 50–51.

9. J. B. Cecil and W. G. Howard, "A Fully Monolithic 6-Bit Multiplying D/A Converter," Int. Solid State Ckts. Conf., *Digest Tech. Papers,* **14** (Feb., 1971): 52–53.

10. J. J. Pastoriza, H. Krabbe and A. A. Molinari, "A High Performance Monolithic D/A Converter Circuit," Int. Solid State Ckts. Conf., *Digest Tech. Papers,* **13** (Feb., 1970): 122–123.

11. J. J. Pastoriza and W. R. Spofford, "Designing with μDAC Monolithic D/A Converters," Analog Devices, Inc., App. Note, Sept., 1970.

12. M. B. Rudin, R. O'Day and R. Jenkins ,"System/Circuit and Device Considerations in the Design and Development of a D/A and A/D Integrated Circuit Family," Int. Solid State Ckts. Conf., *Digest Tech. Papers,* **10** (Feb., 1967): 16–17.

13. J. Graeme, "Monolithic D/A Improves Conversion Times," *EDN Magazine,* March 15, 1971, pp. 39–41.

14. Ref. 1, pp. 156–162.

15. Ref. 1, pp. 355–365.

16. Ref. 2, pp. 247–274.

11 Analog Circuits for Special Applications

PART I Micropower Circuits

In a number of specialized circuit applications, low power dissipation is a key design requirement. This is particularly true for many circuits used in aerospace telemetry or in biomedical electronics, where only a very limited amount of battery power is available.

When designing integrated circuits for operation at microwatt power levels, one faces a number of design constraints.[1] The most fundamental among these are the degradation of active device performance at low current levels and the need for very high value resistors to provide biasing and amplification. In many cases it is possible to overcome the limitations of low power operation by careful circuit and device design. In the following sections of this chapter some of these design guidelines for monolithic micropower circuits will be outlined.

11.1 DEVICE PERFORMANCE AT MICROPOWER LEVELS

The active devices used in micropower circuits are required to provide an acceptable electrical performance with a few microamperes of bias current. As discussed in Chapter 2, the basic performance characteristics of all active devices are functions of the bias current level. For example, the transconductance g_m of either a bipolar or a field-effect transistor is proportional to the total current through the device. Similarly, the β and the unity gain frequency f_T of a bipolar transistor also decrease rapidly as the bias current is reduced to microampere levels.

The degradation of f_T at low current levels is readily apparent from the hybrid-π model of the bipolar transistor described in Chapter 2. Following the discussion of Section 2.1, f_T can be written from Eqs. (2.15) and (2.22) as

$$f_T = \frac{1}{2\pi}\left(\frac{g_m}{C_t + C_d + C_c}\right) \tag{11.1}$$

where $g_m = I_E/V_T$ is the device transconductance, C_t and C_d are the emitter transition and diffusion capacitances, and C_c is the collector-base depletion capacitance. Diffusion capacitance C_d is proportional to emitter current; therefore, at low current levels $C_d \ll (C_t + C_c)$. With this condition, f_T can be related to the emitter and the collector current levels as

$$f_T \approx \frac{I_E}{2\pi V_T(C_t + C_c)} \approx \frac{I_C}{2\pi V_T(C_t + C_c)} \tag{11.2}$$

where the last term in Eq. (11.2) assumes that $\beta \gg 1$. Thus, a transistor with an f_T of 500 MHz or more at milliampere current levels can only provide an f_T of less than 500 kHz with a collector current level of 1 μA.

The reduction of common-emitter current gain β at low currents is a more complex effect and is harder to predict than the f_T reduction. This is because the low current β is determined by the cumulative effect of a number of independent recombination processes in the transistor base and emitter regions. The effects of the individual recombination processes on the common-emitter current gain can be examined better by rewriting β in terms of the common-base current gain:

$$\beta = \frac{\alpha}{1 - \alpha} \tag{11.3}$$

The denominator of Eq. (11.3) is normally referred to as the "alpha-defect," since it represents the fraction of the emitter current which does not get to the collector but instead appears as the base current. Ideally, if $\alpha = 1$, i.e., zero α-defect, β can be infinite. Some of the dominant recombination effects in the emitter-base regions are relatively independent of current level so that their effect or contribution to the α-defect becomes more significant as the total current level is reduced.

Figure 11.1 shows a schematic diagram of current injection across the base-emitter junction of a planar transistor. Note that the carriers are injected into the base all along the emitter-base junction. Yet, only those

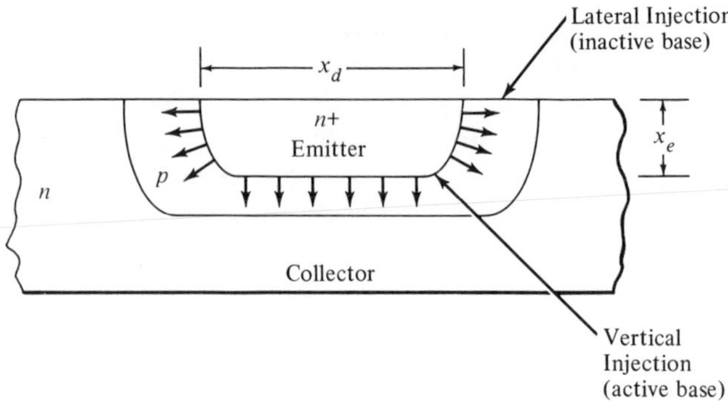

Figure 11.1 Lateral and vertical injection across base-emitter junction.

carriers injected vertically into the "active base" region have a high probability of reaching the collector since they have to traverse a much shorter distance. The laterally injected carriers end up in the "inactive base" region and recombine without reaching the collector. To minimize lateral injection, it is necessary to minimize the emitter sidewall area for a given amount of "active base" area. This can be done by using a circular emitter geometry with a large emitter diameter, X_d to emitter depth X_e ratio.[2]

Other recombination mechanisms which contribute to the base current are the recombination in the emitter-base space-charge region, and the recombination of carriers injected into the emitter from the base. The space-charge recombination has two components: bulk and surface. Bulk recombination in the space-charge region comes about because of the finite lifetime of minority carriers; and, the finite time they have to spend in transit through the emitter-base depletion layer. Surface recombination takes place in the part of the emitter-base junction touching the semiconductor surface. This can be minimized by passivating the semiconductor surface, using silicon-nitride deposition to prevent ionic contamination (see Chapter 1, Section 1.4). The surface recombination can also be reduced by letting the emitter aluminum extend over the dielectric layer covering the emitter-base junction. This metal layer serves as an electrostatic shield and keeps the semiconductor surface potential very nearly at the "flat-band" condition, such that there is no additional widening or narrowing of the space charge region at the surface.

Figure 11.2 shows the photomicrograph of a matched pair of NPN transistors designed for low current operation. Note that both devices use circular emitter structures to minimize lateral injection and have the emitter aluminum contact overlapping the oxide over the base-emitter junction.

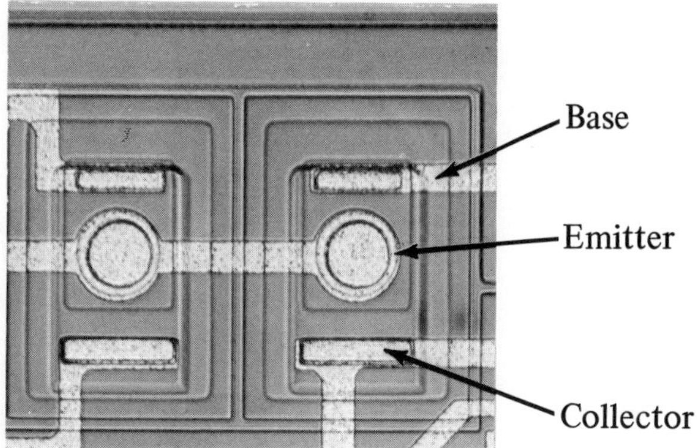

Figure 11.2 Photomicrograph of a matched pair of low-current *npn* transistors. (*Photo: Fairchild.*)

Except for surface recombination, all other modes of recombination effects are inversely proportional to the minority carrier lifetime τ in the bulk of the semiconductor. Therefore, it is necessary to use a high quality, defect-free starting material with a high minority carrier lifetime ($\tau \geqq 1$ μsec) if superior low current performance is required.

In summary, to obtain high β at low current levels, one should[2]

1. Use high bulk-lifetime material
2. Reduce surface recombination (use additional surface passivation if needed)
3. Minimize emitter periphery to area ratio to reduce lateral injection

The last item on the list also implies a design trade-off. The emitter-transition capacitance, which makes up most of C_π, is proportional to the emitter area. Thus, increasing the emitter area to reduce lateral injection effects causes the already low f_T of the device to decrease further.

Figure 11.3 shows the β versus I_C characteristics of a double-diffused *npn* transistor optimized for low current operation.[2] The current gain of conventional lateral *pnp* and substrate *pnp* devices are also shown in the figure. Note that, even though the current gain of the *pnp* devices is quite low compared to the *npn*, the β falloff at low currents is not any worse for the *pnp* devices. This is because the β falloff with current is primarily determined by the parasitic recombination effects, such as surface recombination, and not by the bulk recombination in the active base region.

At very low current levels, high frequency performance of bipolar transistors is determined by the emitter-base transition and the base-collector depletion capacitances, and not by the base transit time. Therefore, for

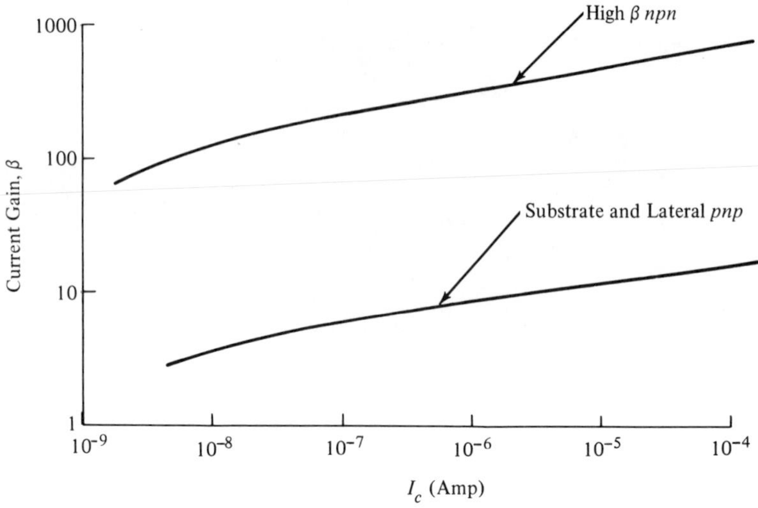

Figure 11.3 Current gain as a function of collector current (after Ref. 2).

micropower operation the frequency response of the lateral or the substrate *pnp* devices are not much worse than for the *npn*. Therefore, in the design of micropower circuits, the *pnp* devices can be used more freely than at conventional power levels.

11.2 RESISTORS IN MICROPOWER CIRCUITS

In order to get voltage gains from transistors operating with low collector currents, large value load resistors are required. For example, the voltage gain of a simple common-emitter stage with a resistive load R_L is

$$A_v = -g_m R_L = \frac{-I_C R_L}{V_T} \tag{11.4}$$

To obtain a voltage gain of 40 from such a stage with 1 μA of collector current requires at least 1 MΩ of resistance. Using conventional base-diffused resistors having a sheet resistivity of about 200 Ω/sq, such a resistor alone would require approximately 3000 sq mils of chip area which is prohibitively large. The need for high value resistors can be in part satisfied using active devices as loads, as described in Chapter 5. This technique is suitable to operational amplifier circuits where high open loop gain is the primary concern. However, it is not readily usable for other classes of circuits where the frequency and bias dependence of the dynamic impedance of active loads can be a problem. Thus, in almost any micropower circuit design there is a need for high value resistors.

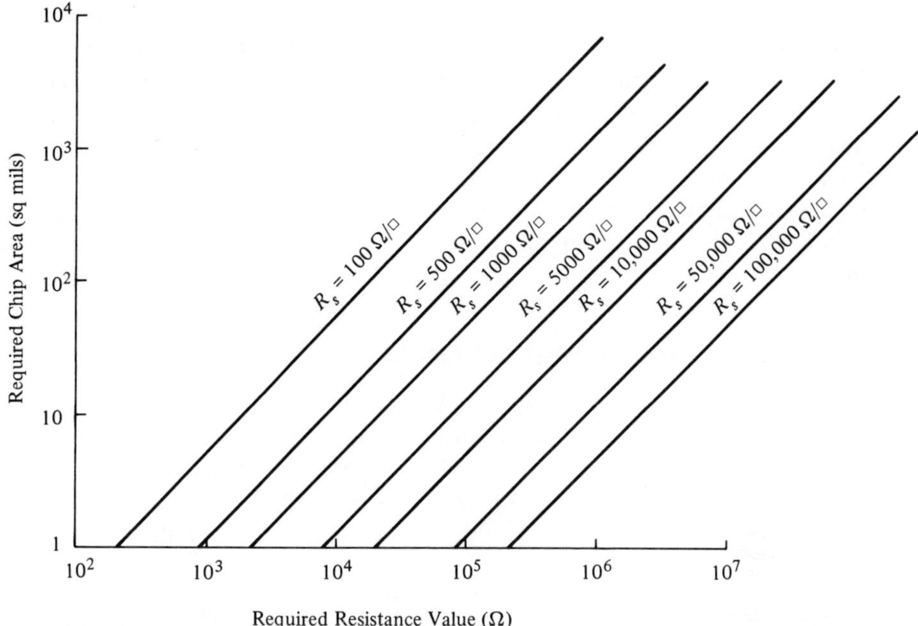

Figure 11.4 Approximate substrate area vs. total required resistor value for different sheet resistivities (0.4 mil wide resistor structure is assumed with 0.8 mil spacing between adjacent resistors).

Figure 11.4 shows the apprixomate substrate area required for a given resistance value, for various values of sheet resistance R_s. In this case a resistor geometry made up of parallel resistor strips 0.4 mil wide, with 0.8 mil spacing, is considered. As indicated by the figure, to obtain resistor values in the megaohm range within practical chip dimensions, sheet resistivity values of 5 kΩ/sq or more are required.

In monolithic circuits the most readily available high value resistors are the pinch-resistor and the bulk-resistor structures discussed in Chapter 3. In each case, these resistors offer a sheet resistance of 10 kΩ/sq or more; however, they also exhibit a strong positive temperature coefficint of $\geqq$4000 ppm/°C and have large manufacturing tolerances. In the case of base-diffused pinch resistors, the maximum voltage across the resistor also is restricted to a few volts because of linearity and breakdown considerations. In micropower circuits, low voltage operation is often required, along with low current; therefore, the breakdown limitations of pinch resistors do not necessary present a problem.

If relatively tightly controlled resistors are required in the sheet resistivity range of 5 kΩ/sq or greater, these can be provided at the expense of process complexity by using ion-implanted or thin-film deposited resistors. The ion-implanted resistors, discussed briefly in Chapter 3, can cover a resistivity range up to 10 kΩ/sq with a temperature coefficient range of ±500 ppm/°C.[3] Among thin-film deposited resistors, silicon-chromium films have been successfully used in a number of monolithic micropower circuits. These resistors have a typical sheet resistivity range of 5 to 20 kΩ/sq, with a temperature coefficient of −400 to −1000 ppm/°C.[2]

11.3 MONOLITHIC DESIGN EXAMPLES

In most micropower circuit applications, the low current and low voltage requirements are present simultaneously. In other words, the circuit is required to operate with only a few microamperes of current from a low battery voltage (typically of the order of 1 to 3 V). The lower limit on the current drain is set by any one or a combination of the following two effects:

1. Degradation of device performance (β and f_T) at low current levels
2. Junction leakage currents

The effects of low current level on device parameters were examined in Section 11.1. The junction leakages, which increase quite rapidly with temperature, provide a second limitation. To provide reliable control of current levels within the device, the collector current should be maintained at a level at least an order of magnitude higher than the leakage currents, i.e.,

$$(I_C)_{\min} \geqq 10\, I_{CEO} \tag{11.5}$$

where I_{CEO} is the collector-base leakage current with the base open-circuited. In high quality *npn* transistors of small or medium geometry (see Fig. 11.2) the maximum value of I_{CEO} is less than 1 nA for temperatures $\leqq$125°C, which implies a lower limit for $(I_C)_{\min}$ of approximately 10 nA.

The lower limit on the operating voltage is normally set by the transistor V_{BE} drops. Thus, in a micropower circuit involving bipolar transistors, the supply voltage V_{CC} has to be $> V_{BE}$ for the circuit to function as an amplifier. This fundamental restriction is not present in unipolar devices such as field-effect transistors.

The high input and output impedance levels of active devices at low current levels can often make it possible to ac couple successive micropower stages by using monolithic coupling capacitors, particularly for radio frequency applications. Figure 11.5 shows a self-biased micropower "gain

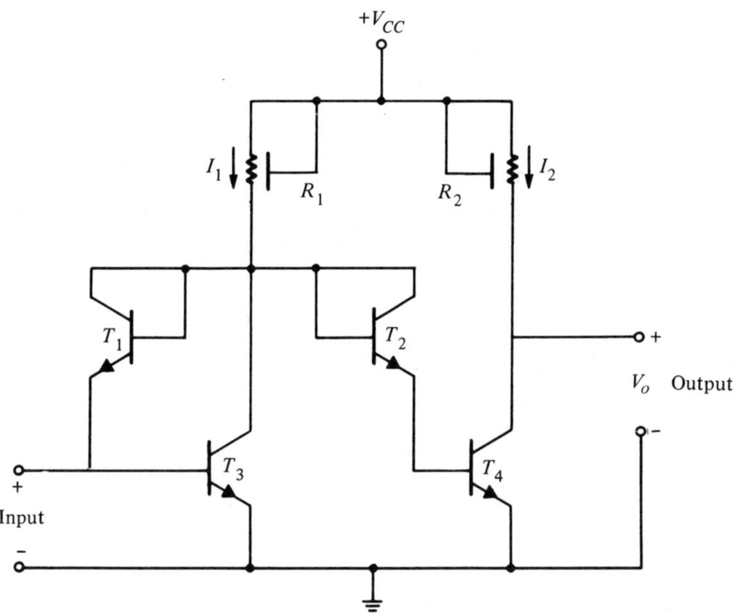

Figure 11.5 A self-biased micropower gain cell.[4]

cell" structure which is suitable for such an application.[4] This particular circuit was developed as a multifunction building block for an implantable biomedical telemetry system. It can operate either as an R-F amplifier or as an AM detector by proper choice of the bias resistor values. The circuit is designed to operate from a single 1.35 V mercury battery. At micropower levels, the input and the output impedances of the circuit are sufficiently high to facilitate ac coupling with a 20 pF monolithic capacitor for radio frequencies.

Resistors R_1 and R_2 of Fig. 11.5 are formed as base-pinch types with a sheet resistivity of $\approx$40 kΩ/sq. This high sheet resistivity of the pinch resistors is obtained by using bipolar transistors with very narrow base width, similar to the punch-through devices described in Chapter 2. At $V_{CC} = 1.35$ V, the punch-through breakdown voltage does not present a problem. If all the transistors are well matched, and if the pinch resistor ratio R_1/R_2 is well controlled, the circuit self-biases itself to keep the entire gain stage in active region. In the circuit, transistors (T_1, T_2) and (T_3, T_4) conduct the same dc current among themselves. Therefore, if the devices are well matched:

$$V_{BE_1} = V_{BE_2} \qquad \text{and} \qquad V_{BE_3} = V_{BE_4}$$

The output dc level V_o can be automatically adjusted to $V_{CC}/2$, inde-

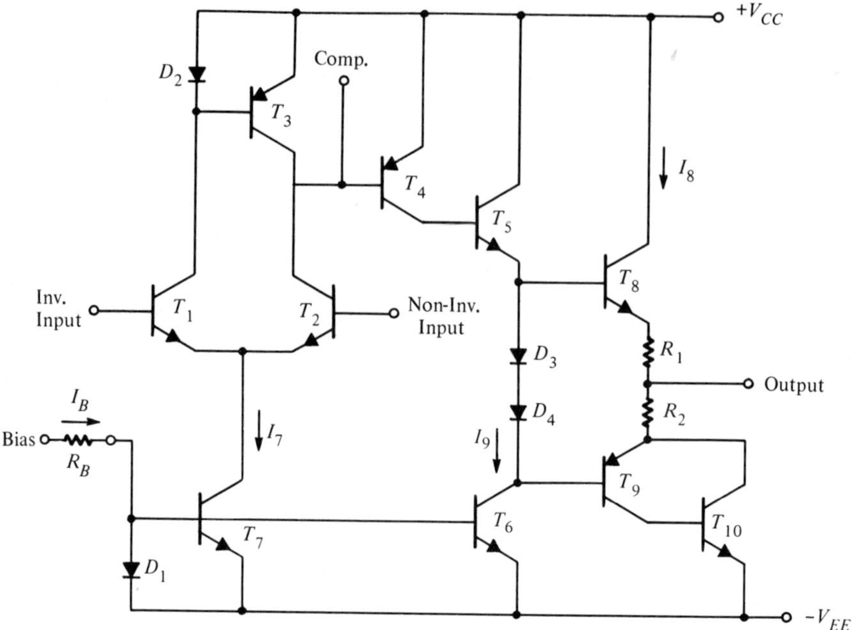

Figure 11.6 A micropower operational amplifier.[5]

pendent of the absolute values of the pinch resistors, by setting the pinch-resistor ratio to be:

$$\frac{R_1}{R_2} = 2\left(1 - \frac{V_{BE_1} + V_{BE_3}}{V_{CC}}\right) \tag{11.6}$$

Since the pinch-resistor ratios can be controlled quite accurately (to within ± 10 percent), the bias levels within the circuit become relatively insensitive to absolute tolerances. Using this circuit as an R-F amplifier, voltage gain of 24 dB has been obtained at 500 kHz, with a total power dissipation of less than 11 μW at $V_{CC} = 1.34$ V.[4]

One of the most commonly used micropower analog circuits is the operational amplifier. Figure 11.6 shows a circuit diagram of a practical micropower operational amplifier.[5] By imaginative design, this particular circuit performs the operational amplifier function without requiring any resistors. Resistors R_1 and R_2 are included only to provide output short-circuit protection. Resistor R_B is external to the monolithic chip and is used to set the bias levels within the circuit. Assuming that the *npn* β is large, quiescent currents I_7 and I_9 would be related to the externally supplied bias current I_B as

$$I_B/a_1 = I_7/a_7 = I_9/a_6 \tag{11.7}$$

where a_1, a_7, and a_6 are the respective emitter areas of D_1, T_7, and T_6.

The input stage of the micropower amplifier uses the diode D_2 and the transistor T_3 as a set of active loads (see Fig. 5.7) to obtain a large voltage gain and to convert the differential input signal to a single-ended output. This signal is further amplified by the second gain stage made up of T_4 and T_5. To conserve quiescent power, a class-B output stage is used similar to that shown in Fig. 5.13. For relatively low values of $R_L (R_L < 10\ \text{k}\Omega)$, the voltage gain of the circuit depends on the external load resistor used and can be approximated as:

$$A_v \approx \beta_p \beta_n^2 g_{m_1} R_L \tag{11.8}$$

where g_m is the transconductance of T_1, and β_p and β_n are the current gains of the *pnp* and the *npn* devices. Neglecting the transistor saturation voltages, the circuit of Fig. 11.6 can operate with a total supply voltage of $\geqq 4\ V_{BE}$ and provides a peak to peak output swing of:

$$(V_o)_{pp} = (V_{CC} - V_{EE}) - 3\ V_{BE} \tag{11.9}$$

For ± 3 V operation the operational amplifier provides a voltage gain of about 80 dB into a 10 kΩ load, with a power dissipation of $\approx 300\ \mu$W.[5]

In the design of micropower operational amplifiers, it is sometimes desirable to set the bias levels within the circuit, internally; and independently of the power supply changes. Figure 11.7 shows the circuit configuration for a micropower operational amplifier stage using internal biasing. Note that, in this case a *pnp* rather than an *npn* input stage is used. At low current levels, the high input impedance requirement can be satisfied with moderate values of *pnp* current gain. For example, for a *pnp* β of 10, the input impedance of the circuit is approximately 0.5 MΩ with 1 μA of collector current. Therefore, at micropower levels, lateral *pnp* devices can be used as input devices without degrading circuit performance.[6]

The input stage of Fig. 11.7 is internally biased by the current source network comprised of transistors T_7 through T_{10}. Following the discussion of Section 4.1 (see Eqs. 4.13 and 4.14), the current levels I_1, I_o, and I_2 are interrelated as:

$$I_1 = \frac{V_T}{R_1} \ln \frac{I_o}{I_1} \tag{11.10}$$

and

$$I_2 = \frac{V_T}{R_2} \ln \frac{I_o}{I_2} \tag{11.11}$$

Assuming that R_3 and R_4 are sufficiently large (see Eqs. 4.11 and 4.12),

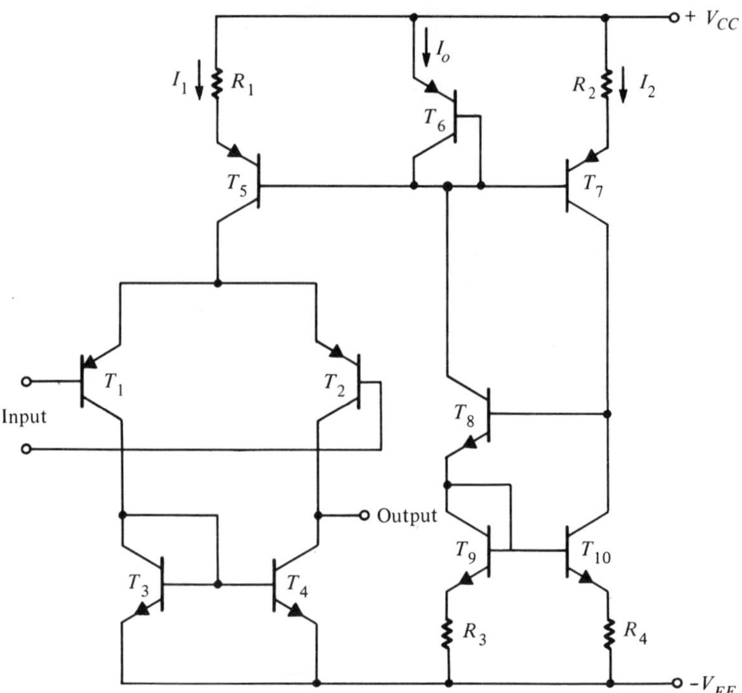

Figure 11.7 An internally biased input stage for micropower operational amplifier.[6]

I_o and I_2 become related as:

$$I_o/I_2 \approx R_4/R_3 \tag{11.12}$$

Letting $R_1 = R_2$, currents I_1 and I_2 are equal from Eqs. (11.10) and (11.11). Then from Eqs. (11.10) and 11.12) one obtains:

$$I_1 \approx \frac{V_T}{R_1} \ln \frac{R_3}{R_4} \tag{11.13}$$

which is independent of the supply voltage and set only by the thermal voltage and the resistor ratios. Since V_T is small, large resistor values are not needed to set the bias. Note that this bias network requires a total supply voltage of only two V_{BE} and one V_{CE} drop to be operative.

Figure 11.8 shows an alternate micropower operational amplifier circuit which incorporates some of salient features of conventional operational amplifiers, such as output short-circuit protection. The bias levels of the circuit are set by an external resistor, connector to the emitter of T_3. This arrangement is somewhat similar to that described in connection with Fig. 11.6. The input stage is a conventional differential gain stage with matched

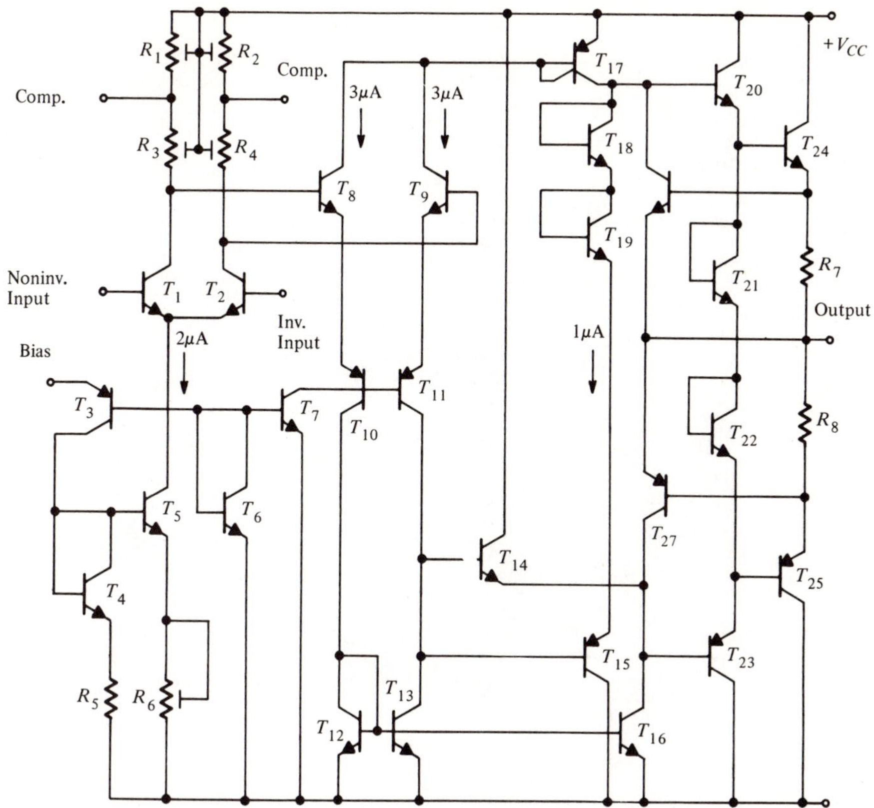

Figure 11.8 A micropower operational amplifier (Signetics SE 533).

resistive loads comprised of pinch resistors $(R_1 + R_3)$ and $(R_2 + R_4)$. The total value of these load resistors is approximately 1 MΩ. Input stage is biased at approximately 1 μA of collector current which provides only a small voltage drop across the pinched load resistors, and thus avoids breakdown problems. Frequency compensation is provided by connecting a capacitor across the collector loads of the input stage.

The second stage circuit configuration is similar to that described in Fig. 5.8, with T_{12} and T_{13} functioning as active loads. T_{14} and T_{15} are buffer emitter follower stages which drive the class-B output stage. To reduce the output impedance, the output stage is driven by the complementary Darlington pairs (T_{20}, T_{24}) and (T_{23}, T_{25}), where the output *pnp* transistor T_{25} is a substrate-*pnp*. Transistors T_{26} and T_{27} are normally off, and are used only for output short-circuit protection (see Fig. 5.14).

The topological layout for the micropower operational amplifier of Fig. 11.8 is given in Fig. 11.9. The corresponding circuit terminals are also

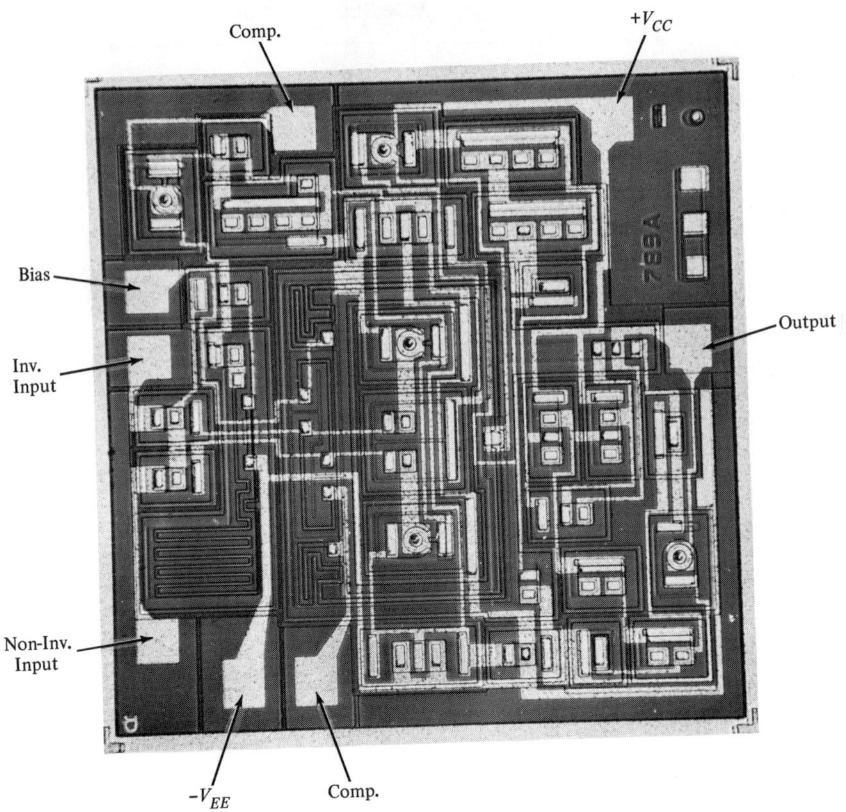

Figure 11.9 Topological layout of the micropower operational amplifier of Figure 11.8. (*Photo: Signetics.*)

identified in the figure. Note that some of the basic features of micropower device design described in Section 11.1 are also evident in the Figure: transistor emitter areas are made large to reduce lateral emission; and the aluminum interconnection layer is extended over the base-emitter junction areas of the devices to reduce surface recombination. The circuit is designed to operate over a power supply range of ±3 to ±5 V and has the following typical performance characteristics:

Open loop gain	90 dB
Input offset voltage	0.5 mV
Input bias current	1 nA
Input impedance	60 MΩ
Common mode rejection	100 dB
Power dissipation:	
±3 V	<100 μW
±15 V	<1 mW

PART II Radiation Resistant Circuits

Semiconductor device parameters are sensitive to various kinds of radiation. Exposure of a semiconductor device to a radiation environment results in a transient or permanent degradation of certain device parameters such as current gains, saturation resistances, or junction leakages. These degradation mechanisms often make the conventional integrated circuits not suitable for usage in a radiation environment.

The recent advances in space exploration, nuclear reactor engineering, and weapons technology have brought about an extensive need for monolithic circuits which can operate under radiation environment with a high degree of reliability. In many cases the resistance of a monolithic circuit to radiation damage can be significantly increased, or the circuit can be "radiation-hardened" by proper choice of the structures, process technology, and careful circuit design. In this section the basic effects of radiation on integrated circuits will be examined and special device and circuit design techniques will be outlined for the design of radiation-hardened analog circuits.

11.4 EFFECTS OF RADIATION ON INTEGRATED CIRCUITS

In the design of radiation-hardened integrated circuits, two basic types of radiation are of primary concern: (1) ionizing radiation, i.e., X-rays and gamma rays, and (2) neutron radiation. Each of these types has a different effect on the semiconductor device and circuit performance.

Ionizing Radiation

The main form of ionizing radiation for integrated circuits is gamma rays. The quantity of ionizing radiation to which a device is exposed is described in terms of energy deposited per unit volume of exposed material. In the case of silicon, the most commonly used unit of ionizing radiation is the "rad," which is the dose of absorbed radiation equal to 100 ergs of energy per gram of silicon.

Exposure to gamma rays results in excess carrier generation (i.e., production of hole-electron pairs in silicon) as well as in heat generation. Of these two effects, the former is by far the most important since it interferes with the electrical operation of the devices. Gamma-induced ionization can interfere with circuit operation by one of three mechanisms:[7] photocurrent generation; latch-up; or degradation of surface properties. The general effects of ionizing radiation on semiconductor components have been studied extensively in the literature.[8-10] Only the basic results will be summarized in this section.

Photocurrent generation is the most important effect of ionizing radiation. A pulse of ionizing radiation incident on a *p-n* junction will generate a photocurrent as a result of hole-electron pair generation in the junction depletion layer. These carriers then drift or diffuse across the depletion layer and constitute a photocurrent transient which lasts during the radiation pulse and then decays to zero. The decay time of the photocurrent, after the radiation pulse is over, is determined by the minority carrier lifetime τ in the semiconductor material. The peak value of the generated current is proportional to the junction area and also increases as the radiation dose rate (in rads/sec) is increased. Even though the photocurrent generation is a transient effect, it can disrupt circuit operation temporarily; and if the current is not limited by the external circuitry, can result in a permanent burn-out. Therefore, the peak photocurrent generated for a given dose rate of radiation is one of the key design parameters for photocurrent protection. Emperical results indicate that this value is typically of the order of 25 nA/cm² of junction area per rad/sec of dose rate.[9] Since the photocurrent generation is proportional to the junction area, in a conventional integrated circuit the main contributions to the total photocurrent come

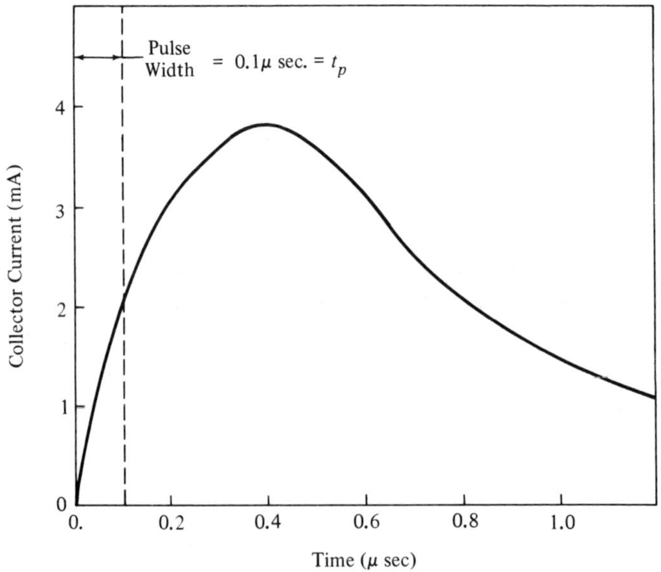

Figure 11.10 Typical transistor response to an ionizing radiation pulse.[11] Test conditions: radiation pulse width = 0.5 μsec; dosage = 0.5 rad.

from the collector-isolation and the collector-base junctions, since these make up most of the junction area available in monolithic devices. Emitter-base junctions usually occupy a much smaller area than the rest; therefore, their contribution to photocurrent generation is negligible.

In the case of a transistor, the photocurrent generated in the base-collector junction also becomes multiplied by the β of the transistor and thus results in a much higher current level. This effect is called "secondary" photocurrent generation. Figure 11.10 shows the photocurrent response of typical *npn* transistor as a result of a 0.5 rad radiation pulse of 0.1 μsec duration.[11] In this case the transistor is biased in off condition and the current conduction is due solely to the primary and the secondary photocurrent generation. Note that the current peak lags the pulse, and the decay time is significantly longer (due to the minority carrier lifetime) than the rise time.

In conventional junction-isolated circuits, ionizing radiation can also create latch-up by causing the collector-substrate junction to conduct during the radiation transient. Since the substrate, collector, base and the emitter junction form a four-layer *p-n-p-n* structure, the current flow across the collector-isolation junction can turn on the parasitic *pnp* transistor to the substrate (see Fig. 2.2). and cause the entire transistor to latch-up in a conducting mode. Latch-up can be eliminated by using dielectric isolation instead of conventional junction isolation.

Both photocurrent generation and latch-up are reversible processes if the total current through the device can be limited by external means. However, ionizing radiation can also create a permanent degradation by generating surface states at the semiconductor-oxide interface.[12] The electron-ion pairs produced in the passivating oxide layer tend to separate, leaving a net positive charge at the silicon-oxide interface. These charges can cause an inversion of lightly doped *p*-regions at the surface. Surface recombination rate also increases causing the low current β of transistors to decrease. Surface degradation induced by ionizing radiation also limits the usefulness of MOS devices in radiation environment. Most bipolar circuits can operate with exposure up to 10^6 rad/sec; however, surface degradation effects in general limit the MOS devices to ionizing radiation doses of $<10^4$ rad/sec.

Neutron Radiation

During neutron radiation the collisions between the neutrons and the silicon atoms result in permanent damage to the silicon lattice. As a result of these collisions, some of the silicon atoms are displaced from their lattice sites. This type is called "neutron-displacement" damage. The silicon atoms displaced from their lattice sites act as recombination and scattering centers and result in a decrease of the minority carrier lifetime τ and the carrier mobility. Since the neutron damage is cumulative, the neutron radiation dose is normally measured in terms of the total neutron fluence incident on a unit surface area of the device.

The generation of recombination centers due to neutron-displacement damage results in a reduction of minority lifetime. Since many of the transistor parameters such as β, reverse leakage current I_{CO}, and storage time τ_s depend on the minority carrier lifetime, the overall device performance can be strongly effected by neutron radiation.

The common emitter current gain β is one of the most important device parameters effected by neutron radiation. It can be shown[7,13] that the post-irradiation value of β can be related to its pre-irradiation value β_o, and the total neutron fluence ϕ by an expression of the form,

$$\frac{1}{\beta} = \frac{1}{\beta_o} + \frac{\tau_B \phi}{\kappa} \tag{11.14}$$

where τ_B = base transit time
ϕ = neutron fluence in neutrons/cm^2
κ = experimentally determined damage constant

The constant of proportionality κ is a function of conductivity type, resistivity, injection level, neutron energy, and temperature. Tabulated values of κ are available in the literature for a wide range of test conditions.[13] For most *npn* transistor structures encountered in monolithic circuits, the values of κ are in the range of (10^6) to (3×10^6) neutrons-sec/cm^2.

The base transit time τ_B is inversely proportional to the square of the base width ω_B. Therefore, transistor structures which have large base widths, such as the lateral and the substrate *pnp* devices are more susceptible to neutron damage than the conventional *npn* transistors. Since f_T of the transistor is closely related to the base transit time, τ_B, high f_T transistors in general have their current gain more resistant to radiation damage than those having lower f_T.[7] Figure 11.11 shows some typical β degradation characteristics for typical integrated circuit transistors under neutron radiation.[14]

The lattice atoms displaced by neutron radiation also serve as scattering centers and reduce the carrier mobility μ, or increase the resistivity ρ of the bulk semiconductor material. Experimental results indicate that an order-of-magnitude estimate of the resistivity increase can be obtained by assuming that approximately one carrier is removed per neutron/cm^2 of fluence. Figure 11.12 shows the expected change in resistivity ρ for various

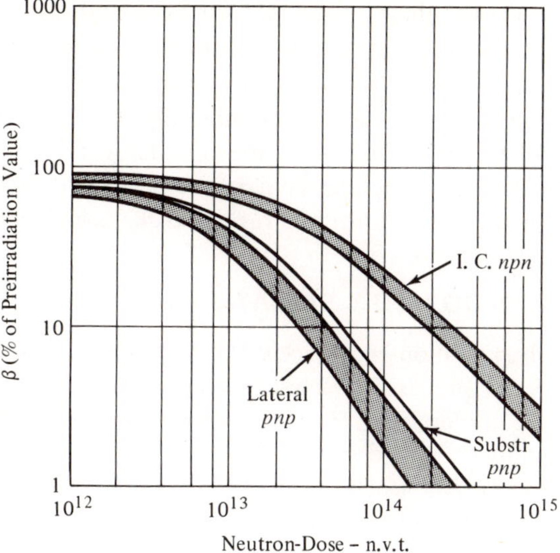

Figure 11.11 Beta degradation of integrated circuit transistors under neutron radiation.[14]

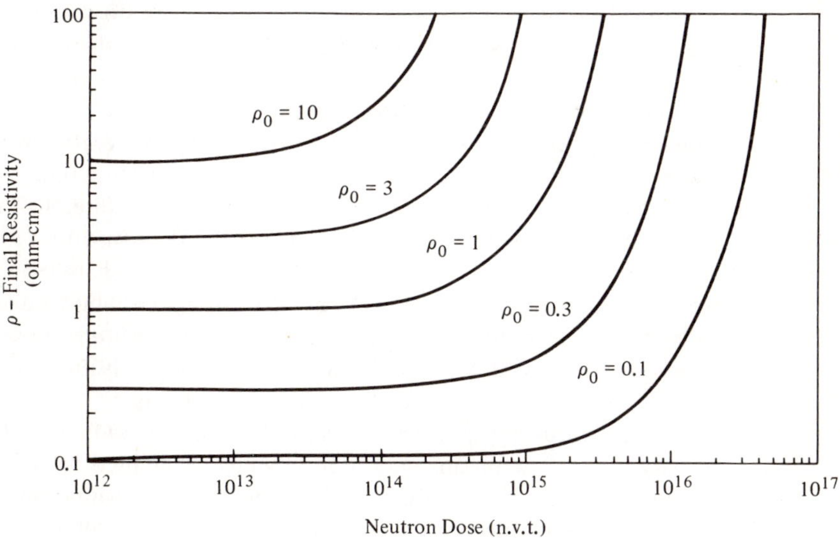

Figure 11.12 Bulk resistivity vs. neutron dosage for *N*-type silicon.[7]

values of pre-irradiation resistivity ρ_o, for n-type silicon.[7] As indicated by the Figure, for most resistivity ranges encountered in monolithic structures, neutron-damage effects on resistivity do not provide a significant limitation on the radiation hardness of analog circuits since the active devices degrade at much lower doses of radiation (see Fig. 11.11). As the resistivity increases due to neutron damage, other device parameters which depend on resistivity, such as the junction breakdown and reverse-leakage currents, also increase.

11.5 TECHNOLOGY FOR RADIATION-HARDENED CIRCUITS

To improve the radiation resistance of a monolithic circuit, it is necessary to employ special processing technology as well as additional device and circuit design considerations. The technology aspects of radiation-hardening will be briefly examined in this section.

Elimination of large isolation junctions is the first step in improving the resistance of a monolithic circuit to ionizing radiation. This can be done by using dielectrically isolated device structures, as described in Chapter 1 (see Figs. 1.14 and 1.15). In this manner, collector-substrate photocurrents and latch-up effects can be eliminated. The use of dielectric isolation also makes it possible to fabricate complementary *pnp* transistors having narrow base widths (see Fig. 2.18) which offer much higher neutron radiation resistance than the lateral or substrate *pnp* devices. In addition to conventional dielectric isolation techniques, other isolation schemes such as air-isolated beam-lead or "silicon-on-sapphire" structures which do not utilize reverse-biased junctions can also be employed in the fabrication of radiation-hardened circuits.

Conventional base-diffused resistors are not impervious to radiation. The *p-n* junction outlining the resistor becomes a virtual short circuit under high level ionizing radiation, due to photocurrent generation; and the sheet resistivity can change with neutron fluence. Therefore, to improve the radiation resistance of a circuit, it is necessary to use thin-film resistors in place of diffused resistors. Thin film resistors such as Ni-Cr, tantalum, or silicon-chromium which are impervious to radiation effects are normally utilized in radiation-resistant circuits as a substitute for diffused resistors. In noncritical parts of the circuit dielectrically isolated bulk resistors can also be used in place of base-diffused resistors.

During the photocurrent transients caused by ionizing radiation, the current density in the aluminum metallization must be limited to a safe value to avoid immediate burn-out due to joule heating, or gradual failure due to electromigration effects (see Section 6.5). For reliability under normal (nonradiative) operating conditions, the current density in the aluminum conductors should be $\leq 10^5$ A/cm^2; and the transient current density due to ionizing radiation should be maintained $\leqq 10^6$ A/cm^2 to avoid electromigration effects. Thus, in radiation-resistant circuits, it is customary to use wider and thicker aluminum interconnections than would be used in conventional integrated circuits. Additional passivating glass deposition over the aluminum interconnections protects the circuit from scratches during assembly and improves the integrity of the aluminum interconnections. Therefore, this "glassivation" step is also used as a part of the radiation-hardening technology.

Due to the specialized nature of their applications, radiation-hardened integrated circuits always have to exhibit a high degree of reliability along with radiation resistance. One of the recognized failure mechanisms in conventional integrated circuits using gold wire bonds to aluminum interconnections is the formation of brittle intermetallic compounds at the gold-aluminum interface, which weakens the integrity of the wire bond. A typical example of this is the so called "purple-plague" in gold-aluminum interconnection systems. To avoid this problem in radiation hardened circuits, ultrasonic-bonded aluminum wires are used in place of gold wires.

Soft X-rays or gamma rays also cause internal heating of the circuit package. This is due to X-rays or gamma-rays giving up their energy to the package materials with which they interact. This joule heating caused by the radiation is dependent on the atomic weight of the material, and increases with increasing weight. Therefore the use of high atomic-number materials, such as gold, in the circuit package should be minimized or avoided for improved radiation hardness.

11.6 CIRCUIT AND DEVICE DESIGN FOR RADIATION ENVIRONMENT

Radiation-hardened analog circuit design is a highly specialized subject since in almost all cases the specifics of the actual design depend very strongly on the particular type, dose, and energy of the radiation environment. Extensive testing and evaluation is required to obtain useful hardness and reliability data, and in most cases the results are applicable only to a specific case (such as a particular circuit configuration) and cannot be generalized. However, based on the radiation effects discussed in Section 11.4, a number of device and circuit design guidelines can be established as follows:

(A) Device Design:

1. Minimize junction areas: since photocurrent generation is proportional to junction area, all junction areas and particularly the base-collector junction area should be minimized.

2. Use narrow-base transistors to reduce β degradation due to neutron damage.
3. Maintain high initial β so that a higher neutron dosage can be tolerated before β is reduced to a minimum allowable level .
4. Maintain high surface concentration in all *p*-type regions to avoid inversion due to surface states generated by ionizing radiation.
5. Avoid using MOS transistors whenever possible since these devices have a lower radiation tolerance than bipolar or JFET devices.
6. Use thin-film resistors instead of base-diffused resistors.
7. Use relatively low bulk-resistivity material to reduce resistivity increase due to neutron damage.

(B) Circuit Design:

1. Use photocurrent compensation to reduce secondary photocurrent generation. Photocurrent generation is proportional to junction area. Thus, photocurrent generated in the base-collector junction and flowing into the base region can be shunted to ground by connecting an identical diode from the base to ground. This is shown schematically in Fig. 11.13, for a common-emitter gain stage. Transistor T_2 has the same base-collector area as T_1. Thus the primary photocurrent I_p flowing into the base of T_1 is equal to the photocurrent I_D in the base-collector junction of T_2. Thus, most of I_p is shunted to ground, and only a small fraction of it is multiplied by β to result in secondary photocurrent generation. Photocurrent compensation diodes can provide up to an order of magnitude reduction in secondary photocurrent generation.

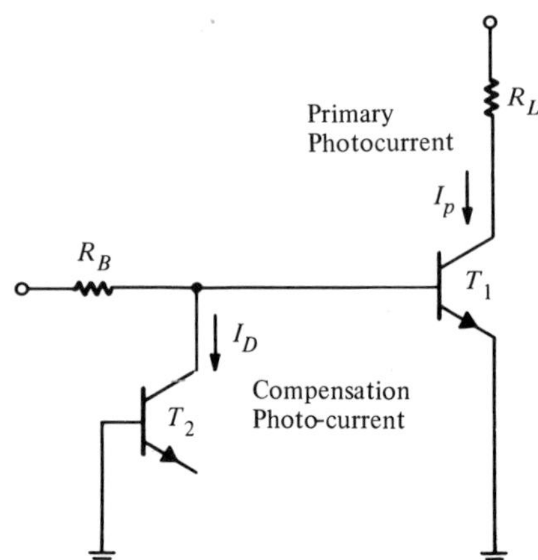

Figure 11.13 Example of photocurrent compensation in a common-emitter gain stage.

2. Use current-limiting resistors in each circuit branch between the power supplies and the dc ground to limit photocurrent surges to a safe level.
3. Use low active-to-passive component ratio. This is a departure from the basic integrated circuit design philosophy where a large number of active devices is readily available. However, under radiation environment, active devices in general offer a much lower radiation resistance than passive components.
4. When wide-base devices such as lateral or substrate *pnp*s must be used, use compound *pnp-npn* configuration (see Section 4.5) to minimize circuit dependence on the *pnp* current gain.
5. Maintain low quiescent power dissipation to increase transient power-handling capability during photocurrent surges. However, device operation at micropower levels should be avoided since the β degradation due to radiation damage is much more pronounced at micropower operation than at conventional power levels.

The above set of design postullates are by no means exhaustive; and are intended only as general design guidlines. The particular emphasis to be placed on each of these guidelines would depend on the specifics of the particular design. In many cases a compromise is necessary between the circuit performance and the radiation resistance. A typical example of this is the design of an operational amplifier output stage where the use of current-limiting resistors results in reduced output swing.

11.7 CIRCUIT DESIGN EXAMPLE

Figure 11.14 shows the circuit diagram of a radiation-hardened monolithic operational amplifier.[15] This particular circuit is based on the well known μA709 operational amplifier design (see Section 5.6 and Fig. 5.17) which has been modified for operation in radiation environment. The basic design and operation of the 709 operational amplifier has been covered in detail in Section 5.6, and will not be repeated here. Only the additional features provided in the circuit of Fig. 11.14 for radiation-hardening purposes will be examined.

The added features of the circuit of Fig. 11.14 can be best examined by comparing it with the original 709 design shown in Fig. 5.17(*a*). In the radiation-hardened circuit diode connected transistors T_3 and T_4 at the input are used as clamping diodes to prevent latch-up due to large positive common-mode input swings. Resistors R_3, R_4, R_5, R_7, R_{14}, R_{22}, R_{23}, and R_{24} are not present in the original design, but are included in the radiation-hardened version to provide photocurrent limiting. These protection re-

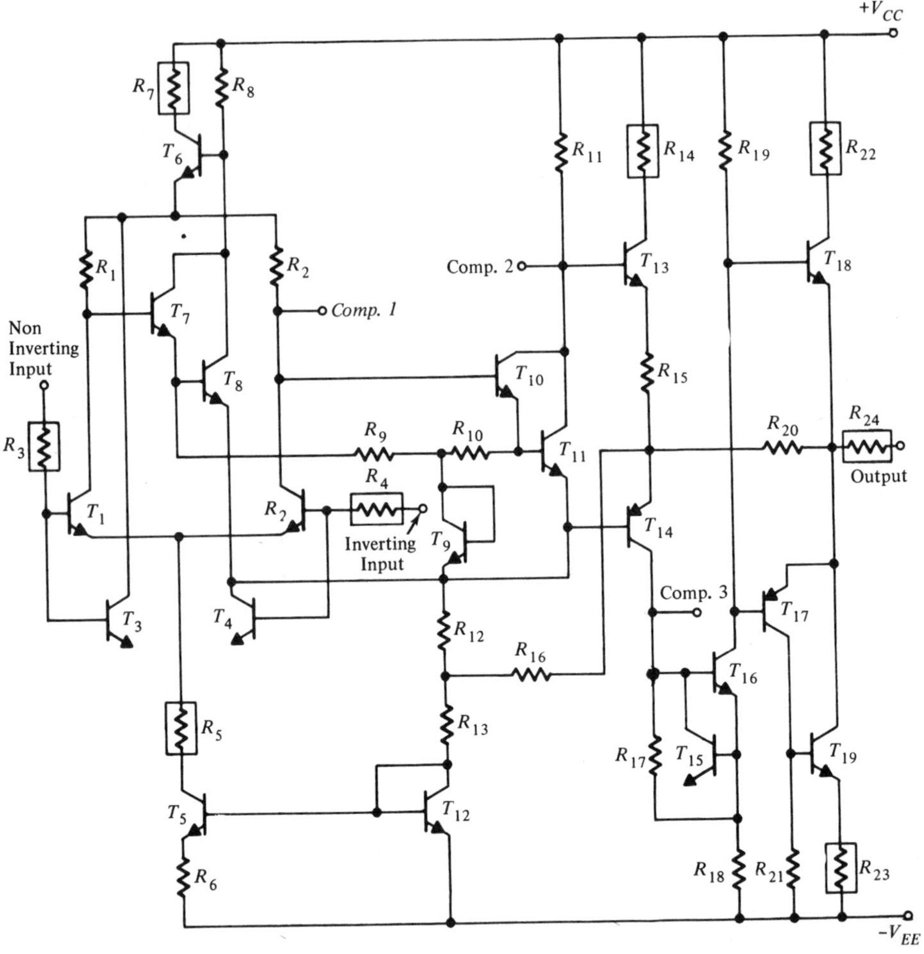

Figure 11.14 Circuit diagram of a radiation-hardened operational amplifier (Fairchild μA 744).

sistors are shown within boxes in the circuit diagram to separate them from the functional components.

T_{15} serves as a photocurrent compensation diode for the output driver stage T_{16}, in a manner illustrated in Fig. 11.13. The lateral *pnp* T_{14} is retained in the circuit since it only requires a $\beta \geqq 0.5$ to perform its circuit function.

Another basic difference between the circuit of Fig. 11.14 and the conventional 709 design is the negative-going portion of the output stage. The substrate *pnp* (T_{13}) of Fig. 5.7 is replaced by the composite connection of the later *pnp* (T_{17}) and the *npn* (T_{19}). This is because the circuit of Fig.

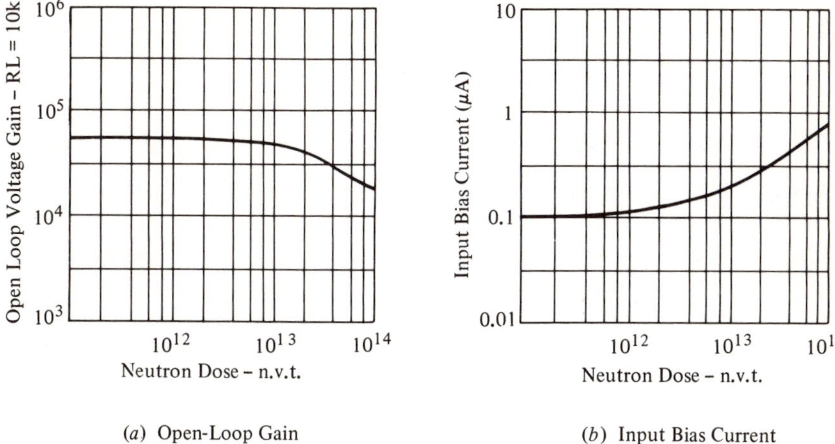

Figure 11.15 Effect of neutron radiation on electrical characteristics of the circuit of Figure 11.14.

11.14 is fabricated using dielectric isolation which does not provide a substrate-*pnp* structure. Also note that the composite connection of T_{17} and T_{19} makes the circuit operation relatively insensitive to the lateral *pnp* β. To enhance its radiation tolerance, the circuit is fabricated using dielectric isolation technology and silicon-chrome thin-film resistors.

The electrical performance characteristics of the circuit of Fig. 11.14 are very similar to those listed in Table 5.1 in connection with the 709 operational amplifier. The two key parameters of the circuit which are effected by the neutron radiation use the open-loop gain and the input bias current. The dependence of these parameters on the neutron radiation fluence is shown in Fig. 11.15.

PART III High Voltage Circuits

In conventional monolithic circuits, breakdown considerations normally limit the operation of the circuit to total supply voltages of 50 V or less. However, there is a rapidly increasing need for analog integrated circuits which can operate at supply voltages of 100 V or more. Some specific applications of such circuits are in driving analog display devices such as the TV picture tube, or in servo or amplifier systems which can be operated directly from the ac power lines.

By proper device and circuit design techniques, it is possible to extend the voltage capability of monolithic circuits to 100 V or more. Some of these design guidelines and techniques will be outlined in the following sections.

11.8 HIGH VOLTAGE DEVICES

Junction breakdown voltages impose a fundamental limitation on the high voltage capability of monolithic devices. Figure 2.9 of Chapter 2 shows the avalanche breakdown voltage for a *p-n* step junction as a function of the impurity concentration on the lighter doped side of the junction. As discussed in Section 2.1, the breakdown voltage vs. impurity concentration characteristics can be approximated by an expression of the form,[16]

$$BV \approx (2.7)(10^{12})N^{-2/3}\ \text{V} \tag{11.15}$$

where BV is the avalanche breakdown voltage, and N is the impurity concentration in atom/cm^3 on the lighter doped side of the junction. However, the data of Fig. 2.9 and Eq. 11.15 refer only to the avalanche breakdown in the bulk of the silicon. The discontinuity of the silicon lattice at the device surface as well as the electrostatic charge trapped at the Si-SiO_2 interface can modify the electrostatic field distribution within the junction at or near the device surface. This can lead to a narrowing down of the junction depletion layer at the surface as shown schematically in Fig. 11.16(*a*). It results in a localized "surface-breakdown" at a lower voltage than the bulk breakdown predicted by Eq. 11.15.

The surface-breakdown effect can be greatly reduced or eliminated by using an electrostatic shield or "field-plate" over the junction surface.[17] Such a field-plate can be formed by depositing a metal layer on the oxide layer directly above the junction, as shown in Fig. 11.17(*b*). When con-

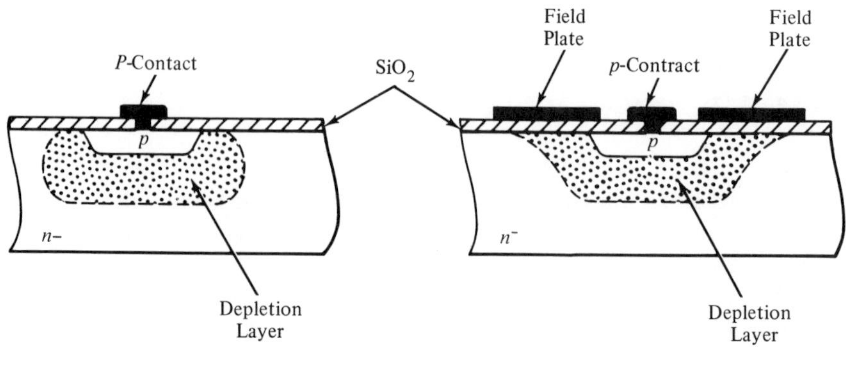

(*a*) Without Field-Plate (*b*) With Field-Plate

Figure 11.16 Effect of field-plate on *P-N* junction depletion layer.

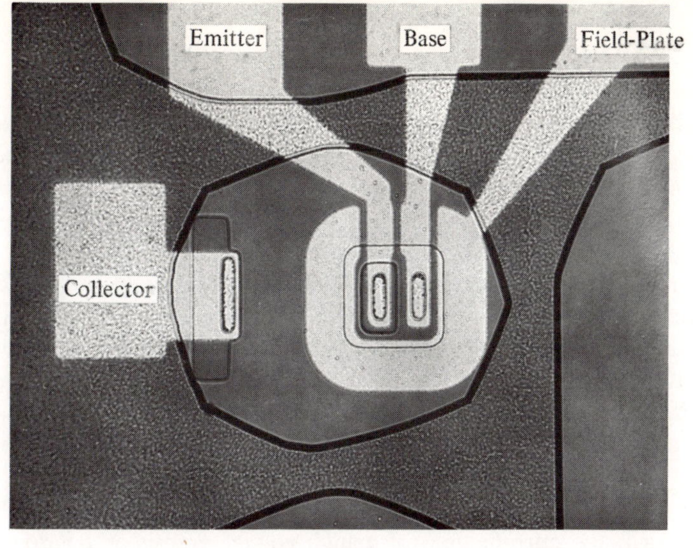

(*a*)

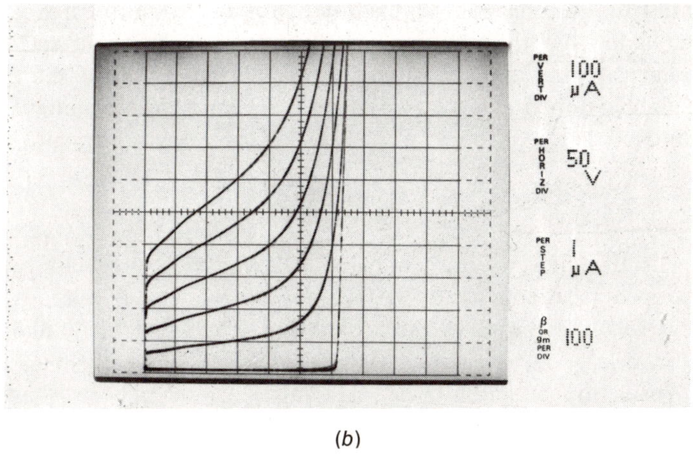

(*b*)

Figure 11.17 Photomicrograph and I-V characteristics of a dielectrically isolated high voltage transistor. (*Photo: Signetics.*)

nected to a more negative potential than the *n*-side of the *p-n* junction, this field-plate tends to spread out the depletion layer near the surface, and thus avoids surface breakdown. In most applications the field-plate is connected directly to the *p*-contact by allowing the contact metal to overlap the junction on the surface. It can be shown that the oxide thickness below the field plate is noncritical for most applications.

In a bipolar integrated circuit, the two junctions which have to sustain the maximum reverse voltage are the base-collector and the collector-isolation junctions. By proper circuit design, the collector-base voltage level and voltage swings can be kept to a level which is below that of the supply voltage. However, the collector-isolation junction is at all times forced to sustain the full supply voltage as a reverse bias across it.

The collector-emitter breakdown BV_{CEO} of a transistor is always smaller than the base-collector breakdown BV_{CBO}, due to the carrier multiplication effects in the transistor base region. As discussed in Section 2.1, collector-emitter breakdown is related to the transistor β and the base-collector breakdown as

$$BV_{CEO} = \frac{BV_{CBO}}{\sqrt[m]{\beta + 1}} \tag{11.16}$$

where $m = 4$ for an *npn* transistor. Therefore, for a given collector-base breakdown, the current gain of a high voltage transistor has to be maintained relatively low (typically $\beta < 80$) to maintain an acceptable BV_{CEO}.

To avoid punch-through breakdown in a device structure, sufficient space must be allocated for the junction depletion region to expand without touching a low resistivity region. The junction depletion layer thickness X_d is related to the applied voltage V and the impurity concentration N on the lighter doped side of the junction. For a step-junction, it can be approximated as

$$X_d = \sqrt{\frac{2\epsilon V}{qN}} \tag{11.17}$$

where ϵ is the dielectric constant and q is the electronic charge. For a lightly doped junction, X_d can be of the order of tens of microns at high voltage levels. Therefore in the design and the layout of a high voltage transistor, a larger space must be allocated for the collector-base and the collector-isolation depletion layer spreading. Since the collector base depletion layer must not touch the n^+ buried layer at the collector-substrate interface, a deep n-type pocket is required for the collector. This calls for a large epitaxial layer thickness (typically of the order of 20 to 30 μ). As the epitaxial layer thickness is increased, the time for p-type isolation diffusion increases and causes additional out-diffusion of n^+ buried layer into the epitaxial region and reduces the effective thickness of the high resistivity collector region. Therefore, this puts a practical limit to high resistivity collector pocket depth that can be obtained with conventional planar epitaxial devices, and limits the maximum collector-base breakdown to approximately 120 V.

Some of the problems of junction isolation in high voltage applications can be avoided by using dielectric-isolation technology described in Chapter 1 (see Fig. 1.15). With this approach the pocket resistivity is set by the choice of starting material, and the pocket depth is determined by the initial oxide-cut width prior to etching of isolation moats (see Eq. 1.18 and Fig. 1.16). Figure 11.17 shows the lateral geometry and the I-V characteristics of a high voltage *npn* transistor fabricated using dielectric isolation technology. The device has a collector resistivity of 30 ohm-cm and an isolation pocket depth of approximately 50 μ, and exhibits a breakdown voltage of >250 V with a current gain of 90. Note that an aluminum field-plate is used over the base-collector junction to eliminate surface breakdown. In the normal operation of the transistor the field-plate would be shorted to either the base or the emitter of the device, since the dc potential difference between the base and the emitter is only a fraction of a volt. Note that the collector n^+ contact is also moved a fair distance from the base-collector junction to provide sufficient room for the depletion layer.

High voltage JFET devices or pinch resistors can also be fabricated using dielectric isolation. Figure 11.18 shows the structural diagram of a high voltage *n*-channel FET which exhibits a low pinch-off voltage and a high drain-gate or drain-source breakdown. The device operates as an FET by pinching off the channel against the dielectric isolation layer which functions as an inert back-gate. The depth of the pocket in the vicinity of the source is kept low in order to maintain a low pinch-off voltage. The pocket depth at the drain end of the channel is increased to provide space for the gate-drain depletion layer. Note that a field plate is also utilized over the gate-drain junction to reduce surface-breakdown effects. This field-plate can be connected to either the gate or the source in the actual operation of the device.

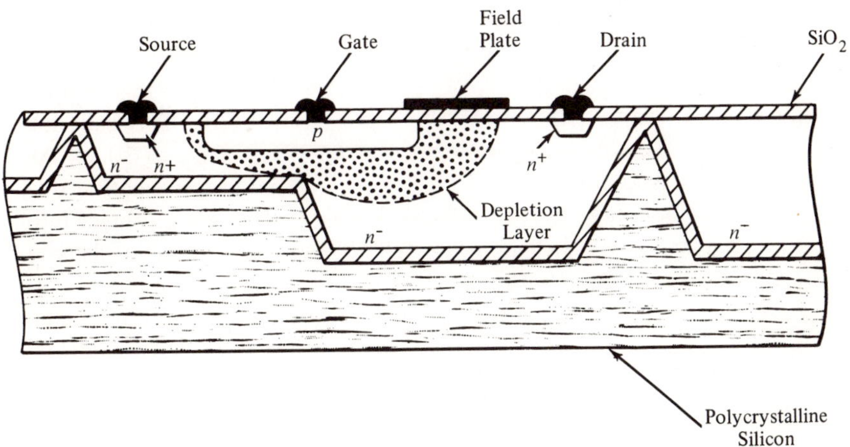

Figure 11.18 A high voltage field-effect transistor using dielectric isolation.[17]

The current drain in a high voltage circuit must be kept to a minimum, to maintain the quiescent power dissipation at a reasonable level. This often requires the use of high value resistors in the circuit. These high-value resistors can be obtained using either thin-film deposited resistors or the "bulk" resistor structures described in Chapter 3. In the case of junction isolated circuits epitaxial pinch resistors can also be used. In the case of dielectric-isolated circuits, the high voltage JFET structure of Fig. 11.18 can be utilized as an n-type pinch resistor by shorting the gate and the source together.

In the fabrication of high voltage circuits, one normally uses a thicker layer of surface dielectric than would be used in conventional integrated circuits. This is done to avoid the breakdown of the oxide layer between the aluminum interconnections and the underlying circuit. As a precautionary measure, it is also customary to place each of the bonding pads of the circuit into a separate isolation pocket so that the circuit can still be functional even if the oxide layer immediately beneath the metal bonding pad may be defective.

The high positive potential in the aluminum conductors also tends to cause surface inversion in the p-type regions directly below it. This effect can be minimized by maintaining a relatively high surface concentration (typically $>(5)(10^{17})$atom/cm^3) as well as by maintaining a relatively thick surface oxide.

The basic guidelines for the device design and layout for high voltage circuit can be briefly summarized as follows:

1. Use field-plate over base-collector or drain-gate junctions.
2. Use high-resistivity collector regions.
3. Use deep collector pockets to allow for depletion layer spreading between the collector-base junction and the n^+ buried layer.
4. Allow space for depletion layer spreading in the lateral dimensions of the device, particularly between the base and the n^+ collector contact, and base and isolation regions.
5. Use a thick oxide layer.
6. Place each bonding pad in a separate isolation pocket.
7. Maintain relatively high surface concentration in p-type regions to avoid surface inversion.

11.9 CIRCUIT DESIGN

In designing high voltage circuits the circuit designer faces two fundamental limitations: (1) voltage breakdown of devices, and (2) power dissipation considerations. Within these boundary conditions it is still possible to improve the circuit performance and high voltage capability to imaginative design approaches.

Proper internal biasing of high voltage circuits often requires the use of high value internal voltage sources. The reverse breakdown voltage of the base-emitter diode, which is of the order of 6 to 9 V is usually too low for such applications and cascading a large number of these diodes to obtain a 50 to 60 V reference is wasteful of chip area. An alternate approach which can be useful to simulate a high value avalanche diode for level-shift and internal voltage reference applications is the circuit of Fig. 11.19. This circuit has already been discussed in Chapter 4 (see Fig. 4.9) and will be briefly reviewed once more. If the current levels in the circuit are properly chosen such that transistor T_1 is conducting, then its collector voltage V_A is an up-scaled version of the voltage level V_B at the base node, i.e.,

$$V_A = V_B\left(1 + \frac{R_1}{R_2}\right) = (V_z + V_{BE})\left(1 + \frac{R_1}{R_2}\right) \tag{11.18}$$

Thus, by proper choice of R_1 and R_2, the circuit of Fig. 11.19 provides a convenient high voltage reference which can be used for biasing or level-shifting in a monolithic high voltage circuit.

In many applications it is possible to extend the high-voltage capability

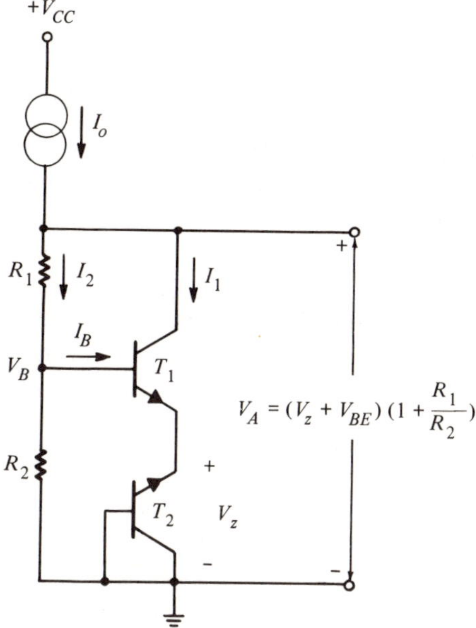

Figure 11.19 A voltage reference circuit for high voltage operation.

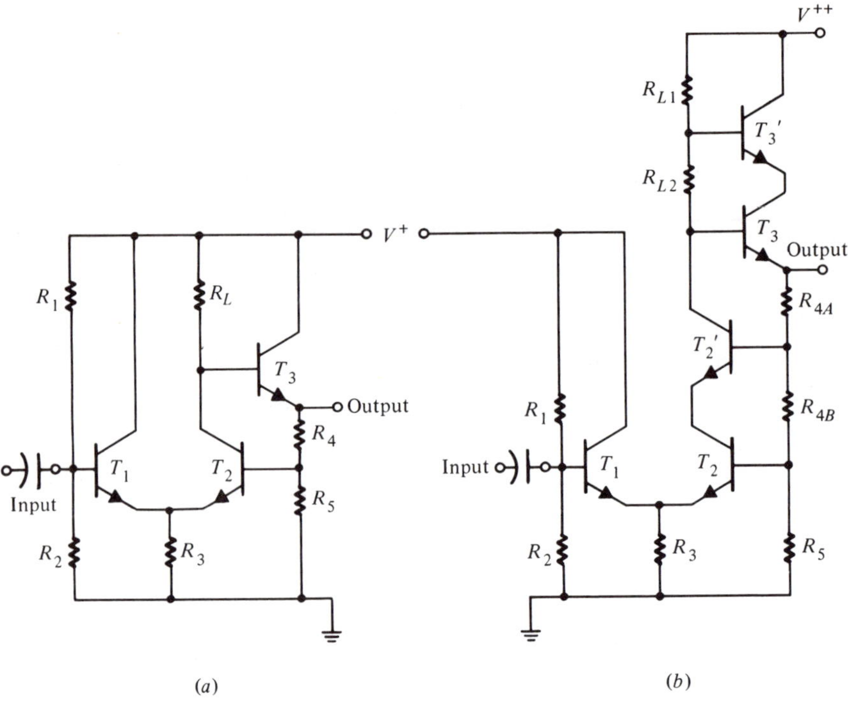

Figure 11.20 A differential shunt-series feedback amplifier: (*a*) conventionl design; (*b*) high voltage design using voltage-sharing principle.[18]

of the circuit by using "voltage-sharing" between the transistors in the circuit such that no single device sees the full supply voltage across it. This can be done by connecting the devices in series, in a "cascode" configuration. Figure 11.20 shows a simplified schematic for high-voltage amplifier circiut using this principle.[18] The circuit of Fig. 11.20(*a*) shows the conventional (low voltage) design for the basic circuit. The resistors R_4 and R_5 form the shunt-series feedback network which sets the voltage gain of the circuit as

$$A_V \approx \frac{R_4 + R_5}{R_5} \tag{11.19}$$

The circuit of Fig. 11.20(*b*) is the high voltage version of the same circuit designed with the "voltage sharing" approach. For simplicity, it is assumed that a low voltage supply V^+ is available to bias the input stage in addition to the high voltage supply V^{++}. In this circuit transistor pairs (T_2, T_2') and (T_3, T_3') effectively function in the same manner as the single devices T_2 and T_3 of the conventional design. Yet, all times, the net voltage drop is shared across two devices. In this manner the circuit can provide a peak-to-peak swing approaching V^{++}, with the net voltage across any transistor being less than $(V^{++}/2)$ at all times. Thus, using this principle it is possible to design circuits for 100 V operation transistors with only a 50 V breakdown. Note that in the high voltage circuit the load and the feedback resistors are appropriately tapped to set the bias levels and the net voltage drops across the series transistors T_2' and T_3'. Neglecting the V_{BE} drops, choosing

$$R_{L1}/R_{L2} = R_{4A}/R_{4B} \approx 1 \tag{11.20}$$

automatically sets the voltage drops across (T_2 and T_2') or (T_3 and T_3') to be equal at all times. As compared to the conventional design, the overall gain of the high voltage circuit is approximated as

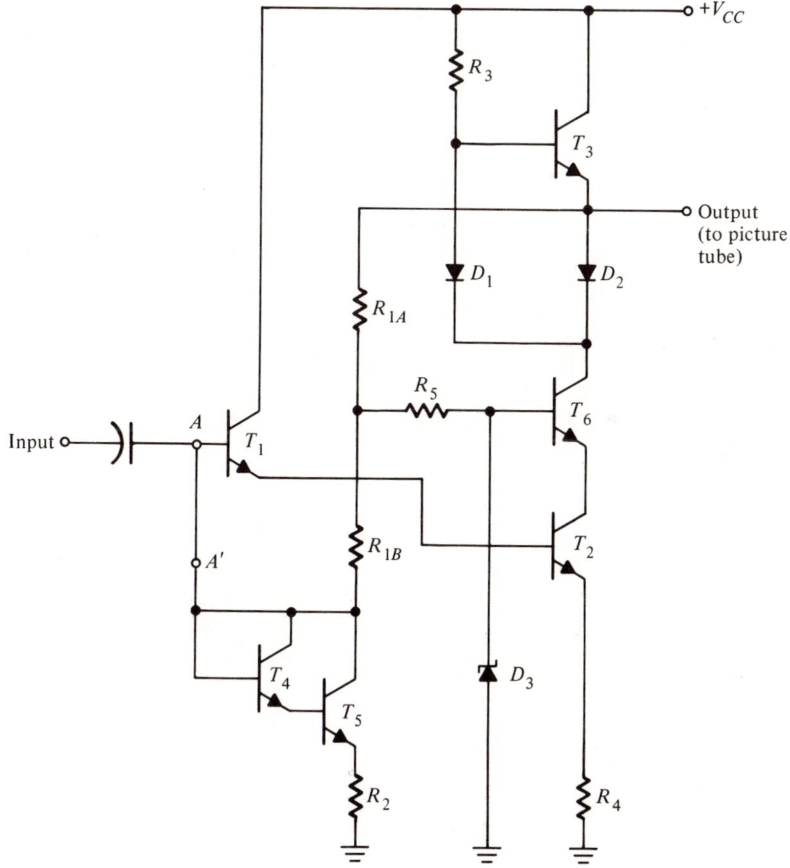

Figure 11.21 A high voltage video amplifier.[17]

$$A_v \approx \frac{R_{4A} + R_{4B} + R_5}{R_5} \tag{11.21}$$

Figure 11.21 shows an alternate circuit design for high voltage and high frequency applications. This particular circuit is designed as a video amplifier for a black and white television set. The amplifier output directly drives the cathode or the grid of the picture tube; it can operate with power supply voltages as high as 250 V and deliver a peak-to-peak output swing of 100 V for frequencies up to 4 MHz. The circuit uses a class-B "totem-pole" output stage (see Section 5.4) to minimize quiescent power dissipation. A shunt feedback network formed by R_{1A}, R_{1B}, and R_2 sets the gain of the circuit as

$$A_v \approx -\frac{R_{1A} + R_{1B}}{R_{s'}} \tag{11.22}$$

where R_s' is the parallel combination of the source resistance and R_2. The diode connected transistors T_4 and T_5 set the dc bias for the input transistor T_1. D_3 is a base-emitter avalanche diode which sets the dc bias level for T_6. In this case, the "cascode" connection of T_2 and T_6 is used for im-

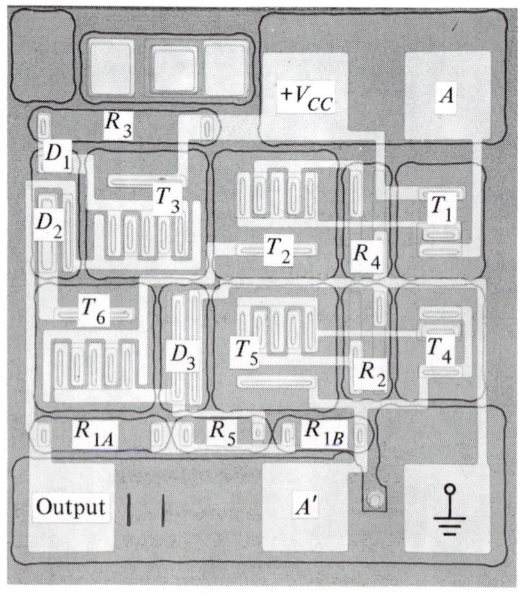

Figure 11.22 Circuit layout for the high voltage video amplifier (chip size: 40 × 45 mils). (*Photo: Signetics.*)

proved frequency response, rather than for voltage-sharing purposes as described earlier.

The circuit is designed using dielectric-isolation technology and uses the high-voltage transistors which have electrical characteristics similar to that shown in Fig. 11.17. The photomicrograph of the finished circuit chip is given in Fig. 11.22. The parts of the circuit corresponding to the circuit schematic of Fig. 11.21 are also identified on the chip layout.

REFERENCES

Part I

1. J. D. Meindl, *Micropower Circuits,* Wiley, New York, 1969.
2. C. A. Bittman, G. H. Wilson, R. J. Whittier and R. K. Waits, "Technology for the Design of Low-Power Circuits," *IEEE J. Solid State Ckts.,* **SC–5**(1) (Feb., 1970): 29–37.
3. H. H. Stellrecht, D. S. Perloff and J. T. Kerr, "Precision Ladder Networks Using Ion-Implanted Resistors," presented at WESCON 1971, Aug., 1971.
4. P. H. Hudson and J. D. Meindl, "A Monolithic Micropower Command Receiver," Int. Solid State Ckts. Conf., *Digest Tech. Papers,* **14** (Feb., 1971): 108–109.
5. W. R. Harden and M. J. Hellstrom, "A Triple-Channel Micropower Operational Amplifier," *IEEE J. Solid State Ckts.,* **SC–4**(4) (Aug., 1969): 236–240.
6. A. Tuszynski and J. H. Bohorquez, "Micropower Techniques in Differential Amplifiers," *NEREM 1970 Record,* Nov., 1970, pp. 26–27.

Part II

7. J. P. Spratt, G. L. Schnable and J. D. Standever, "Impact of Radiation Environment on Integrated Circuit Technology," *IEEE J. Solid State Ckts.,* **SC–5**(1) (Feb., 1970): 14–24.
8. J. L. Wirth and S. C. Rogers, "The Transient Response of Transistors and Diodes to Ionizing Radiation," *IEEE Trans. Nuclear Sci.,* **NS–11** (Nov., 1964): 24–38.
9. G. C. Messenger, "Radiation Effects on Microcircuits," *IEEE Trans. Nuclear Sci.,* **NS–13** (Dec., 1966): 141–159.
10. W. C. Bowman, R. S. Caldwell, A. H. Johnston and K. E. Kells, "Transient Radiation Response Mechanisms in Microelectronics," *IEEE Trans. Nuclear Sci.,* **NS–13** (Dec., 1966): 309–315.
11. "A Monolithic Radiation-Resistant Operational Amplifier," Fairchild Semiconductor Application Brief No. 122, June, 1969.

12. J. P. Mitchell and D. K. Wilson, "Surface Effects of Radiation on Semiconductor Devices," *Bell System Tech. J.*, **46** (Jan., 1967): 1–80.
13. *TREE Handbook,* 2d ed., R. K. Thatcher (ed.), Batelle Memorial Institute, Columbus, Ohio, Aug., 1967.
14. J. D. Lieux, "Radiation Testing of Linear Microcircuits," Fairchild Semiconductor Technical Paper No. 40, Dec., 1966.
15. K. R. Stafford and D. W Oberlin, "A Monolithic Radiation-Hardened Operational Amplifier," *Solid State Technol,* May, 1970, pp. 67–72.

Part III

16. S. K. Ghandhi, *The Theory and Practice of Microelectronics,* Wiley, New York, 1968, pp. 268–273.
17. H. R. Camenzind, B. Polata and J. Kocsis, "ICs Break Through the Voltage Barrier," *Electronics,* March, 1969, pp. 90–95.
18. G. W. Haines and R. Genesi, "An Integrated High-Swing Video Amplifier," Int. Solid State Ckts. Conf., *Digest Tech. Papers,* (Feb., 1969): 94–95.

TABLE OF CONSTANTS

Boltzmann's constant $k = 8.63 \times 10^{-5}$ ev/°K
Electron charge $q = 1.6 \times 10^{-19}$ coulomb
$q = 4.8 \times 10^{-10}$ esu
Free space permittivity $\epsilon_0 = 8.85 \times 10^{-14}$ farad/cm
Planck's constant $h = 6.63 \times 10^{-27}$ erg-sec
Electron mass $m = 9.11 \times 10^{-28}$ g
1 ev $= 1.6 \times 10^{-12}$ erg
$kT = 0.026$ ev ($T = 300$°K)
$kT/q = 0.026$ volt ($T = 300$°K)
1 mil $= 10^{-3}$ in.
1 $\mu = 10^{-4}$ cm
1 $\mu = 10^4$ A
1 in. $= 2.54$ cm
1 mil$^2 = 6.45 \times 10^{-6}$ cm^2

INDEX